Electromagnetic Radiation

The Manchester Physics Series

General Editors

F. MANDL : R. J. ELLISON : D. J. SANDIFORD

Physics Department, Faculty of Science,
University of Manchester

Properties of Matter:	B. H. Flowers and E. Mendoza
Optics:	F. G. Smith and J. H. Thomson
Statistical Physics:	F. Mandl
Solid State Physics:	H. E. Hall
Electromagnetism:	I. S. Grant and W. R. Phillips
Atomic Physics:	J. C. Willmott
Electronics:	J. M. Calvert and M. A. H. McCausland
Electromagnetic Radiation:	F. H. Read

ELECTROMAGNETIC RADIATION

F. H. Read

Department of Physics,
Faculty of Science,
University of Manchester

John Wiley & Sons

CHICHESTER NEW YORK BRISBANE TORONTO

***British Library Cataloguing in Publication Data*:**

Read, Frank Henry
Electromagnetic radiation.—(Manchester
physics series).

1. Electromagnetic waves
I. Title II. Series
539.2 QC661 79-41484

ISBN 0 471 27718 5 (Cloth)
ISBN 0 471 27714 2 (Paper)

Printed by J. W. Arrowsmith Ltd., Bristol, England

Editors' Preface to the Manchester Physics Series

In devising physics syllabuses for undergraduate courses, the staff of Manchester University Physics Department have experienced great difficulty in finding suitable textbooks to recommend to students; many teachers at other universities apparently share this experience. Most books contain much more material than a student has time to assimilate and are so arranged that it is only rarely possible to select sections or chapters to define a self-contained, balanced syllabus. From this situation grew the idea of the Manchester Physics Series.

The books of the Manchester Physics Series correspond to our lecture courses with about fifty per cent additional material. To achieve this we have been very selective in the choice of topics to be included. The emphasis is on the basic physics together with some instructive, stimulating and useful applications. Since the treatment of particular topics varies greatly between different universities, we have tried to organize the material so that it is possible to select courses of different length and difficulty and to emphasize different applications. For this purpose we have encouraged authors to use flow diagrams showing the logical connection of different chapters and to put some topics into starred sections or subsections. These cover more advanced and alternative material, and are not required for the understanding of later parts of each volume.

Since the books of the Manchester Physics Series were planned as an integrated course, the series gives a balanced account of a large part of undergraduate physics. The level of sophistication varies. *Properties of*

Matter is for first year, *Solid State Physics* and *Electromagnetic Radiation* for the third. The other volumes are intermediate, allowing considerable flexibility in use. *Atomic Physics* and *Optics* are suitable for the first or second year, *Statistical Physics* for the second or third. The remaining two volumes of the series, *Electromagnetism* and *Electronics* cover material which is spread over several years in the Manchester course (this is reflected in their greater lengths), starting from first year level and progressing to material suitable for second and third year courses.

Although planned as a series, the books have been written in such a way that each volume is self-contained and can be used independently of the others. Several of them are suitable for use outside physics courses; for example *Electronics* is used by electronic engineers, *Properties of Matter* by chemists and metallurgists. While the series has been written for undergraduates at an English university, it is equally suitable for American university courses beyond the Freshman year. Each author's preface gives detailed information about the prerequisite material for his volume.

Preliminary editions of these books have been tried out at Manchester University and circulated widely to teachers at other universities, so that much feedback has been provided. We are extremely grateful to the many students and colleagues, at Manchester and elsewhere, who through criticisms, suggestions and stimulating discussions helped to improve the presentation and approach of the final version of these books. Our particular thanks go to the authors, for all the work they have done, for the many new ideas they have contributed, and for discussing patiently, and frequently accepting, our many suggestions and requests. We would also like to thank the publishers, John Wiley and Sons, who have been most helpful in every way, including the financing of the preliminary editions.

Physics Department F. MANDL
Faculty of Science R. J. ELLISON
Manchester University D. J. SANDIFORD

Author's Preface

This book is based on a course of lectures of the same title, given by the author at the University of Manchester. The lectures have been attended by third year undergraduate physics students, together with a few first year postgraduate students, for whom the lectures have formed part of their M.Sc. or Diploma courses. To make the book more unified and complete, some introductory second year material has been added.

The subject of electromagnetic radiation offers a unique combination of interests. The foremost of these is the enormous range of frequencies in the electromagnetic spectrum, from the radio waves of frequency less than 10 kHz that are used for submarine communications, to the γ-rays of frequency greater than 10^{24} Hz that are observed as part of the cosmic-ray flux—a total span of more than 20 decades! Associated with this range is the exceptionally large number of techniques by which electromagnetic radiation is generated and detected. At one end of the spectrum the techniques are exclusively based on wave properties, while at the other end only the photon nature of the radiation is relevant. Over much of the middle part of the spectrum the wave and particle natures of light both have a role to play, and the understanding of this duality is surely one of the high points in the development of physics. Indeed, one of the most rewarding aspects of the subject is the way in which so many phenomena can be perfectly well understood in terms of either waves or particles. In addition to all these interests there is the laser, a unique light source, fascinating for its theory as well as its many practical uses.

In the present book these elements are fused together into a single comprehensive treatment, covering all the important aspects of the generation, propagation, and detection of electromagnetic radiation. The level is appropriate to a third year undergraduate degree course in physics, at either a British or a North American university.

The book starts with a brief survey of the more important historical landmarks. This survey is included not only because of the intrinsic interest of the historical quest, but also in the belief that a passing acquaintanceship with the history of the subject, covering up to the present time, will give the student a useful and interesting perspective for his later studies. In Chapter 2 we start with a brief reminder of Maxwell's equations, and then go on to their consequences, the existence of classical electromagnetic waves and the way in which they propagate in free space and in matter. In Chapter 3 classical methods are pursued further, to show how electromagnetic radiation can be generated by various forms of accelerating charge. Here we deal with topics as diverse as the Hertzian dipole, synchrotron radiation, bremsstrahlung and Cerenkov radiation, linking them together through classical ideas. In the second part of the chapter refraction processes are discussed, from the refraction of radio waves by the ionosphere to the refraction of visible light by crystals, again giving a coherent presentation to the whole range. In Chapter 4 we discuss the quantum nature of light, starting with the vital contribution made by Planck. This leads us to Einstein's work on stimulated emission, and to the connection with quantum mechanics. The relationships between the classical and quantal results are emphasized throughout.

In Chapter 5 we change direction and consider the coherence of light, firstly from a classical and then a quantal point of view. The important experiments of Hanbury-Brown and Twiss, and the practical uses of coherence measurements, are discussed. Chapters 6 and 7 are concerned with the laser. We deal firstly with the theory of the laser and then use the first laser, the ruby laser of Maiman, as an example. Then we go on to discuss some of the more important types of laser, such as the dye laser and the solid-state laser. We finish this part of the book by considering the properties and uses of laser light, from holography to cutting and welding. The physical mechanisms by which light is scattered and absorbed are the subject of Chapter 8. Here we deal with topics as diverse as the scattering of light by the sun's corona to the mechanism of the solar cell, again achieving a unity of treatment. Finally, in Chapter 9, we consider the various means by which electromagnetic radiation can be detected and measured, from the bolometer used at infra-red frequencies to the nuclear detectors used for γ-rays.

The complete book is intended to cover a two-semester course of about 40 lectures, but it is also possible to select various sub-sets of the material to cover shorter one-semester courses. The flow diagram on the inside of the front cover shows the interdependences of the various chapters, and examples of suitable sub-sets are given below the diagram.

A wide selection of problems is included at the end of each chapter, many of them designed to amplify specific points. The answers to many of the problems are either given at the end of the book, or are contained (or hinted at) in the questions themselves. SI units are used throughout the book.

I should like to thank the many students who by their questions and comments on the early drafts of the book have slowly but surely improved its presentation. My sincere thanks go also to the colleagues in the physics department (in particular Drs. T. A. King, H. Baker, and R. Booth) for their help with particular chapters, and especially to the three colleagues, Drs. R. J. Ellison, I. S. Grant, and F. Mandl for their help and advice on the lay-out and content of the whole book. I am grateful also to Dr. J. Marsh of the Department of History of Science and Technology (UMIST) for help with the historical details, to Miss Lesley Burns for typing the penultimate and final manuscripts so accurately and quickly, to Mr. John Rowcroft for preparing the initial drawings of many of the figures, and finally to my wife for her patience and understanding during the writing of the text.

FRANK H. READ

Manchester, August, 1979

Contents

CHAPTER 1

Introduction

1.1 THE ELECTROMAGNETIC SPECTRUM

Isaac Newton was the first to show that light beams of different colour are essentially different in some respect. His curiosity was aroused on setting up a demonstration of 'the celebrated Phaenomena of Colours'. He used a glass prism to refract a beam of sun-light coming through a small hole in his window shutters, and saw on the opposite wall not the coloured circular form that he had been led to expect, but an oblong-shaped spectrum, red at the bottom and violet at the top. Further experiments showed that the various colours are immutable. To quote his own words

> 'When one sort of Ray hath been well parted from those of other kinds, it hath afterwards obstinately retained its colours, not withstanding my utmost endeavours to change it'.

We know now of course that the different colours are characterized by different frequencies, which clearly makes their inter-changeability impossible.

What Newton could not possibly have imagined is the enormous range of 'colours' of light that we now know exists. This range extends from radio waves of frequency less than 10 kHz (used for submarine communications) to γ-rays of frequency greater than 10^{24} Hz (observed as part of the cosmic ray flux), a total span of more than 20 decades! The part of this spectrum observed by Newton is indeed a very small part of the total, covering a span of frequencies of less than a factor 2.

During the course of this book the reader will become familiar with much of this spectrum, but before this happens it is interesting to see the whole of it, in Fig. 1.1. The end limits of each type of radiation are only approximately defined, and in general there is some overlap between neighbouring types, as shown in the figure. Three different, but equivalent, characterizations of 'colour' are given in the figure, namely frequency, energy, and wavelength. The relationships between these is given on the inside back cover of the book, along with other numerical information that we shall need. The most familiar part of the spectrum is of course the part that we can *see*, from deep red at a wavelength of 720 nm to violet at a wavelength of 400 nm.

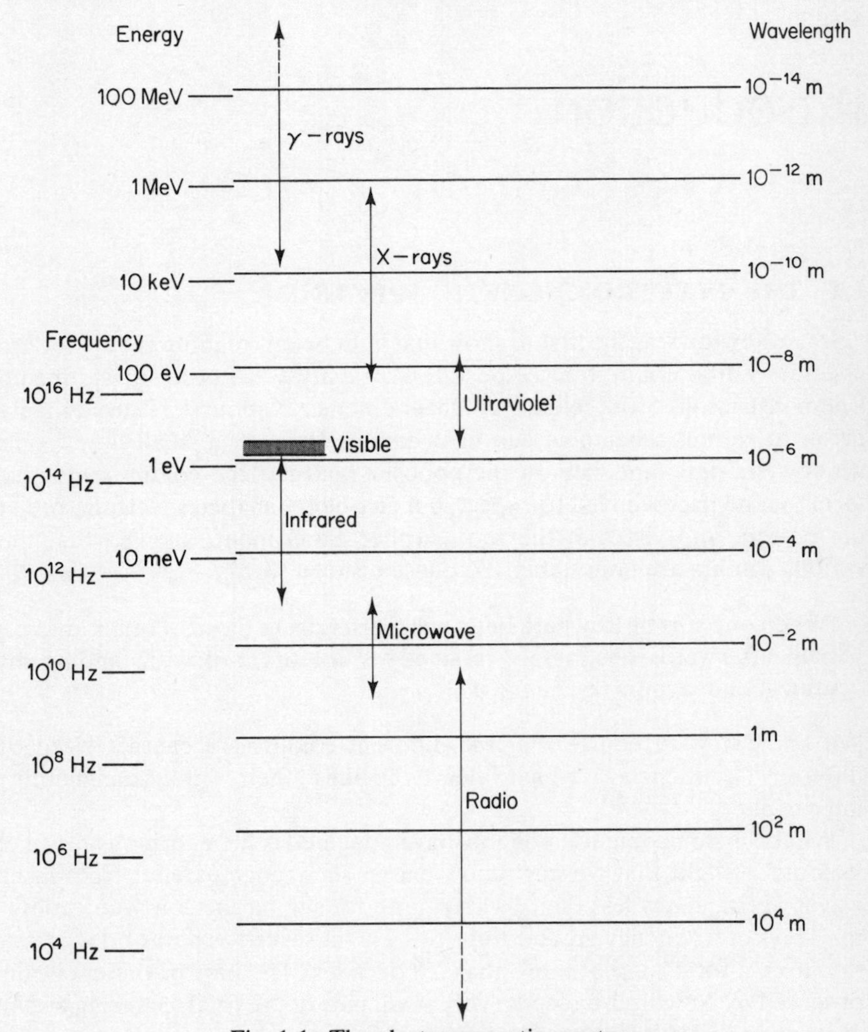

Fig. 1.1. The electromagnetic spectrum.

1.2 THE HISTORY OF LIGHT WAVES AND PHOTONS

Two aspects of light which are even more fascinating than the range of its spectrum are the difficult question of its fundamental nature, and the many controversies to which this has given rise. The questions 'What is light?', and later, 'Does light consist of waves or photons?' have surely attracted more answers and occupied more of the intellectual energies of more of the greatest scientists of the past three centuries than any other subject in physics. Although we shall deal with this aspect, and attempt to answer the question of the nature of light, there will unfortunately be space only for scant mention of the details of the historical quest. To partially remedy this we give now a summary of the more important dates, names, and discoveries, with a few comments on their significance. The comments are necessarily brief, and many of them assume a knowledge of what is in the later chapters, but it is hoped nevertheless that the reader will be able at this stage to gain some impression of the fascination and depth of the subject on which he or she is about to embark. The dates have been divided for convenience into seven 'periods', each representing a significant step forward in the subject.

First period *The corpuscular and wave models are considered, the latter ultimately being accepted*

1621 Snell discovers experimentally the law governing the refraction of light as it passes through an interface between two media.

1637 Descartes publishes an explanation of this law in terms of a corpuscular theory of light, and has to suppose that the corpuscules travel more quickly in dense media than in air.

1657 Fermat enunciates his Principle of Least Time, in the form 'Nature always acts by the shortest course', and uses it to deduce the laws of reflection and refraction. The experimental results are explained if light travels more slowly in dense media than in air.

1665 Hooke publishes a description of his observations of the colours produced when white light falls on a thin film (in fact this phenomenon had previously been studied by Boyle), and his observation that light is not propagated in straight lines but bends through small angles into the geometrical shadow of an object (this phenomenon had also been studied previously, by Grimaldi, who named it diffraction). He proposes the hypothesis that light consists of vibrations of an aether, and proceeds to explain refraction in terms of the deflection of a wave-front. The wave-front travels more slowly in dense media than in air. He does not associate differences in colours with differences in the frequency of vibration or in other inherent properties of light.

1669 Bartholinus describes the properties of Iceland spar including that of producing two images of one object (double refraction).

1671 Newton suggests, after studying the separation of colours by refraction in a prism, that beams of different colours are essentially different in some respect. He rejects Hooke's hypothesis, because it does not seem to be able to account for the rectilinear propagation of light, amongst other reasons. Instead he offers a wide range of models, without committing himself ('let every man here take his fancy'), but later authors select the corpuscular model.

1675 Newton proposes the 'hypothesis' that light particles in certain circumstances generate wave motions in the aether. He uses this to explain 'Newton's rings', the colour of thin films and diffraction at a straight edge.

1675 Roemer observes that the times of eclipse of a satellite of Jupiter depend on the relative positions of Jupiter and the Earth, and so discovers that light has a finite velocity.

1678 Huygens develops a wave theory in which each point of a wave-front is regarded as a source of secondary wavelets. This gives the laws of reflection and refraction, provides a construction for calculating the relative position of the images in double refraction, and is consistent with rectilinear propagation. However, it does not attempt to account for interference or diffraction.

1704 Newton publishes the first edition of his *Opticks*. The light particles and associated aether waves of the 'hypothesis' are now replaced by particles which suffer *periodic* 'fits'. In later editions the light particles are also considered to have 'sides' thereby accounting, at least in part, for double refraction.

1728 Bradley reports that the apparent direction of a star depends on the direction of the component of the earth's velocity transverse to the line of sight. This is interpreted in terms of light corpuscules of finite velocity.

1800 Herschel discovers infra-red radiation, and infers that radiant heat has essentially the same nature as visible light.

1801 Young explains Newton's rings in terms of the interference between different wave trains. This marks the beginning of the general acceptance of the wave theory of light.

1808 Malus discovers polarization by reflection.

1816 Fresnel uses the wave theory to calculate diffraction patterns that agree with experimental observations.

1817 Young suggests that certain polarization phenomena can be explained by supposing that the light vibrations are transverse to the direction of propagation.

1860 Kirchhoff states his theorem that the ratio w of the emissivity of a surface to its absorptivity depends only on the wavelength and temperature, and not on the nature of the surface. The theoretical struggle to find the form of $w(\lambda, T)$, and of the associated energy density $u(\lambda, T)$ of thermal radiation, eventually leads to Planck's law and the discovery of the photon.

Second period The electromagnetic nature of light is established

1864 Maxwell reads a paper to the Royal Society, entitled 'A dynamical theory of the electromagnetic field'. He finds that the velocity of the electromagnetic waves has a numerical value close to that found by astronomers for the velocity of light, and so infers that light is an electromagnetic disturbance. This marks the beginning of the subject of electromagnetic radiation (and also the starting point for the material of this book).

1879 Stefan finds empirically that the total energy density of thermal radiation is proportional to T^4, thus showing for the first time the connection between thermal radiation energy and temperature.

1887 Hertz discovers long wavelength electromagnetic radiation, by finding that a spark can be induced across a short gap in a circuit when it is placed near to a discharging induction coil. In the following year he shows that the disturbance is propagated with the velocity c. He also discovers the photoelectric effect by noting that sparks are induced more easily when the electrodes of the gap are illuminated by ultraviolet light.

1893 Wien discovers that the function $u(\lambda, T)$ giving the radiation energy density at wavelength λ and temperature T has the form $T^5 f(\lambda T)$.

1895 Röntgen notices that paper painted with barium platino-cyanide fluoresces when near a Crookes tube, and thus discovers X-rays.

1896 Wien finds that the energy density function has the form $F(\lambda, T) \propto \lambda^{-5} \exp(-b/\lambda T)$, which agrees with experimental results except at long wavelengths.

1900 Villard discovers γ-rays.

1900 Lord Rayleigh calculates the energy density of radiation, assuming equipartition of energy, finding $u(\lambda, T) = 8\pi kT\lambda^{-4}$, but it is realized that this cannot be correct at short wavelengths.

Third period The quantum hypothesis is proposed but is not widely accepted

1900 Planck reads a paper before the German Physical Society, in which he gives the correct radiation law, $u(\lambda, T) = 8\pi ch\lambda^{-5} [\exp(hc/\lambda kT) - 1]^{-1}$. This is the first indication that the 'resonators' in the walls enclosing the radiation have quantized energies $\varepsilon = h\nu$. Planck did not believe that the radiation field itself is quantized.

1902 Lenard discovers that there is a threshold frequency for the photoelectric effect and that the kinetic energy of the photoelectrons is independent of the light intensity.

1905 Einstein publishes a paper in *Annalen der Physik* (in the same volume that contains his celebrated papers on the special theory of relativity and on Brownian motion), in which he uses thermodynamic reasoning to show that the radiation field is quantized in units $h\nu$. He derives the photoelectric equation and explains the main features of the experimental observations. This marks the birth of the photon.

1909 Einstein considers the energy and momentum fluctuations in a radiation field, and notices that the theoretical expressions seem to imply the co-existence of waves and particles in the field.

1911 Barkla identifies two series of characteristic secondary X-rays produced in circumstances analogous to fluorescence of light.

1912 Friedrich and Knipping, following a suggestion of Laue, find that X-rays are diffracted by a copper sulphate crystal.

1914 Einstein's photoelectric equation is verified experimentally by Richardson and Crompton, and also by Hughes, but the validity of the equation is not taken as proof of the validity of the light quantum hypothesis.

1915 Douane and Hunt find experimentally that the relationship between the potential V across an X-ray tube and the maximum frequency ν_{max} of the X-rays is $h\nu_{max} = eV$. This is given the status of a law, and is not seen as a trivial consequence of the existence of the photon. In 1918 Rutherford says of the law 'There is at present no physical explanation possible of this remarkable connection between energy and frequency'.

1917 Einstein publishes a paper in which he points out the existence of stimulated emission, derives what are now known as the Einstein A and B coefficients, re-derives Planck's radiation law, and shows that photons have momentum $h\nu/c$.

1922 Bohr expresses the commonly held opinion of the time when he says in his Nobel lecture that 'in spite of its heuristic value, Einstein's hypothesis of light quanta, which is quite irreconcilable with the so-called interference phenomena, is not able to throw light on the nature of radiation'.

Fourth period *The existence of quanta of radiation is established, but the concept of wave-particle duality presents difficulties*

1923 Compton publishes a paper in *The Physical Review* in which his

X-ray scattering results are interpreted in terms of the scattering of quanta of radiation by free electrons, with the quanta possessing energy $h\nu$ and momentum $h\nu/c$. He had previously tried every other conceivable explanation. This marks the beginning of the general acceptance of the photon hypothesis.

1923 Debye publishes a paper in *Physikalische Zeitschrift* in which he independently suggests a quantum theory of scattering to explain Compton's results.

1924 Bose and Einstein independently develop the statistics, now known as Bose–Einstein statistics, that are obeyed by photons and certain other particles.

1924 de Broglie argues in his doctoral thesis that a wavelength $\lambda = h/p$ can be associated with particles of momentum p. This is confirmed experimentally three years later by Davisson and Germer, who record the diffraction patterns formed when slow electrons are scattered from a single crystal of nickel. This shows that light is not the only type of radiation exhibiting both a wave and a particle nature.

Fifth period The concept of wave-particle duality is understood

1925 Heisenberg discovers matrix-mechanics.

1925 Born and Jordan develop the mechanics and apply it to the harmonic oscillator. They also discuss the quantization of the electromagnetic field.

1926 Schrödinger discovers his equation by developing de Broglie's ideas, and thus initiates wave-mechanics. Later in the same year he shows that matrix-mechanics and wave-mechanics are equivalent.

1926 The quantum of radiation is named the 'photon' by G. N. Lewis.

1927 Heisenberg develops the Uncertainty Principle and in the following year Bohr introduces the Complementarity Principle. These principles give a clearer understanding of the nature of wave-particle duality.

1927 Dirac, and later Jordan and Pauli, develop the quantum theory of the pure radiation field.

Sixth period The techniques of quantum electrodynamics are developed

1929 Heisenberg and Pauli develop the general theory of quantum electrodynamics.

1947 Lamb and Retherford discover experimentally that the $2s_{1/2}$ and $2p_{1/2}$ states of atomic hydrogen are slightly different in energy. The difference is now known as the Lamb shift. The theoretical interpretation in terms of the interaction of electrons with the radiation field followed soon afterwards.

1949 Kastler suggests, and later carries out, optical double resonance and optical pumping experiments, thus opening the way to many two-photon and multi-photon studies.

Seventh period The maser and laser are invented, and the subject of quantum optics is started

1954 Basov and Prokhorov publish a detailed theoretical exploration of the maser, and independently Gordon, Zeigler, and Townes publish details of the construction and operation of the ammonia maser, following the original idea of Townes.

1956 Hanbury-Brown and Twiss demonstrate an intensity correlation between coherence beams of light.

1960 Maiman announces the successful operation of a ruby laser.

1960 Javan makes the first helium-neon laser.

1966 Sorokin and Lankard publish details of a tunable laser, the dye laser. This makes possible a wide variety of experiments in atomic and molecular physics.

Even this is not the end of the story. There is still intense activity in the laser field, leading to new types of lasers and new uses of laser light. On a more fundamental level, unification of the electromagnetic and weak interactions has been achieved theoretically and verified experimentally. Who can doubt that the subject may yet hold further surprises in store?

CHAPTER 2

Maxwell's equations and electromagnetic waves

Maxwell's equations are a mathematically concise statement of the laws of electricity and magnetism that had been established by the painstaking work of several great scientists during the first half of the 19th century. Maxwell's own brilliant contribution was to add to one of the equations a term which could not have been observed experimentally with the apparatus available at that time, and without which electromagnetic radiation would not exist. The purpose of the present chapter is to present the equations and to discuss the types of wave to which they give rise.

Before the time of Maxwell it had not been at all obvious that electric and magnetic fields had anything to do with the propagation of light and heat. The connection was made first and most convincingly by Michael Faraday, although he was able to express the idea only in qualitative terms. He was helped in making the connection by his preference for thinking of electric and magnetic phenomena in terms of models rather than mathematical equations. In particular he made much use of a vivid and powerful model in which electric and magnetic fields are regarded as containing *lines of force* which are in tension along their length and which mutually repel each other sideways. With this model he was able to understand for example the forces of attraction or repulsion between two charged bodies: the lines of electric force set up a system of stress in the region between (and surrounding) the two bodies, leading to a resultant attractive or repulsive force between the bodies themselves. Since all media, even vacua, contain these lines of force, it was natural to go further and suggest that light might consist of transverse vibrations propagating along the lines of force, in analogy with a wave travelling along a stretching string. Faraday was, in this sense, the first to suggest that light has an electromagnetic nature. He was also the first to establish experimentally a connection between magnetism and light, when he found that in some circumstances magnetic fields can change the plane of polarization of a beam of light.

Despite the usefulness of Faraday's models, it was not until the laws of electricity and magnetism were expressed in an exact and concise mathematical form that progress was made towards understanding the details of light propagation. We start therefore with a brief reminder of the set of equations representing the laws of classical electromagnetism.

2.1 MAXWELL'S EQUATIONS

An obvious starting point is the equations themselves. In empty space they take the form

$$\operatorname{div} \mathbf{E} = 0 \tag{2.1}$$

$$\operatorname{div} \mathbf{B} = 0 \tag{2.2}$$

$$\operatorname{curl} \mathbf{E} = -\frac{\partial \mathbf{B}}{\partial t} \tag{2.3}$$

$$\operatorname{curl} \mathbf{B} = \mu_0 \varepsilon_0 \frac{\partial \mathbf{E}}{\partial t} \tag{2.4}$$

where $\mathbf{E}$ and $\mathbf{B}$ are the electric and magnetic fields, and μ_0 and ε_0 are the absolute permeability and absolute permittivity of free space respectively. The numerical value of μ_0 is by definition

$$\mu_0 = 4\pi\ 10^7 \text{ henry metre}^{-1}.$$

The numerical value of ε_0 is then found from experimental measurements to be

$$\varepsilon_0 = 8.854 \times 10^{-12} (\approx 1/36\pi \times 10^9) \text{ farad metre}^{-1}.$$

The product $\mu_0\varepsilon_0$ has the dimensions of (velocity)$^{-2}$; its value is

$$\mu_0\varepsilon_0 = 1/c^2,$$

where

$$c = 2.998 \times 10^8 \text{ m/s}.$$

We shall see in the next section that this velocity is the velocity of light in free space.

The first of the free-space equations arises because of the absence of any sources of the electric field (i.e. charges) in empty space. The second arises in a similar way because point sources of the magnetic field (i.e. free magnetic poles, sometimes called magnetic monopoles) do not exist; the equation applies therefore even in the presence of matter. In the case of static fields the third of the equations becomes curl $\mathbf{E} = 0$. This is a consequence of the fact that static electric flux lines begin and end on electric charges, and therefore cannot form closed loops. In the time-dependent case the third equation is simply Faraday's law of electromagnetic induction.

Maxwell's own contribution to these equations is the term on the right-hand side of the fourth equation. One thing that we notice straight away about this term is that it gives an overall symmetry to the roles of **E** and **B** in the four equations. The symmetry is not spoiled by the fact that the fourth equation contains the factor $\mu_0\varepsilon_0$ while the third does not, which is simply a result of the system of units that we are using. Another point about the term is that it is too small to have been experimentally observable with the apparatus available to the early investigators. Thus although Faraday was able to discover the phenomenon of electromagnetic induction, he failed to observe the analogous phenomenon of magneto-electric induction.

In the presence of matter, and of currents and charges, Maxwell's equations are

$$\operatorname{div} \mathbf{D} = \rho_f \tag{2.5}$$

$$\operatorname{div} \mathbf{B} = 0 \tag{2.6}$$

$$\operatorname{curl} \mathbf{E} = -\frac{\partial \mathbf{B}}{\partial t} \tag{2.7}$$

$$\operatorname{curl} \mathbf{H} = \frac{\partial \mathbf{D}}{\partial t} + \mathbf{j}_f \tag{2.8}$$

where ρ_f and $\mathbf{j}_f$ refer to free charge and current densities, **D** is the electric displacement and **H** is the magnetic intensity.

The origin of the first of these equations is illustrated by considering a dielectric medium composed of polarizable molecules. Each molecule acquires an average dipole moment **p** proportional to the local electric field $\mathbf{E}^{\text{local}}$ applied to it. If there are N such molecules per unit volume the macroscopic polarization **P** is then given by

$$\mathbf{P} = N\mathbf{p} = N\alpha\varepsilon_0\mathbf{E}^{\text{local}}, \tag{2.9}$$

where α is the molecular *polarizability*. The local electric field acting on a molecule is the sum of the externally applied field **E** and an extra field caused by the surrounding polarized molecules, but for isotropic media this extra component is itself proportional to **E** (unless the magnitude of **E** is very large), and so we may write

$$\mathbf{P} = \chi \varepsilon_0 \mathbf{E}. \tag{2.10}$$

This equation defines the *electric susceptibility* χ of the dielectric. The electric displacement is now defined by

$$\mathbf{D} = \varepsilon_0 \mathbf{E} + \mathbf{P} = (1+\chi)\varepsilon_0 \mathbf{E} = \varepsilon\varepsilon_0 \mathbf{E}, \tag{2.11}$$

where ε is the *relative permittivity* of the dielectric. The external field **E** is essentially caused by the density ρ_f of external free charges, while the polarization **P** has its origins in the density ρ_v of polarization charges. The total charge density ρ is of course the sum of ρ_f and ρ_v. By considering the divergences of **E** and **P** is can be shown that the divergence of **D** depends only on the free charges, as given by Eq. (2.5).

The second of Maxwell's equations remains the same as in the free-space case, as mentioned before. The third of the equations is also unaffected by the presence of matter, again because of the absence of free magnetic poles. If such poles were to exist Maxwell's equations would be completely symmetric, with a pole density term in the second and a magnetic current term in the third.

The fourth equation contains the magnetic intensity in place of the magnetic field which appears in the equivalent free-space Eq. (2.4). In the absence of magnetic materials **H** is defined simply as

$$\mathbf{H} = \mathbf{B}/\mu_0. \tag{2.12}$$

In the presence of magnetic materials the applied field **B** induces a density $\mathbf{j}_M$ of magnetization currents, and hence a magnetization **M** in the material. This is analogous to the production of electric polarization by an electric field. In the case of non-ferromagnetic materials the relationship between **B**, **M**, and **H** is

$$\mathbf{B} = \mu_0 \mathbf{H} + \mathbf{M} = \mu\mu_0 \mathbf{H}, \tag{2.13}$$

where μ is the *relative permeability* of the medium. With this definition of **B** the fourth of Maxwell's equations can be written as

$$\text{curl } \mathbf{B} = \mu\mu_0 \mathbf{j}_f \tag{2.14}$$

for static fields. This is one form of Ampère's law.

Finally we come to the term $\partial\mathbf{D}/\partial t$ in the fourth of the equations. This term is called the *displacement current*. It is similar to the term on the right-hand side of the equivalent free-space equation, but is modified because of the possible presence of dielectric materials. In the free-space case the displacement current is $\varepsilon_0\,\partial\mathbf{E}/\partial t$.

Maxwell showed that the displacement current must exist by considering the conservation of free electric charge. Suppose that a volume V has a surface S and that it contains a total charge Q. Suppose also that the charge density within the volume is ρ_f and that the current density flowing through the surface is $\mathbf{j}_f$. Now if the charge is to be conserved, the total current flowing out through the surface must equal the rate of decrease of the total charge within the surface,

$$\int_S \mathbf{j}_f \cdot d\mathbf{S} = -\frac{dQ}{dt} = -\frac{d}{dt}\int_V \rho_f \, d\tau.$$

Using the divergence theorem (A.16) to change the left-hand side,

$$\int_S \mathbf{j}_f \cdot d\mathbf{S} = \int_V \operatorname{div} \mathbf{j}_f \, d\tau,$$

we obtain the relationship

$$-\frac{\partial \rho_f}{\partial t} = \operatorname{div} \mathbf{j}_f. \tag{2.15}$$

This is the *equation of continuity* for the free charges. By using the first of Maxwell's equations this can be written as

$$\frac{\partial}{\partial t}(\operatorname{div} \mathbf{D}) + \operatorname{div} \mathbf{j}_f = 0,$$

and by reversing the order of the operators $\partial/\partial t$ and div this becomes

$$\operatorname{div}\left(\frac{\partial \mathbf{D}}{\partial t} + \mathbf{j}_f\right) = 0. \tag{2.16}$$

The remaining step is to take the divergence of both sides of Eq. (2.8). Since div curl $\mathbf{F}$ is zero for any vector $\mathbf{F}$ (Eq. (A.14)), we again obtain Eq. (2.16). We see from this that the equation of continuity would not be satisfied if the displacement current were not included in Eq. (2.8).

2.2 THE EXISTENCE OF ELECTROMAGNETIC WAVES

There are no conceptual difficulties in understanding how sound waves travel through the air or how displacement waves travel along a stretched string, or indeed how any wave motion propagates through an elastic medium. A disturbance or displacement at any one point in the medium produces a force that acts on neighbouring points and causes the disturbance to be propagated throughout the medium. Also the neighbouring and more distant points react back on the initially disturbed points, usually after some time delay, thus causing the displacements with time of all the points in the medium to be inter-dependent.

The concept of propagation in the absence of a medium is much more difficult. How can a displacement $\mathbf{x}$ or field $\mathbf{E}$ propagate itself if it has nothing on which to act and if there is no medium which can react back on it?

The answer to this question lies, in the case of electromagnetic waves, in the existence of the two fields $\mathbf{E}$ and $\mathbf{B}$, and in the inter-play between them. We can see from Maxwell's free space equations that a time-dependent electric field gives rise to a magnetic field, and if this is itself time-dependent it in turn gives rise to an electric field. This return loop gives a type of 'elasticity' to the electric field, and hence gives the possibility of propagation of the electric field. This same elasticity exists for the accompanying magnetic field, and the two fields are so closely related that the propagation can only be described as that of an *electromagnetic* field. As discussed at the beginning of the previous chapter, Faraday appears to have been the first to have suggested the possibility that the propagation of light is connected with electric or magnetic fields, but it required the systematization and concise description by Maxwell of the laws of electromagnetism, together with his identification of the displacement current $\partial\mathbf{D}/\partial t$, before this suggestion could be confirmed, as we shall now see.

The propagation of electromagnetic waves is treated in a purely classical way in the present chapter. It is assumed therefore that matter is continuous, with no atomic structure, and that there are no restrictions on the energy of the electromagnetic field. The essentially atomistic and quantal problems of the generation, detection, absorption, and quantization of the field will be the subjects of later chapters. The discussion is also still limited to media having relative permeabilities μ and permittivities ε that are isotropic and linear.

Wave equations

Starting with the four free-space equations, we should like to eliminate $\mathbf{E}$ or $\mathbf{B}$ and obtain an equation in one of the fields only. This requires three steps. First we take the curl of both sides of Eq. (2.3),

$$\text{curl curl }\mathbf{E} = -\text{curl}\,\frac{\partial \mathbf{B}}{\partial t}.$$

Then we reverse the order in which the space and time derivatives operate,

$$-\text{curl}\,\frac{\partial \mathbf{B}}{\partial t} = -\frac{\partial}{\partial t}(\text{curl }\mathbf{B}).$$

For the third step we use Eq. (2.4) to obtain

$$\text{curl curl } \mathbf{E} = -\varepsilon_0 \mu_0 \frac{\partial^2 \mathbf{E}}{\partial t^2}. \tag{2.17}$$

The left-hand side of this can now be simplified by using the identity (Eq. A.15)

$$\text{curl curl } \mathbf{E} = \text{grad div } \mathbf{E} - \nabla^2 \mathbf{E}.$$

The first term on the right-hand size of this is zero, from Eq. (2.1), and the second term is easily expressed in terms in second order partial derivatives (see Eqs. (A.11, 12)). Finally we have the equation

$$\boxed{\nabla^2 \mathbf{E} = \frac{1}{c^2} \frac{\partial^2 \mathbf{E}}{\partial t^2}} \tag{2.18}$$

We can establish in a similar way that

$$\boxed{\nabla^2 \mathbf{B} = \frac{1}{c^2} \frac{\partial^2 \mathbf{B}}{\partial t^2}} \tag{2.19}$$

These are the *wave equations* for the electromagnetic field. They are the simplest, indeed the only, equations having as solutions undamped vector waves travelling in three dimensions at any frequency and with constant velocity.

Plane waves

To see the significance of these equations, we start with a simple example of an electromagnetic field in which the direction of **E** is along the x axis only, and the magnitude of **E** depends on the time t and the displacement z only,

$$\mathbf{E} = \mathbf{i}_x E_x(z, t).$$

The z direction is clearly the direction of propagation of this field (assuming that we shall find that it does propagate). This type of wave is called a *plane wave.* By substituting this form of **E** in Eq. (2.18) (and also using Eq. (A.11) for ∇^2) we obtain

$$\mathbf{i}_x \frac{\partial^2 E_x}{\partial z^2} = \mathbf{i}_x \frac{1}{c^2} \frac{\partial^2 E_x}{\partial t^2}. \tag{2.20}$$

A possible solution of this is

$$E_x(z, t) = f(z - ct), \tag{2.21}$$

where f is any twice-differentiable function. To see this we denote $\mathrm{d}f(u)/\mathrm{d}u$ by $f'(u)$, and write

$$\frac{\partial E_x}{\partial t} = \frac{\mathrm{d}f(z-ct)}{\mathrm{d}(z-ct)} \frac{\partial (z-ct)}{\partial t} = -cf'(z-ct),$$

$$\frac{\partial^2 E_x}{\partial t^2} = c^2 f''(z-ct),$$

$$\frac{\partial^2 E_x}{\partial z^2} = f''(z-ct).$$

Now the function $f(z-ct)$ has the same value at the time $t+\tau$ and the position $z+c\tau$ that it has at the time t and the position z. Figure 2.1 shows an

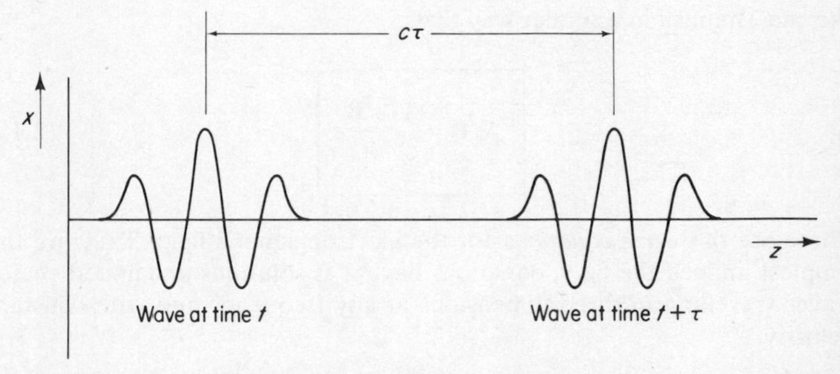

Fig. 2.1. A localized wave-packet $E_x(z-ct)$, travelling in the $+z$ direction with the velocity c.

example of an electric field of this type. Clearly it represents a wave packet which travels in the $+z$ direction with the velocity c.

An equally valid solution of Eq. (2.20) is

$$E_x(z, t) = g(z + ct), \tag{2.22}$$

where g is again any twice-differentiable function. This corresponds to a wave packet travelling in the negative z direction, but still with the velocity c. No other type of solution of (2.20) exists, and so the general solution can be written as a linear combination of the types (2.21) and (2.22), namely

$$E_x(z, t) = f(z - ct) + g(z + ct). \tag{2.23}$$

Using Eq. (2.3) we can see that the associated magnetic field is given by

$$\frac{\partial B_y}{\partial t} = -\frac{\partial E_x}{\partial z} = -f'(z-ct) - g'(z+ct),$$

which can be integrated to yield

$$B_y(z, t) = \frac{1}{c}[f(z-ct) - g(z+ct)]. \tag{2.24}$$

The magnetic field therefore also consists of two waves moving in opposite directions with the velocity c.

The velocity of light

Before considering other solutions of the wave equations (2.18) and (2.19) we should comment at this stage on the value of the velocity c. The precise measurement of this velocity has been the subject of many prolonged and exacting experimental investigations, and in earlier times any comprehensive treatment of the subject of electromagnetic radiation would have included a discussion of the measurement of the velocity of light in free space. Now, however, the numerical value of this velocity is close to being a defined, rather than a measured, quantity. The reason for this is that the units of length and time are both defined in terms of atomic properties, namely the wavelength of a transition in ^{86}Kr (605.7802105 nm) and the frequency of a transition in ^{133}Cs (9192.63177 MHz) respectively. Because the wavelength and frequency of a given transition are related by

$$\lambda \nu = c \tag{2.25}$$

a measurement of c amounts only to a direct or indirect comparison of either the wavelengths or the frequencies of the krypton and caesium transitions. The *1973 Comité Consultatif pour la Définition du Mètre* have therefore expressed the hope that the presently accepted mean value of the velocity of light in free space, namely

$$c = 299\,792\,458 \text{ ms}^{-1}, \tag{2.26}$$

will be preserved, with any further small adjustments being absorbed into a redefinition of the metre. For most practical purposes it is of course sufficiently accurate to use the value 3×10^8 m/s for c.

Sinusoidal fields

A particular form of $f(z-ct)$, describing a continuous oscillation at a single frequency ν and wavelength λ, and having an amplitude E_0, is

$$E_x = E_0 \cos\left[2\pi\left(\frac{z}{\lambda} - \nu t\right)\right]. \tag{2.27}$$

The accompanying magnetic field can be deduced from Eq. (2.24). It is

$$B_y = \frac{1}{c} E_0 \cos\left[2\pi\left(\frac{z}{\lambda} - \nu t\right)\right]. \tag{2.28}$$

This electromagnetic wave is illustrated in Fig. 2.2. We shall usually find it more convenient to express this type of solution in a slightly different form, by using the *wavenumber* k, defined to be

$$k = 2\pi/\lambda, \tag{2.29}$$

in place of λ, and by using the *angular frequency* ω, where

$$\omega = 2\pi\nu, \tag{2.30}$$

in place of ν. The quantity ω is also known as the angular velocity or the cyclic frequency. We shall often refer to it simply as the frequency, when there is no ambiguity with the frequency ν. The velocity c is related to ω and k by

$$c = \nu\lambda = \omega/k. \tag{2.31}$$

Two other convenient modifications to Eq. (2.27) are to reverse the sign of the argument and to add a phase angle ϕ. Then we have

$$\mathbf{E} = \mathbf{i}_x E_0 \cos(\omega t - kz + \phi). \tag{2.32}$$

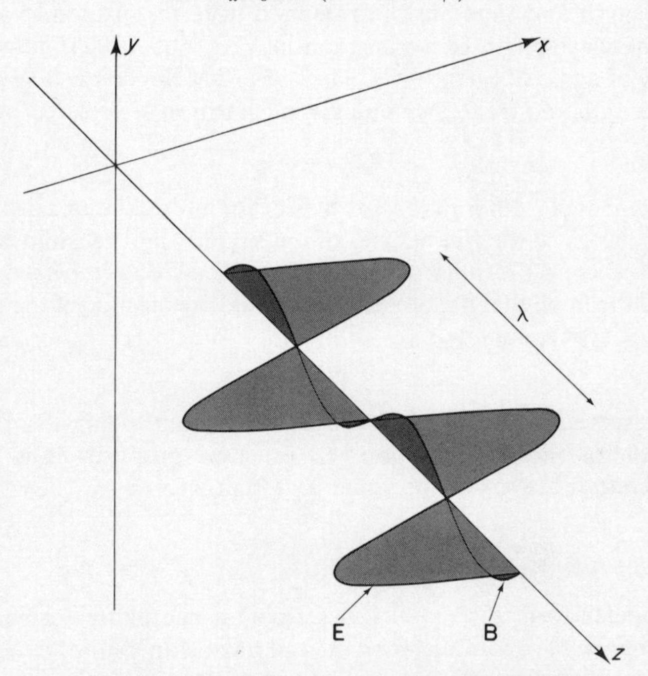

Fig. 2.2. The electromagnetic wave given by Eqs. (2.26) and (2.27).

The corresponding expression for waves travelling in the negative z direction is

$$\mathbf{E}=\mathbf{i}_x E_0 \cos(\omega t+kz+\phi). \tag{2.33}$$

Electric and magnetic fields are of course real, observable quantities, but the formulas involving them are often easier to handle if the fields are expressed in terms of complex quantities. In practice this means for example that for the purpose of calculation the electric field (2.32) is treated as the complex field

$$\mathbf{E}^{\text{complex}}=\mathbf{i}_x E_0\,\mathrm{e}^{\mathrm{j}(\omega t-kz+\phi)},$$

while at the end of the calculation it is remembered that the physically significant quantity is the real part of this,

$$\mathbf{E}=\mathrm{Re}\,(\mathbf{E}^{\text{complex}})=\mathbf{i}_x E_0 \cos(\omega t-kz+\phi).$$

The label 'complex' will be omitted when it is clear from the context which form of $\mathbf{E}$ is being used.

To see how waves travelling in other directions may be represented let us consider the electric field

$$\mathbf{E}(\mathbf{r},t)=\mathbf{E}_0\,\mathrm{e}^{\mathrm{j}(\omega t-\mathbf{k}\cdot\mathbf{r}+\phi)}. \tag{2.34}$$

Now the equation

$$\mathbf{k}\cdot\mathbf{r}=\text{constant}=kb, \quad \text{say,}$$

is the equation of a plane normal to the direction $\mathbf{k}$. The perpendicular distance from the origin to the plane is b. We see from the form of Eq. (2.34) that at a given time t the field $\mathbf{E}(\mathbf{r},t)$ is the same at all points in a plane of constant $\mathbf{k}\cdot\mathbf{r}$. Furthermore $\mathbf{E}$ has the same value at the time t in the plane at a distance b from the origin, as it has at the later time $t+\tau$ in the plane at a distance $b+c\tau$ from the origin, since

$$\omega(t+\tau)-k(b+c\tau)=\omega t-kb.$$

In other words Eq. (2.34) represents a plane wave travelling in the direction $\mathbf{k}$ with velocity c. The vector $\mathbf{k}$ is referred to as the *wave vector*. For the wave to be transverse $\mathbf{E}_0$ must of course be perpendicular to $\mathbf{k}$. As a check we can notice that Eqs. (2.32) and (2.33) are particular forms of Eq. (2.34), with $\mathbf{k}=+\mathbf{i}_z k$ and $-\mathbf{i}_z k$ respectively.

Polarized waves

An important feature of all the wave solutions that we have discussed so far is that the electric field exists in one transverse direction only. This type of wave is described as being *linearly polarized*. For example the waves represented by Eqs. (2.32) and (2.33) are linearly polarized in the x

direction (or they can be described as being linearly polarized in the xz plane). A more general direction of linear polarization is given by

$$\mathbf{E} = (\mathbf{i}_x E_{0x} + \mathbf{i}_y E_{0y})\, \mathrm{e}^{\mathrm{j}(\omega t - kz + \phi)}$$
$$= (\mathbf{i}_x \cos\theta + \mathbf{i}_y \sin\theta)(E_{0x}^2 + E_{0y}^2)^{1/2}\, \mathrm{e}^{\mathrm{j}(\omega t - kz + \phi)}, \tag{2.35}$$

where

$$\theta = \tan^{-1}(E_{0y}/E_{0x}),$$

and E_{0x} and E_{0y} are assumed to be real.

Linear polarization is not the only form of polarization that a plane wave may possess. Suppose for example that a wave travelling in the z direction has E_x and E_y components of the same magnitude, but with the E_y component having a phase $\pi/2$ greater than that of the E_x component,

$$\mathbf{E} = \mathbf{i}_x E_0\, \mathrm{e}^{\mathrm{j}(\omega t - kz + \phi)} + \mathbf{i}_y E_0\, \mathrm{e}^{\mathrm{j}(\omega t - kz + \phi + \pi/2)}.$$

This has the real part

$$\mathbf{E} = \mathbf{i}_x E_0 \cos(\omega t - kz + \phi) - \mathbf{i}_y E_0 \sin(\omega t - kz + \phi). \tag{2.36}$$

We see from this expression that $\mathbf{E}$ is of constant magnitude E_0 and lies in the (x, y) plane with direction given by $E_y/E_x = -\tan(\omega t - kz + \phi)$, i.e. rotating in the (x, y) plane. This rotation of the direction of polarization is illustrated in Fig. 2.3. Plane waves which behave in this way are said to be *circularly polarized.* If we were to observe the electric field vector by looking into the oncoming beam (that is, looking in the negative z direction, which is the perspective shown in Fig. 2.3) at a fixed z position, we would see that the tip

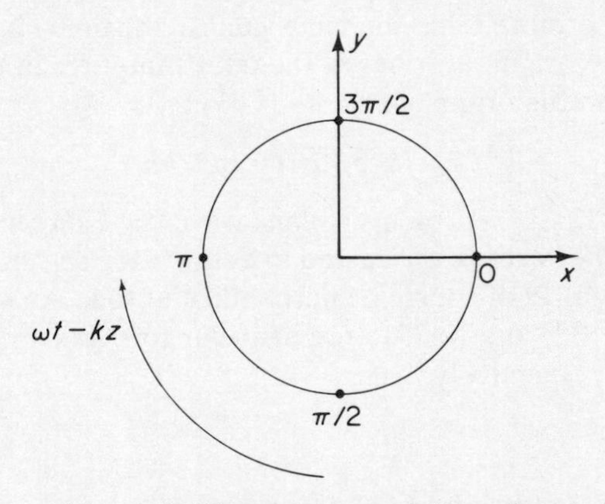

Fig. 2.3. The rotation of the plane of polarization of the circularly polarized plane wave represented by Eq. (2.36). The values of $\omega t - kz$ are indicated (and the phase ϕ is taken to be zero).

of the vector rotates in a clockwise sense as time increases. This particular wave is described as having *left-handed* circular polarization, and the opposite sense of rotation would of course be described as *right-handed.* This is a modern definition of the direction of rotation, and is the opposite of that often used in the past; it conforms with the convention concerning the direction of polarization of photons as we shall see in Section 2.6.

As well as considering the direction of **E** as a function of time at a fixed value of z, it is also instructive to 'freeze' the beam at a fixed time and to consider its dependence on z. This is illustrated in Fig. 2.4, which shows the frozen shape of a right-hand circularly polarized wave.

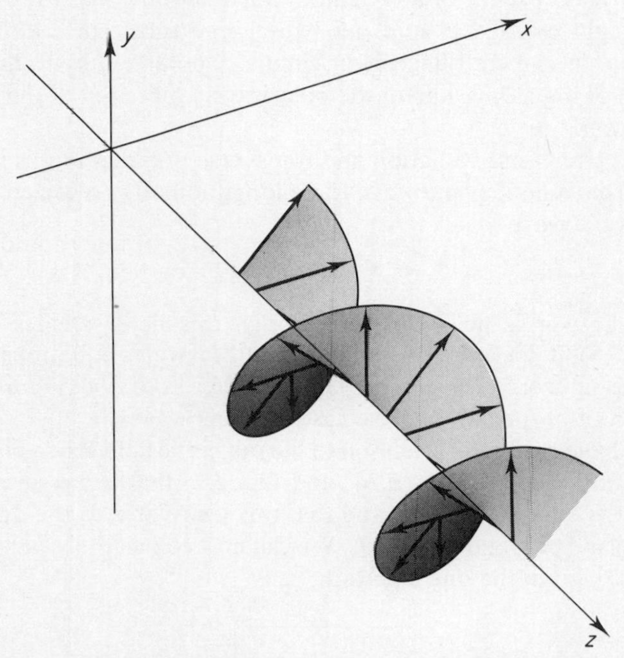

Fig. 2.4. The electric vector of a right-hand circularly polarized wave, at a fixed time t.

A more general form of polarization occurs when E_x and E_y have different amplitudes and when their relative phase is arbitrary but constant,

$$\mathbf{E} = \mathbf{i}_x E_{0x} \cos(\omega t - kz + \phi_x) + \mathbf{i}_y E_{0y} \cos(\omega t - kz + \phi_y).$$

The path traced out by the tip of the electric vector is then an ellipse, and waves of this sort are described as being *elliptically polarized.* Circular and linear polarization are clearly two special cases of elliptical polarization.

Yet another type of plane wave is that in which the plane of polarization changes randomly with time. This is called an *unpolarized* or *randomly polarized* plane wave. The changes in the plane of polarization are caused by random changes in the phases, and usually also the amplitudes, of the E_x and E_y components, and so we can write the electric vector in the form

$$\mathbf{E} = \mathbf{i}_x E_{0x}(t) \cos[\omega t - kz + \phi_x(t)] + \mathbf{i}_y E_{0y}(t) \cos[\omega t - kz + \phi_y(t)]. \tag{2.37}$$

An observer trying to measure the polarization of this type of wave would obtain a null result if the time taken to make the measurements is long compared with the time in which the amplitudes and phases vary (see Chapter 5). This absence of any definite polarization exists for example in the visible light emitted from a neon discharge tube, since although the individual atoms can emit linearly or circularly polarized light, the relative orientations of the atoms, and of the polarization direction of the light they emit, is random.

As well as transverse radiation and transverse polarization, is it possible for an electromagnetic plane wave to be longitudinally polarized? In other words, does the wave

$$\mathbf{E} = \mathbf{i}_z E_0 \cos(\omega t - kz)$$

exist? The answer is no, because although this field satisfies the wave equation (2.18), it does not satisfy the first of Maxwell's equations, (2.1): its divergence is not zero. The electric field $\mathbf{E}$ is therefore *always transverse* to the direction of propagation in the case of plane waves.

So far we have paid little attention to the magnetic field $\mathbf{B}$ in a plane wave, apart from noticing (see Eq. (2.24) and Fig. 2.2) that it has an amplitude E_0/c, that it is in phase with $\mathbf{E}$, and that it is transverse to the direction of propagation and perpendicular to $\mathbf{E}$. We can now collect these facts together and express them in the one equation

$$\boxed{\mathbf{B} = \frac{1}{c}\mathbf{i}_k \wedge \mathbf{E}} \tag{2.38}$$

where $\mathbf{i}_k$ is a unit vector in the direction of propagation.

2.3 FLOW OF ENERGY IN PLANE WAVES IN FREE SPACE

A propagating electromagnetic wave carries energy which can in principle be totally absorbed and converted into other forms of energy. To find the amount of energy available we consider plane linearly polarized radiation

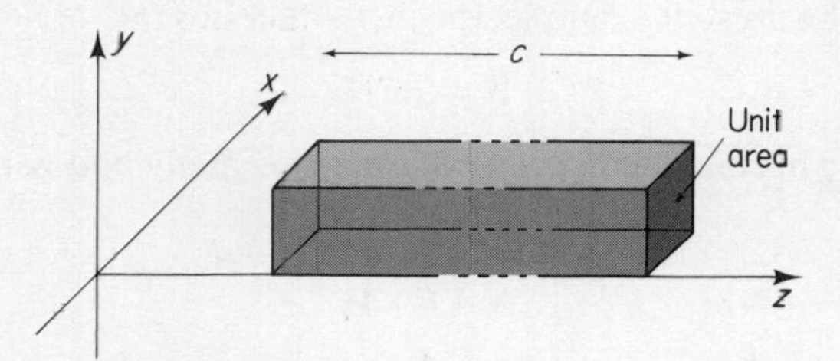

Fig. 2.5. A box of unit cross-sectional area and length c, containing electromagnetic radiation propagating in the z direction.

flowing through the box shown in Fig. 2.5, with

$$\mathbf{E}=\mathbf{i}_x E_0 \cos(\omega t-kz), \qquad \mathbf{B}=\mathbf{i}_y \frac{E_0}{c}\cos(\omega t-kz).$$

The energy $\mathrm{d}U$ contained in an infinitesimal volume $\mathrm{d}\tau$ of an electromagnetic field is given by

$$\mathrm{d}U=\tfrac{1}{2}(\mathbf{E}\cdot\mathbf{D}+\mathbf{B}\cdot\mathbf{H})\mathrm{d}\tau, \tag{2.39}$$

and therefore the total energy present in the box shown in Fig. 2.5 at time t is

$$\begin{aligned}U&=\tfrac{1}{2}\int_{x=0}^{1}\int_{y=0}^{1}\int_{z=0}^{c}\left(\varepsilon_0 E_0^2+\frac{1}{c^2\mu_0}E_0^2\right)\cos^2(\omega t-kz)\,\mathrm{d}x\,\mathrm{d}y\,\mathrm{d}z\\&=\varepsilon_0 E_0^2\left[\frac{z}{2}-\frac{\sin[2(\omega t-kz)]}{4k}\right]_{z=0}^{c}\\&=\tfrac{1}{2}\varepsilon_0 E_0^2 c-\frac{\varepsilon_0 E_0^2}{4k}\{\sin[2(\omega t-kz)]-\sin 2\omega t\}.\end{aligned}$$

The time average of the second term of this is zero (and in any case this term is negligibly small compared with the first term because k^{-1} $(=\lambda/2\pi)$ is very much smaller than the length c). Therefore the average energy in the box is

$$\bar{U}=\tfrac{1}{2}\varepsilon_0 E_0^2 c=\tfrac{1}{2}E_0 H_0.$$

This energy flows through each end face of the box in unit time and therefore it represents, for a plane wave, the mean flow of energy per unit time and per unit area.

The flow of energy is not smooth because the energy is localized near the antinodes shown in Fig. 2.2. The rate at which energy actually flows through an end face at the time t is given by the instantaneous value of the product EH at the end face. Since the flow of energy is a vector quantity, lying in the

z direction in the present example, the instantaneous rate of flow of energy is

$$\mathbf{N} = \mathbf{i}_z EH$$

per unit area. This may be expressed more succinctly and generally as

$$\boxed{\mathbf{N} = \mathbf{E} \wedge \mathbf{H}} \tag{2.40}$$

The important quantity $\mathbf{N}$ is known as *Poynting's vector.* We have shown here that it represents the rate of flow of energy in a plane wave, but Eq. (2.40) applies generally to the energy flow of electromagnetic radiation (e.g. for spherical waves). In terms of the electric field strength alone, the time-averaged flow of power per unit area is

$$\bar{\mathbf{N}} = \varepsilon_0 c \overline{E^2} \mathbf{i}_k = \tfrac{1}{2} \varepsilon_0 c E_{\max}^2 \mathbf{i}_k. \tag{2.41}$$

2.4 PROPAGATION OF PLANE WAVES IN DIELECTRIC AND CONDUCTING MEDIA

Propagation in dielectric media

In the case of a non-magnetic and non-conducting dielectric medium the wave equations (2.18) and (2.19) become

$$\nabla^2 \mathbf{E} = \frac{1}{v^2} \frac{\partial^2 \mathbf{E}}{\partial t^2} \tag{2.42}$$

$$\nabla^2 \mathbf{B} = \frac{1}{v^2} \frac{\partial^2 \mathbf{B}}{\partial t^2} \tag{2.43}$$

where the propagation velocity v is now given by

$$v = (\varepsilon \varepsilon_0 \mu_0)^{-1/2} = c(\varepsilon)^{-1/2}.$$

The quantity

$$\boxed{n = \varepsilon^{1/2} = \frac{c}{v}} \tag{2.44}$$

is known as the *refractive index* of the medium.

The wave equation (2.42) possesses plane wave solutions such as those given by Eqs. (2.35), (2.36), and (2.37), but the relationship between the angular velocity and wavenumber is now

$$\omega/k = v = c/n, \tag{2.45}$$

and the wavelength in the dielectric is

$$\lambda = \frac{2\pi}{k} = \frac{\lambda_0}{n}, \tag{2.46}$$

where λ_0 is the wavelength in vacuum. The associated magnetic field is given by the analogue of Eq. (2.38), namely

$$\mathbf{B} = \frac{1}{v}\mathbf{i}_k \wedge \mathbf{E} = \frac{n}{c}\mathbf{i}_i \wedge \mathbf{E}, \tag{2.47}$$

and Poynting's vector is still given by Eq. (2.40).

Propagation in conducting media

On the other hand the behaviour of electromagnetic waves in conducting media is entirely different from that in free space. The difference arises because propagation in free space (or in a dielectric medium) depends on the existence of the displacement current, whereas in conducting media the propagation characteristics are determined by the much larger free conduction currents.

The treatment may be simplified by assuming that the conducting medium has a constant *conductivity* σ, defined by

$$\mathbf{j}_f = \sigma\mathbf{E}. \tag{2.48}$$

Equations (2.5), (2.6), and (2.7) are then still the appropriate forms for the first three of Maxwell's equations, but the fourth becomes

$$\text{curl}\,\mathbf{H} = \frac{\partial \mathbf{D}}{\partial t} + \sigma\mathbf{E}.$$

Equation (2.17) then changes to

$$\text{curl curl}\,\mathbf{E} = -\mu\mu_0\frac{\partial}{\partial t}\left(\frac{\partial \mathbf{D}}{\partial t} + \sigma\mathbf{E}\right),$$

and the wave equation (2.18) becomes

$$\nabla^2\mathbf{E} = \varepsilon\varepsilon_0\mu\mu_0\frac{\partial^2\mathbf{E}}{\partial t^2} + \mu\mu_0\sigma\frac{\partial \mathbf{E}}{\partial t}. \tag{2.49}$$

The extra, second term on the right-hand side of this equation is usually much larger than the first. We can find the relative magnitude by assuming that $\mathbf{E}$ has the time dependence exp $(\mathrm{j}\omega t)$. Then the ratio of the displacement term to the conduction term is

$$\left(\mathrm{j}\omega\varepsilon\varepsilon_0\mu\mu_0\frac{\partial\mathbf{E}}{\partial t}\right)\Big/\left(\mu\mu_0\sigma\frac{\partial\mathbf{E}}{\partial t}\right)=\frac{\mathrm{j}\omega\varepsilon\varepsilon_0}{\sigma}.$$

For example in the case of copper (for which $\sigma \approx 6\times10^7\ \Omega^{-1}\ \mathrm{m}^{-1}$, and $\varepsilon \approx 1$), this ratio varies from $\sim10^{-12}$ for long radio waves to $\sim10^{-3}$ for visible light. It approaches unity only at wavelengths which are so short ($\lesssim 30$ nm) that our assumption of a uniform conductor (having uniform values of σ and ε) is no longer appropriate (and in this case we have to treat the problem by the methods described in Chapter 4). When our present assumptions are reasonable we can neglect the displacement current, in which case Eq. (2.49) simplifies to

$$\nabla^2\mathbf{E}=\mu\mu_0\sigma\frac{\partial\mathbf{E}}{\partial t}. \tag{2.50}$$

A suitable plane wave solution to this, as can be verified by back-substitution, is

$$\mathbf{E}=\mathbf{E}_0\,\mathrm{e}^{\mathrm{j}(\omega t-z/\delta)}\,\mathrm{e}^{-z/\delta}, \tag{2.51}$$

where

$$\delta=\left(\frac{2}{\mu\mu_0\sigma\omega}\right)^{1/2}. \tag{2.52}$$

This solution is illustrated in Fig. 2.6.

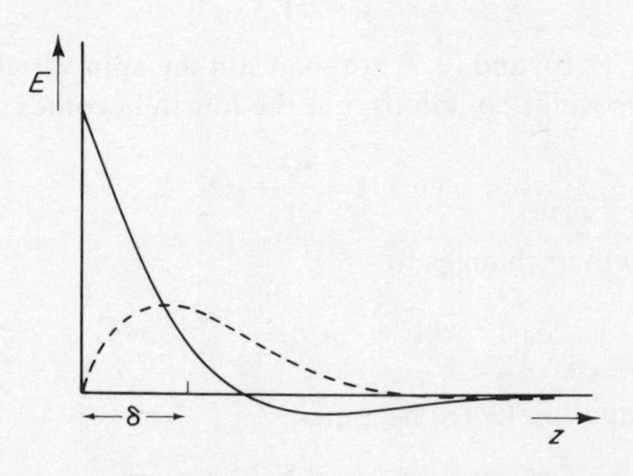

Fig. 2.6. Form of a plane wave inside a conducting medium, showing the strong attenuation with the penetration distance z. The full line shows cos (z/δ) exp $(-z/\delta)$ and the broken line sin (z/δ) exp $(-z/\delta)$.

The quantity δ, having the dimensions of length, appears twice in the solution (2.51). Its first appearance is in the role of the wavelength

$$\lambda = 2\pi\delta = 2\pi\left(\frac{2}{\mu\mu_0\sigma\omega}\right)^{1/2},$$

implying a velocity

$$v = \frac{\omega\lambda}{2\pi} = \left(\frac{2\omega}{\mu\mu_0\sigma}\right)^{1/2}.$$

Note that these values of λ and v are entirely different from their free space values, in contrast to the situations in dielectrics, for which λ and v change only by the refractive index n. For radiation of frequency 1 MHz in copper for example, both λ and v are 6 orders of magnitude smaller than their free space values. This difference is essentially caused by the fact that the conduction current is 12 orders of magnitude greater than the displacement current at this frequency. The second appearance of δ is in the attenuation term $\exp(-z/\delta)$, where δ is now an attenuation length, and is usually known as the *skin depth*. It is the depth or length in which the amplitude of the radiation is reduced by a factor e. For the example used above, copper at 1 MHz, it has the value 0.06 mm.

An interesting example of this is the transmission of an alternating current along a solid cylindrical conductor (such as a power cable, or a connecting wire). If $\delta \ll R$, where R is the radius of the conductor, then the electric field which drives the current penetrates the surface of the conductor as shown in Fig. 2.6. The field, and hence the current itself, exists only in the surface region, to a depth $\sim\delta$. This is known as the *skin effect*. The effective conductivity of a unit length of the conductor can be shown to be

$$\sigma_{\text{eff}} = 2\pi R\delta\sigma,$$

which is the same as the conductivity that would be obtained for a uniform current flowing through a cylindrical shell of thickness δ. The central region of the conductor is therefore essentially inactive, and does not contribute to the overall conductivity.

2.5 REFLECTION AT BOUNDARIES

When an electromagnetic wave crosses a boundary from one region to another its velocity and wavelength change. So, in general, do the magnitudes, phases and directions of its electric and magnetic field components. The purpose of this section is to investigate these boundary effects. We deal only with boundaries between continuous media, and assume that the boundaries are planar (or, more exactly, that their radii of curvature are much greater than the wavelength of the radiation), and that they are sharply discontinuous (in the sense that any irregularities have dimensions much smaller than the wavelength of the radiation). These restrictions are usually unimportant however, and our results are valid for most practical purposes.

Boundary conditions

The four conditions that must be satisfied by the electromagnetic field at a boundary between media 1 and 2 are

$$D_{\perp}^{(2)} = D_{\perp}^{(1)} + \rho_s, \tag{2.53}$$

$$B_{\perp}^{(2)} = B_{\perp}^{(1)}, \tag{2.54}$$

$$E_{\parallel}^{(2)} = E_{\parallel}^{(1)}, \tag{2.55}$$

and

$$H_{\parallel}^{(2)} = H_{\parallel}^{(1)} + j_s. \tag{2.56}$$

Here $F_{\perp}$ signifies the component of $\mathbf{F}$ that is perpendicular to the plane of the boundary, and $F_{\parallel}$ any component lying in the plane of the boundary. These conditions are a direct consequence of Maxwell's equations. We see that $B_{\perp}$ and $E_{\parallel}$ are always continuous across the boundary, $D_{\perp}$ and $H_{\parallel}$ only if the free surface charge density ρ_s and the free surface current density j_s are zero, as will be the case in the applications that we shall consider.

Reflection at normal incidence

A simple example is that of the plane wave

$$\mathbf{E} = \mathbf{i}_x E_0 \, \mathrm{e}^{\mathrm{j}(\omega t - k_1 z)}$$

travelling in the $+z$ direction in medium 1 towards a boundary lying in the x–y plane at $z = 0$. Writing the reflected wave as

$$\mathbf{E}' = \mathbf{i}_x E_0' \, \mathrm{e}^{\mathrm{j}(\omega t + k_1 z)},$$

and the transmitted wave as

$$\mathbf{E}'' = \mathbf{i}_x E_0'' \, \mathrm{e}^{\mathrm{j}(\omega t - k_2 z)},$$

we see that Eq. (2.55) gives the condition

$$E_0 + E_0' = E_0''. \tag{2.57}$$

The associated magnetic fields can be obtained from Eq. (2.47). We find

$$\mathbf{B} = \mu_0 \mathbf{H} = \mathbf{i}_y \frac{n_1 E_0}{c} \, \mathrm{e}^{\mathrm{j}(\omega t - k_1 z)},$$

$$\mathbf{B}' = \mu_0 \mathbf{H}' = -\mathbf{i}_y \frac{n_1 E_0'}{c} \, \mathrm{e}^{\mathrm{j}(\omega t + k_1 z)},$$

and

$$\mathbf{B}'' = \mu_0 \mathbf{H}'' = \mathbf{i}_y \frac{n_2 E_0''}{c} \, \mathrm{e}^{\mathrm{j}(\omega t - k_2 z)},$$

where n_1 and n_2 are the refractive indices of the two media (and we continue to assume that $\mu = 1$ for both media). Equation (2.56) now gives a second condition connecting the electric field strengths, namely

$$n_1(E_0 - E_0') = n_2 E_0''. \tag{2.58}$$

Combining (2.57) and (2.58) we now find

$$\frac{E_0'}{E_0} = -\frac{n_2 - n_1}{n_2 + n_1}, \qquad \frac{E_0''}{E_0} = \frac{2n_1}{n_2 + n_1}.$$

The *reflection coefficient*, the ratio of reflected to incident intensities, is therefore

$$R = \left|\frac{E_0'}{E_0}\right|^2 = \left(\frac{n_2 - n_1}{n_2 + n_1}\right)^2. \tag{2.59}$$

To evaluate the *transmission coefficient*, the ratio of transmitted to incident intensities, we need to remember that the beam intensity is given by Poynting's vector, and depends on the velocity as well as on $|\mathbf{E}|^2$. We find

$$T = \frac{n_2 |E_0''|^2}{n_1 |E_0|^2} = \frac{4 n_1 n_2}{(n_2 + n_1)^2}. \tag{2.60}$$

No power is lost at the boundary, and so

$$T + R = 1.$$

Reflection at non-normal incidence

To find the reflection and transmission coefficients when the incident beam is not normal to the boundary, as in Fig. 2.7, we can write the incident, reflected and transmitted waves as

$$\mathbf{E} = \mathbf{E}_0 \, \mathrm{e}^{\mathrm{j}(\omega t - \mathbf{k}_1 \cdot \mathbf{r})},$$

$$\mathbf{E}' = \mathbf{E}_0' \, \mathrm{e}^{\mathrm{j}(\omega t - \mathbf{k}_1' \cdot \mathbf{r})},$$

and

$$\mathbf{E}'' = \mathbf{E}_0'' \, \mathrm{e}^{\mathrm{j}(\omega t - \mathbf{k}_2'' \cdot \mathbf{r})}.$$

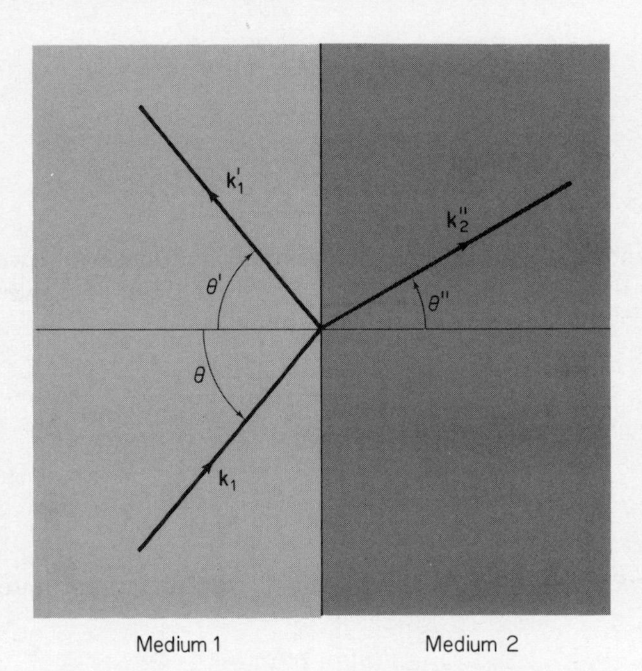

Fig. 2.7. The reflection and refraction of a plane wave at the boundary between two dielectric media.

For the directions shown in the figure,

$$\mathbf{k}_1 = k_1(\mathbf{i}_x \sin\theta + \mathbf{i}_z \cos\theta),$$

$$\mathbf{k}_1' = k_1(\mathbf{i}_x \sin\theta - \mathbf{i}_z \cos\theta),$$

and

$$\mathbf{k}_2'' = k_2(\mathbf{i}_x \sin\theta'' + \mathbf{i}_z \cos\theta'').$$

By taking the angle of reflection (θ') equal to the angle of incidence (θ) we have ensured that $\mathbf{E}$ and $\mathbf{E}'$ have the same dependence on x (so that the boundary conditions can be satisfied simultaneously at all x). To give $\mathbf{E}''$ the same dependence we need

$$k_1 \sin\theta = k_2 \sin\theta,$$

which can be expressed (using Eq. (2.45)) in a more familiar form, namely *Snell's law*

$$n_1 \sin\theta = n_2 \sin\theta''. \qquad (2.61)$$

The relationships between $\mathbf{E}_0$, $\mathbf{E}_0'$, and $\mathbf{E}_0''$ can now be found by substituting these forms of $\mathbf{E}$, $\mathbf{E}'$, and $\mathbf{E}''$, with $t = x = z = 0$ for convenience, into the boundary equations. The result depends on the plane of polarization of the incident wave. When it is polarized in the plane of reflection (the xz plane) we find

$$\frac{E_0'}{E_0} = -\frac{n_2 \cos\theta - n_1 \cos\theta''}{n_2 \cos\theta + n_1 \cos\theta''}, \qquad \frac{E_0''}{E_0} = \frac{2n_1 \cos\theta}{n_2 \cos\theta + n_1 \cos\theta''}. \tag{2.62}$$

When the polarization is perpendicular to the plane of reflection,

$$\frac{E_0'}{E_0} = -\frac{n_2 \cos\theta'' - n_1 \cos\theta}{n_2 \cos\theta'' + n_1 \cos\theta}, \qquad \frac{E_0''}{E_0} = \frac{2n_1 \cos\theta}{n_2 \cos\theta'' + n_1 \cos\theta}. \tag{2.63}$$

These last four relationships are known as *Fresnel's formulas*. Figure 2.8 shows the values of E_0'/E_0 for the two directions of polarization (for the case $n_2/n_1 = 1.5$), together with the corresponding values of the reflection coefficient R. At normal incidence $R_\perp$ and $R_\parallel$ are of course the same, and are given by Eq. (2.59).

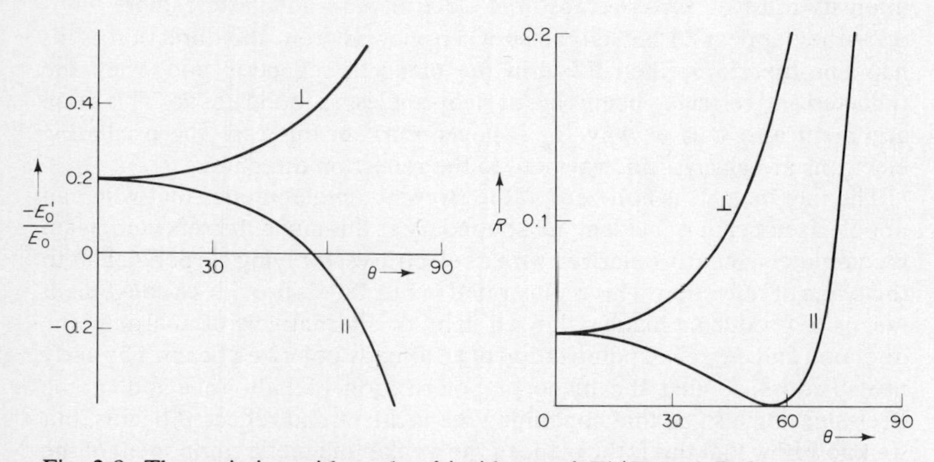

Fig. 2.8. The variation with angle of incidence of E_0'/E_0 and R, for an air–glass interface with $n_2/n_1 = 1.5$. Polarizations parallel and perpendicular to the plane of reflection are denoted by $\parallel$ and $\perp$ respectively.

The Brewster angle

We see from Fig. 2.8 and Eq. (2.62) that R becomes zero when the incident light is in the plane of reflection and when

$$n_2 \cos\theta = n_1 \cos\theta''.$$

Combining this with Snell's law gives

$$\sin 2\theta'' = \sin 2\theta,$$

and hence

$$\theta'' = \frac{\pi}{2} - \theta. \tag{2.64}$$

The reflected ray is therefore at right-angles to the refracted ray. The condition can also be expressed as

$$\tan\theta = \frac{n_2}{n_1}. \tag{2.65}$$

This particular angle of incidence is known as the *Brewster angle*. It is possible to understand physically why $R_{\parallel}$ is zero at this angle by considering what happens when a beam enters medium 2 from a vacuum ($n_1 = 1$). We can suppose that the electric fields cause the electrons of the medium to oscillate, and that these oscillating electrons act as the generators of the reflected and refracted beams. The direction of oscillation is the direction of $\mathbf{E}_0''$ in the medium. Now when the reflected beam is along this direction its intensity must be zero, because the electrons are unable to radiate 'end-on'—they appear to be stationary when viewed from this direction. This happens therefore when $\mathbf{E}_0''$ is in the plane of reflection and when the reflected and refracted beams are at right-angles, as found above. This same argument also tells us why $R_{\perp}$ is never zero: in this case the oscillating electrons are always 'sideways-on' to the reflection direction.

The fact that $R_{\perp}$ is non-zero at the Brewster angle implies that when an unpolarized beam is incident on a boundary at this angle the reflected beam is completely linearly polarized with its electric vector lying perpendicular to the plane of reflection. This is illustrated in Fig. 2.9. It provides a convenient means of producing plane polarized light, or alternatively of analysing the direction and degree of polarization of an already polarized beam. The early investigators defined the plane of polarization of light reflected at the Brewster angle to be that containing the incident and reflected beams, but we know now that this is the plane of the weaker magnetic component of the beam. It has therefore become conventional to redefine the plane of polarization as that of the stronger electric vector.

The difference between $R_{\perp}$ and $R_{\parallel}$ is made use of in Polaroid sunglasses. Much of the light received by the eye when facing the sun is sunlight that has been reflected from approximately horizontal surfaces. Because $R_{\perp} > R_{\parallel}$ the electric vector of this light is predominantly perpendicular to the plane of reflection and hence parallel to the Earth's surface. The sunglasses are therefore designed to filter out this direction of polarization.

Total internal reflection

Fresnel's formulas (2.62 and 2.63) apply equally well when $n_2 > n_1$ and when $n_2 < n_1$, with one important difference: in the latter case, when the

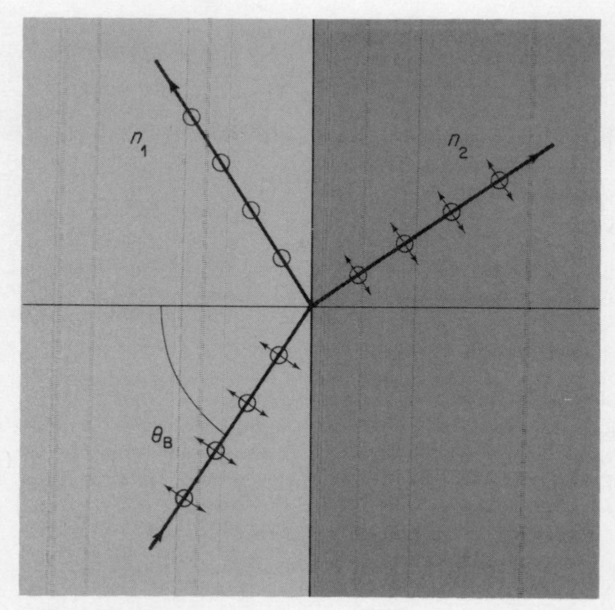

Fig. 2.9. Polarization in the plane of reflection is represented by ↕ and polarization perpendicular to this plane by ○. Only the perpendicular form is reflected at the Brewster angle.

wave is passing from a medium of higher refractive index to one of lower refractive index, no refracted beam exists for angles of incidence greater than the *critical angle* θ_c, given by

$$\sin\theta_c = \frac{n_2}{n_1}. \tag{2.66}$$

This is because, when θ is greater than θ_c, $\sin\theta''$ becomes greater than unity and the angle θ'' becomes an imaginary number. When this happens the electromagnetic field penetrates only a short distance ($\sim\lambda$) into the second medium and the time average of Poynting's vector is zero. As a consequence all the energy in the incident beam appears in the reflected beam. This is known as *total internal reflection*. The phenomenon is illustrated in Fig. 2.10.

Use is made of total internal reflection in the retro-reflector illustrated in Fig. 2.11. In the arrangement shown θ_c must be less than 45°, which means that n_2/n_1 must be greater than $\sqrt{2}$ (see Eq. (2.66)). Another familiar application is in the operation of 'optical fibres', in which light is transmitted down the length of a fibre by successive internal reflections at the walls. This is illustrated in Fig. 2.12. The walls must be smooth to much better than λ if many reflections are required. Bunches of optical fibres can be used as a light

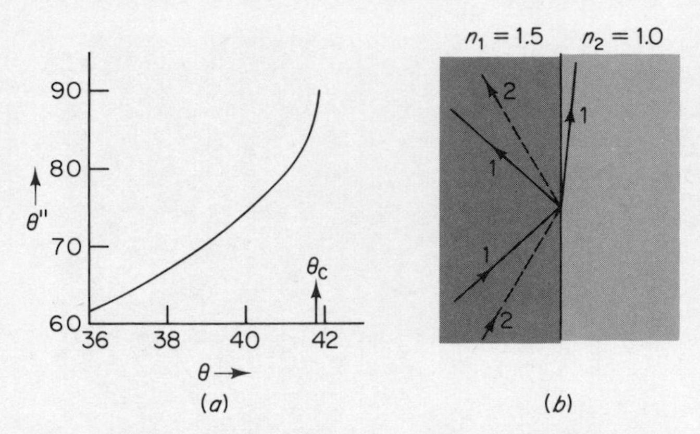

Fig. 2.10. The phenomenon of total internal reflection. Part (a) shows the dependence of the angle of refraction θ'' on the angle of incidence θ, near the critical angle θ_c. The ratio n_1/n_2 has been taken as 1.5. In part (b) the beams marked 1 and 2 have $\theta < \theta_c$ and $\theta > \theta_c$ respectively.

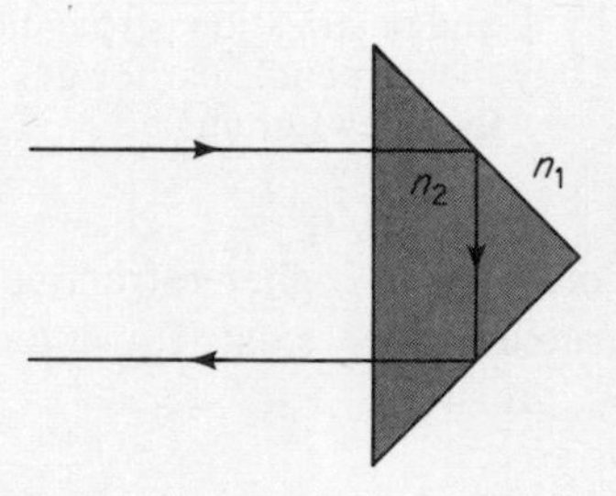

Fig. 2.11. The use of a prism to give complete reflection of an incident beam. $n_2 > n_1\sqrt{2}$.

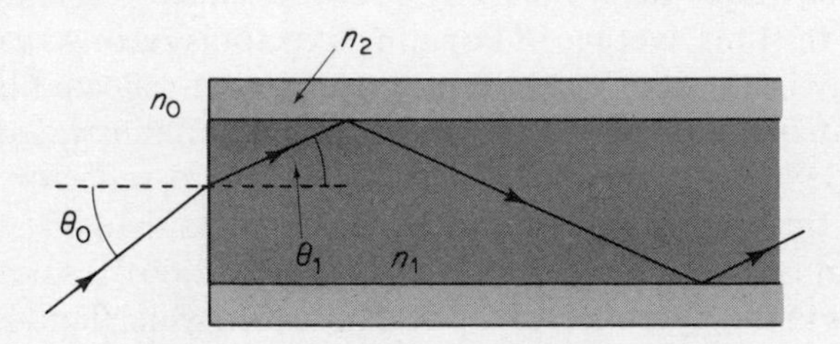

Fig. 2.12. Transmission of a light beam through an optical fibre. The refractive index of the sheath (n_2) is less than that of the core (n_1).

guide to illuminate or view an otherwise inaccessible region (such as the inside of a stomach, for example). If the fibres are too close to each other the nonpropagating electromagnetic field that penetrates a short distance into the outer medium will penetrate the next fibre and so can become a propagating field, causing light to pass between the fibres. This is called *frustrated total internal reflection.* To prevent this happening the individual fibres are usually sheathed by a material of lower refractive index, as shown in the figure (or a graded-index sheath is used). For a ray which passes through the axis of the fibre (assumed to have a cylindrical form), total internal reflection occurs at the sheath when

$$\cos \theta_1 > \frac{n_2}{n_1}$$

(see Eq. (2.66)). Individual fibres can be used also for the transmission of signals. For example quartz fibres of diameter $\sim 100\ \mu$m and having the extraordinarily low attenuation of 1 dB per km—less than city air!—have recently been developed as a possible replacement for conventional telephone lines (see Section 7.2.4).

Yet another use of total internal reflection is in the total reflection of X-rays off the *external* surfaces of crystals, when the angle of incidence is nearly 90° (i.e., for near-grazing incidence). This occurs when the refractive index of the crystal is less than unity (which is certainly possible in the X-ray region, see Section 3.2.2). This type of reflection is often termed *total external reflection.*

Reflection at a boundary between a dielectric and a conductor is entirely different from that at a boundary between two dielectrics. The velocity and wavelength of the light usually change by many orders of magnitude in passing from the dielectric to the conductor, and this large mis-match causes the radiation to be almost totally reflected at such a boundary. For example the reflection coefficient for an air–copper boundary differs from unity by only 10^{-6} at 1 MHz and by 10^{-2} at the infra-red frequency 10^{14} Hz. Conductors are therefore very good reflectors over a wide frequency range. At very high frequencies conduction becomes less important as a contribution to the interaction of electromagnetic radiation with matter, and eventually at frequencies in the optical range or above the correlation between conductivity and reflection coefficient ceases to hold.

2.6 RADIATION PRESSURE

When an electromagnetic wave strikes a boundary between two media it exerts a pressure, called the *radiation pressure*, on the boundary. The physical origin of this can be seen in simple classical terms when the boundary is between a dielectric (or vacuum) and a conductor, as depicted in

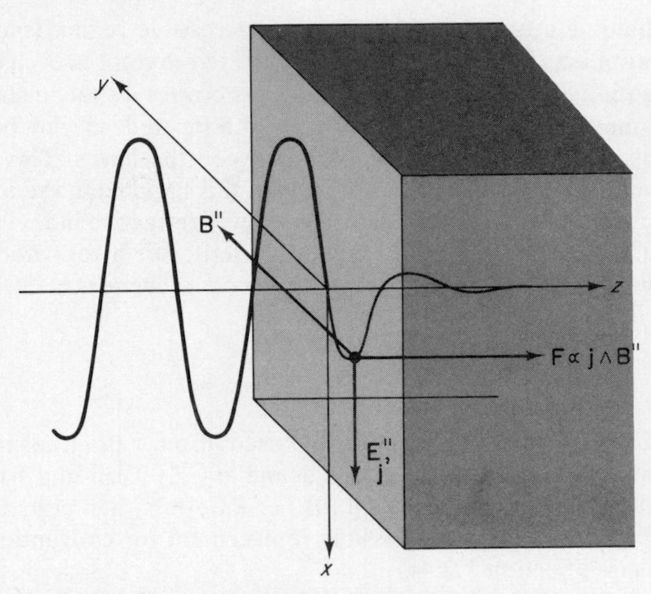

Fig. 2.13. The electric field $\mathbf{E}''$ of the transmitted beam acts on the charge carriers in the conductor to give a current density $\mathbf{j}$, and then the magnetic field $\mathbf{B}''$ reacts with this induced current to give an inwardly directed Lorentz-force $\mathbf{F}$.

Fig. 2.13. The transmitted electric field $\mathbf{E}''$ induces a current density

$$\mathbf{j} = \sigma \mathbf{E}''$$

in the conductor, and the interaction between this and the magnetic field $\mathbf{B}''$ results in an inwardly directed Lorentz force, as shown in the figure. The total force per unit area acting on the conductor can be shown to be

$$\mathbf{F} = \mathbf{i}_z \varepsilon \varepsilon_0 |E_0|^2. \tag{2.67}$$

This is the *radiation pressure* acting on the conductor.

This result can be re-derived and understood in a different way if we anticipate the description in Chapter 4 of light as photons, and regard the incident beam as a stream of photons. For light of frequency ν each photon has an energy

$$E = h\nu,$$

and a momentum

$$p = \frac{E}{c} = \frac{h}{\lambda},$$

where h is Planck's constant. The photons move at the velocity c, and the relationship $p = E/c$ applies of course to any particle moving at this velocity.

In a dielectric medium the photons travel at a velocity less than c, but the relationship $p = h/\lambda$ still applies. The number m of photons falling on the conductor per unit area and unit time is obtained from the time-averaged Poynting's vector,

$$m = \frac{\bar{N}}{h\nu}.$$

Assuming that all the photons are reflected at the surface the total change in momentum is

$$\Delta P = m\frac{2h}{\lambda} = \frac{2\bar{N}}{v} = \varepsilon\varepsilon_0 |E_0|^2,$$

where v is the velocity of propagation in the dielectric. Since this momentum change is the pressure on the surface, we regain Eq. (2.67).

This interpretation also enables us to calculate the radiation pressure when the incident beam is not normal to the surface or when one of the media is not a conductor. Whatever the medium or the direction, a beam of power $\bar{N}$ per unit area carries a momentum

$$P = \frac{\bar{N}}{h\nu} \cdot \frac{h}{\lambda} = \frac{n\bar{N}}{c}$$

per unit area and unit time, where n is the refractive index. To find the net pressure on the boundary one has only to consider the net change in the beam momentum.

We shall see in Chapter 4 that photons have intrinsic angular momenta, of value $h/2\pi$ $(=\hbar)$. In a right-hand circularly polarized beam the angular momentum of all the photons is in the direction of propagation. Therefore a beam of this type carries a total angular momentum

$$\mathbf{J} = \frac{\bar{N}}{h\nu}\hbar\mathbf{i}_k = \frac{\bar{N}}{\omega}\mathbf{i}_k \qquad (2.68)$$

per second and per unit area, where ω is the angular frequency of the radiation and $\mathbf{i}_k$ is a unit vector in the direction of propagation. A left-hand circularly polarized beam would have $\mathbf{J}$ in the direction $-\mathbf{i}_k$. The convention that a right-hand photon has its intrinsic angular momentum in the direction of motion is the usual convention for elementary particles (and this is why the older convention mentioned above for the direction of polarization of electromagnetic waves has been changed). The result (2.68) can also be obtained, but in a much more cumbersome manner, by the methods used above for the linear momentum, for example by considering the torque exerted by the electric field of the beam on electrons in a conductor.

The magnitude of the linear radiation pressure is usually small and inconsequential in laboratory experiments, except in the case of intense laser light (see Chapter 6, and problem 2.7). It is of vital significance however to the balance of forces acting within a star, since it provides an outwardly directed force to help counteract the inwardly directed gravitational attraction (see problem 2.9). Even the small radiation pressure exerted by our own sun at planetary distances may be useful as a means of propelling inter-planetary vehicles. This possibility has been seriously considered by NASA, for example. Figure 2.14 shows the radiation pressure acting on a 'solar sail', which would consist essentially of a thin membrane of large area ($\sim 1\ km^2$). The total force is small, but it can have an appreciable effect if it acts over a long period of time.

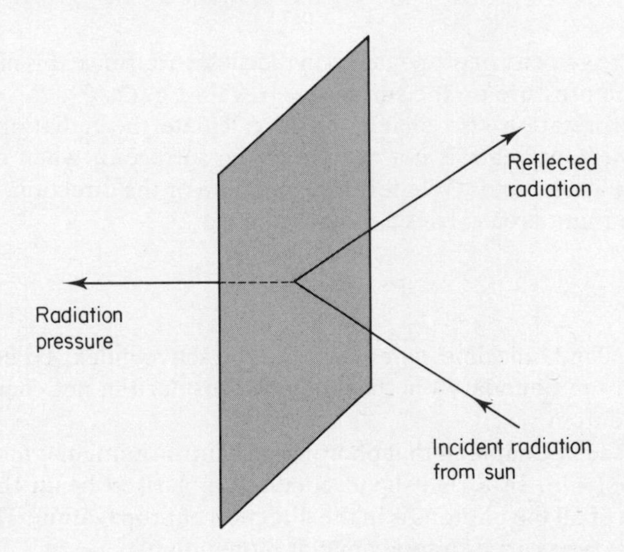

Fig. 2.14. The radiation pressure acting on a solar sail.

PROBLEMS: CHAPTER 2

(Answers to selected questions are given in Appendix E.)

2.1 Faraday's law of electromagnetic induction states that $\oint \mathbf{E}\, \mathrm{d}\mathbf{l} = -\mathrm{d}\phi/\mathrm{d}t$ where $\phi = \int \mathbf{B} \cdot \mathrm{d}\mathbf{S}$. State the analogous law of magneto-electric induction.

2.2 Show how Maxwell's equations would have to be modified if free magnetic poles (often referred to as magnetic monopoles) were to exist. To obtain a consistent set of units you may assume that the magnetic intensity in free space at a distance r from a magnetic pole of strength m is $\mathbf{H} = \mathbf{i}_r m/(4\pi\mu_0 r^2)$.

2.3 Estimate the electric and magnetic field strengths in sunlight at the Earth's surface, and compare these strengths with those of the DC electric and magnetic fields at the Earth's surface (which are of the order of 100 V/m and 5×10^{-5} W/m^2 respectively).

2.4 Show that a laser beam of diameter 1 mm and power 20 kW produces a maximum field strength of 4.4 MV/m.

2.5 The angle of incidence of a beam of light falling on a glass slab of refractive index 1.5 is put equal to the Brewster angle to remove from the reflected beam the component polarized in the plane of reflection. What is the allowed tolerance in the setting of the angle if a reflection coefficient of less than 1% is required for this polarization plane?

2.6 The rate of generation of heat in a long cylindrical wire carrying a current I is equal to I^2R per unit length where R is the resistance per unit length of the wire. Show that this joule heating can be described in terms of Poynting's vector as a flow of energy from the surrounding space.

2.7 A small hollow glass sphere of mass 1 mg is hit by a laser pulse of energy 300 J. Find the change in the velocity of the sphere if it absorbs all the laser light. Find the change in angular velocity of the sphere if it has a moment of intertia 10^{-12} kg m^2 and the laser light is circularly polarized and of wavelength 600 nm.

2.8 Show that the minimum wavelength of radio waves that can be received at a depth of 10 m below the surface of the sea is of the order of 10^5 m. The conductivity of sea water is approximately 4 Ω^{-1} m^{-1}.

2.9 A small black absorbing sphere of density 10^3 kg m^{-3} is situated in the vicinity of the sun. Find the radius of the sphere if the gravitational attraction of the sun is balanced by the pressure of radiation from the sun. The intensity of the sun's radiation at the surface of the Earth's atmosphere is 1350 W/m^2, and the distance of the Earth from the sun is 1.5×10^{11} m.

2.10 A copper rod is 10 m long and 6 mm in diameter. Show that its resistance at 100 Hz, 1 Mz and 10 GHz is approximately 6×10^{-3} Ω, 0.15 Ω and 15 Ω respectively.

CHAPTER 3

Classical treatment of the generation and interaction of electromagnetic waves

3.1 THE GENERATION OF ELECTROMAGNETIC WAVES

The fact that moving charges radiate was discovered experimentally by Heinrich Hertz. The apparatus that he used is shown schematically in Fig. 3.1. In his first experiments he took a copper wire bent into a rectangular shape, with the ends separated by a short air-gap, and connected it by another wire to a point on the discharge circuit of an induction coil. Sparks were seen to pass across the air-gap when the induction coil was operated. Later, he removed the connecting wire and noticed that sparks were still produced in the secondary circuit. He showed also that the velocity of propagation of the disturbance is approximately equal to the velocity of visible light.

It seems that seven years earlier than these discoveries of Hertz, D. E. Hughes had demonstrated a similar effect to a group including the President of the Royal Society, the 'Electrician' to the Post Office, and Sir George Stokes, but these gentlemen were unconvinced, which so discouraged Hughes that he apparently stopped this work and did not publish his results until twenty years later.

Before starting the mathematical treatment of the generation process, it is instructive to consider the simple example illustrated in Fig. 3.2. A charged particle is stationary at point A before time zero, and again after time τ, but between these two times it is moved a short distance from A to B and back again. The particle thus suffers some acceleration and deceleration during the interval from 0 to τ.

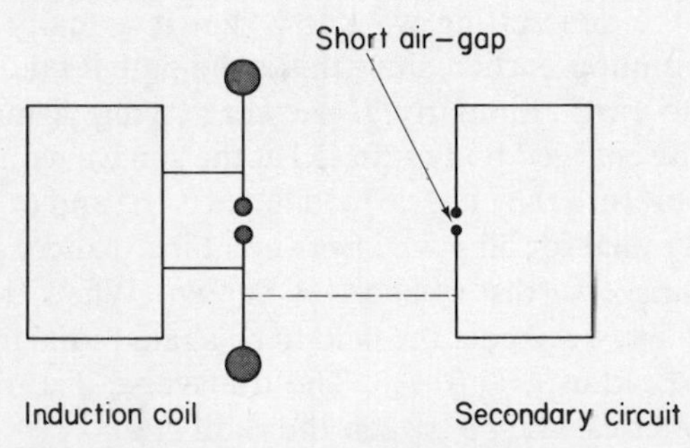

Fig.3.1. Schematic representation of the apparatus used by Hertz. A spark passes across the air gap in the secondary coil when the induction coil is operated. In the first experiments the two circuits were connected by a wire (shown as the broken line), but when this was removed the sparks were still seen.

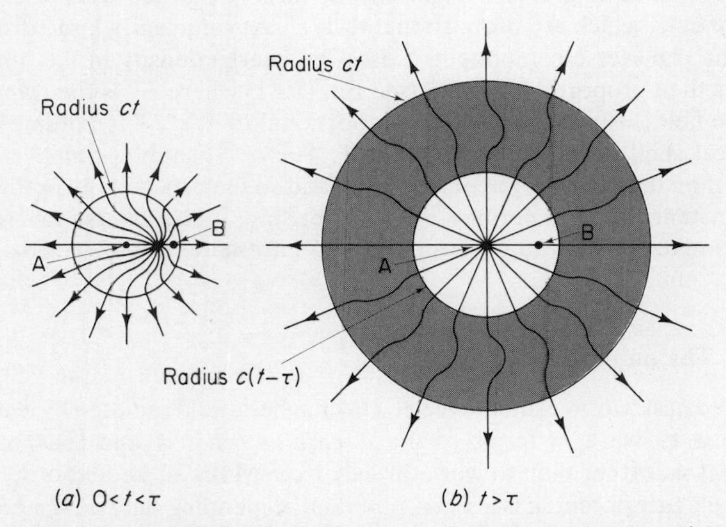

Fig. 3.2. A cross-section of the electric field of a charged particle which is at point A for times $t<0$, moves from A to B and back again between times 0 and τ, and then stays at point A for $t>\tau$. Part (*a*) shows the field at $0<t<\tau$, and part (*b*) at $t>\tau$. The shaded band between the spheres of radius $c(t-\tau)$ and ct contains transverse components, and it propagates outwards with velocity c.

For times $t<0$ the electric field is static and purely radial. After time 0 the field ceases to be radial everywhere, but we still expect it to be radial at distances greater than ct because information about the position of the charged particle cannot travel faster than the velocity of light. For example when we see the sun setting we know that it actually passed below the horizon eight minutes earlier, since this is the time it takes for light to travel from the sun to earth. Similarly, if we were trying to measure the electric field of a moving charged body situated at the sun we would expect the same time delay. Therefore the electric field in parts (a) and (b) of the figure have been drawn as radial for all $r>ct$. Between time 0 and τ the moving charge produces a transverse disturbance, as shown. When the charge is again stationary, at $t>\tau$, we expect the field to be again radial in the vicinity of the charge, up to a radius $r=c(t-\tau)$. The transverse disturbance is therefore limited to a spherical shell between the radii ct and $c(t-\tau)$. We see that it propagates radially outwards with the velocity c. The disturbance is of course an electromagnetic wave, with an associated transverse magnetic field.

As we shall see in Section 3.1 these ideas are substantiated by a rigorous treatment. A very important, and at first sight curious, feature which is not obvious from this qualitative example, is that at large distances the transverse components of the field are proportional to $1/r$. These components therefore have a greater magnitude at large distances than the radial components, which are proportional to $1/r^2$. At sufficiently large distances only the transverse components exist. The energy density in the outgoing wave is then proportional to E_{T}^2 (see Eq. (2.41) where E_{T} is the transverse electric field), which is therefore proportional to $1/r^2$. The volume of the spherical shell occupied by the wave is $4\pi r^2 c\tau$ (assuming that r is much larger than the thickness $c\tau$ of the shell), and so the total energy in the wave is a constant, which is perhaps not so surprising. It is of course this feature which makes possible the transmission of information by electromagnetic waves.

3.1.1 The potentials ϕ and $\mathbf{A}$

The easiest way to evaluate the electromagnetic field radiated by a moving charge is to work in terms of the *electric potential* ϕ and the *magnetic potential* $\mathbf{A}$, rather than to work directly from Maxwell's equations.

The electric potential is a scalar potential, depending only on the position $\mathbf{r}$ and the time t. For a static charge distribution having the density $\rho(\mathbf{r}')$ at the point $\mathbf{r}'$ the electric potential at $\mathbf{r}$ is

$$\phi(\mathbf{r})=\frac{1}{4\pi\varepsilon_0}\int\frac{\rho(\mathbf{r}')}{|\mathbf{r}-\mathbf{r}'|}\,\mathrm{d}\tau'. \tag{3.1}$$

The static electric field is then given by

$$\mathbf{E} = -\text{grad}\,\phi. \tag{3.2}$$

The magnetic potential **A** on the other hand is a vector potential, having three components at each point **r**. For a static system having a current density $\mathbf{j}_f$ it is defined by

$$\mathbf{A}(\mathbf{r}) = \frac{\mu_0}{4\pi} \int \frac{\mathbf{j}_f(\mathbf{r}')}{|\mathbf{r}-\mathbf{r}'|}\, d\tau'. \tag{3.3}$$

The magnetic field is then given by

$$\mathbf{B} = \text{curl}\,\mathbf{A}. \tag{3.4}$$

Equations (3.1) and (3.3) do not apply to time-dependent systems. For example, in the same way that we see on the earth what happened on the sun 8 minutes ago, so the potential at the point **r** due to a charge density at **r**′ depends on the value of this density at the previous time $|\mathbf{r}-\mathbf{r}'|/c$. Figure 3.3

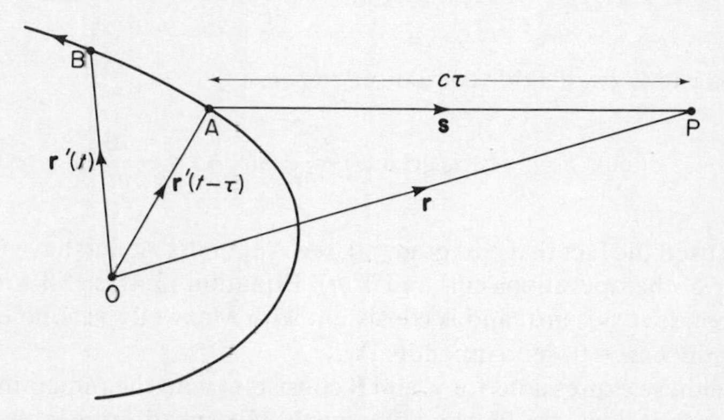

Fig. 3.3. The potential at the point P at time t is determined by the position A of the charge q at the earlier time $t-\tau$, where the distance AP is $c\tau = |\mathbf{r}-\mathbf{r}'(t-\tau)|$. B is the position of q at time t.

shows the same idea applied to a moving point charge q whose position at time t is $\mathbf{r}'(t)$. Here we take the simple point of view that information about the position of a charge q travels with the velocity of light, and that the potential at the position **r** at the time t is related to the position **r**′ of the charge at the earlier time $t-\tau$, i.e. it is determined not by $\mathbf{r}'(t)$ but by $\mathbf{r}'(t-\tau)$ where the distance AP in the figure is given by

$$s = |\mathbf{r}-\mathbf{r}'(t-\tau)| = c\tau. \tag{3.5}$$

We shall meet an example of this equation later.

For a distributed charge of density ρ the potential at position $\mathbf{r}$ and time t due to the part of the charge in the vicinity of $\mathbf{r}'$ depends on the value of ρ at the previous time $t-|\mathbf{r}-\mathbf{r}'|/c$. The potential due to the whole charge is therefore

$$\phi(\mathbf{r}, t)=\frac{1}{4\pi\varepsilon_0}\int\frac{\rho(\mathbf{r}', t-|\mathbf{r}-\mathbf{r}'|/c)}{|\mathbf{r}-\mathbf{r}'|}\,\mathrm{d}\tau'. \tag{3.6}$$

This is called a *retarded potential*, for reasons that are now obvious. The retarded form of the magnetic potential is similarly given by

$$\mathbf{A}(\mathbf{r}, t)=\frac{\mu_0}{4\pi}\int\frac{\mathbf{j}_\mathrm{f}(\mathbf{r}', t-|\mathbf{r}-\mathbf{r}'|/c)}{|\mathbf{r}-\mathbf{r}'|}\,\mathrm{d}\tau'. \tag{3.7}$$

Equation (3.2) also does not apply for time-dependent systems. It must be modified to

$$\mathbf{E}=-\operatorname{grad}\phi-\frac{\partial\mathbf{A}}{\partial t}. \tag{3.8}$$

This satisfies Maxwell's third equation, because

$$\operatorname{curl}\mathbf{E}=-\operatorname{curl}(\operatorname{grad}\phi)-\frac{\partial}{\partial t}(\operatorname{curl}\mathbf{A})=-\frac{\partial\mathbf{B}}{\partial t}$$

(we have used the fact that curl grad $\equiv 0$, see Appendix A, and have reversed the order of the operations curl and $\partial/\partial t$). Equation (3.4) is still correct for time-dependent systems, and is consistent with Maxwell's second equation (because div curl $\equiv 0$, see Appendix A).

But are these expressions for $\mathbf{E}$ and $\mathbf{B}$ consistent with the remaining two of Maxwell's equations (the first and the fourth)? By substituting in the first we obtain

$$(\varepsilon\varepsilon_0)^{-1}\operatorname{div}\mathbf{D}=-\nabla^2\phi-\frac{\partial}{\partial t}\operatorname{div}\mathbf{A}=(\varepsilon\varepsilon_0)^{-1}\rho_\mathrm{f}.$$

If we add $v^{-2}\,\partial^2\phi/\partial t^2$ to both sides, where v is the propagation velocity, this can be rearranged to give

$$\nabla^2\phi-\frac{1}{v^2}\frac{\partial^2\phi}{\partial t^2}=-\frac{\rho_\mathrm{f}}{\varepsilon\varepsilon_0}-\frac{\partial F}{\partial t},$$

where

$$F=\operatorname{div}\mathbf{A}+\frac{1}{v^2}\frac{\partial\phi}{\partial t}.$$

By substituting in the fourth of Maxwell's equations we find in a similar way that

$$\nabla^2\mathbf{A}-\frac{1}{v^2}\frac{\partial^2\mathbf{A}}{\partial t^2}=-\mu\mu_0\mathbf{j}_f+\text{grad}\,\mathbf{F}.$$

These equations are simplified if we put

$$\mathbf{F}=\text{div}\,\mathbf{A}+\frac{1}{v^2}\frac{\partial\phi}{\partial t}=0. \tag{3.9}$$

Then they become

$$\nabla^2\phi-\frac{1}{v^2}\frac{\partial^2\phi}{\partial t^2}=-\frac{\rho_f}{\varepsilon\varepsilon_0} \tag{3.10}$$

and

$$\nabla^2\mathbf{A}-\frac{1}{v^2}\frac{\partial^2\mathbf{A}}{\partial t^2}=-\mu\mu_0\mathbf{j}_f. \tag{3.11}$$

We see that the equations in ϕ and $\mathbf{A}$ are now de-coupled: Eq. (3.10) gives ϕ in terms of the charges only, and Eq. (3.11) gives $\mathbf{A}$ in terms of the currents only. The condition (3.9) which has achieved this is called the *Lorentz condition.* It can always be satisfied because of an inherent arbitrariness in the definitions of ϕ and $\mathbf{A}$.

The arbitrariness is illustrated by making the changes

$$\mathbf{A}\rightarrow\mathbf{A}'=\mathbf{A}+\text{grad}\,\chi \tag{3.12}$$

and

$$\phi\rightarrow\phi'=\phi-\frac{\partial\chi}{\partial t}, \tag{3.13}$$

where χ is an arbitrary function of $\mathbf{r}$ and t. The electric and magnetic fields remain unchanged, because

$$\mathbf{E}'=\mathbf{E}+\text{grad}\,\frac{\partial\chi}{\partial t}-\frac{\partial}{\partial t}\,\text{grad}\,\chi=\mathbf{E}$$

and

$$\mathbf{B}'=\mathbf{B}+\text{curl}\,(\text{grad}\,\chi)=\mathbf{B}.$$

Therefore either the potentials ϕ and $\mathbf{A}$, or the potentials ϕ' and $\mathbf{A}'$ can be used. It is always possible to find a function χ such that ϕ' and $\mathbf{A}'$ satisfy the Lorentz condition. The changes (3.12) and (3.13) are known as a gauge transformation, and potentials satisfying the Lorentz condition are said to belong to the *Lorentz gauge.*

It can be shown that the retarded potentials (3.6) and (3.7) are solutions of the decoupled equations (3.10) and (3.11). The set of Eqs. (3.6), (3.7), (3.9), (3.10), and (3.11) are therefore consistent with each other and with all of Maxwell's equations. They enable ϕ and $\mathbf{A}$ to be calculated from a knowledge of ρ_f and $\mathbf{j}_f$, and then $\mathbf{E}$ and $\mathbf{B}$ to be calculated from ϕ and $\mathbf{A}$. This procedure is usually much easier than attempting to evaluate $\mathbf{E}$ and $\mathbf{B}$ directly from Maxwell's equations without the aid of ϕ and $\mathbf{A}$.

3.1.2 The radiation from an oscillating dipole

A simple but important example of a radiating system is that illustrated in Fig. 3.4. It is an idealized system consisting of a conductor of length l

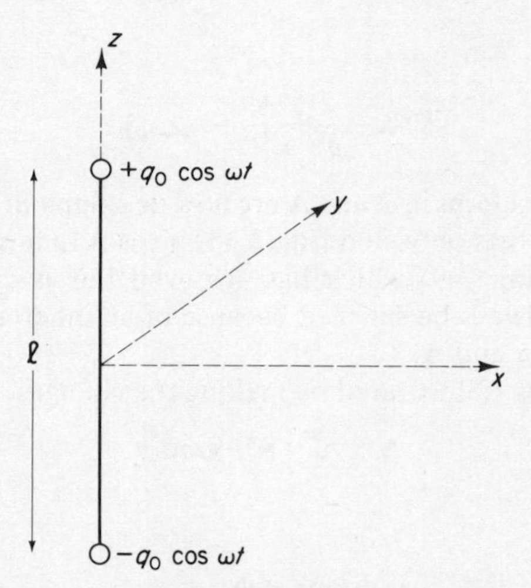

Fig. 3.4. The Hertzian dipole. The length l of the dipole is much smaller than the wavelength of the emitted radiation.

carrying small spheres at each of its ends. The charges $\pm q$ on the spheres vary sinusoidally with time,

$$q = q_0 \cos \omega t.$$

The instantaneous electric dipole moment of the system is therefore

$$\mathbf{p} = \mathbf{i}_z p_0 \cos \omega t,$$

where

$$p_0 = l q_0. \tag{3.14}$$

A particular form of oscillating electric dipole is that in which the length l is much smaller than the wavelength λ of the radiation from the dipole. This is called the *Hertzian dipole*. To simplify the calculations we shall go further and assume that l is vanishingly small, but with the product $lq_0 = p_0$ remaining finite.

The time-dependent charges give rise to a current I in the conductor, where

$$I = \frac{dq}{dt} = -I_0 \sin \omega t$$

and

$$I_0 = \omega q_0. \tag{3.15}$$

Thus the Hertzian dipole contains free charges and free currents, and so Eqs. (3.6) and (3.7) are both possible starting points in a calculation of the radiation field. Equation (3.7) provides the simpler approach for the present problem.

Since the length l is very small, the variable $\mathbf{r}'$ in the integrand of Eq. (3.7) can be set to zero. Integration over the volume element $d\tau'$ can be split up into an integration over the cross-sectional area of the conductor, the important part of which is

$$\int \mathbf{j}_f \, dx \, dy = \mathbf{I} = \mathbf{i}_z I,$$

together with an integration over the z co-ordinate, which gives the length l of the dipole. The result of this is

$$\mathbf{A}(\mathbf{r}, t) = -\mathbf{i}_z \frac{\mu_0 l I_0 \sin [\omega(t - r/c)]}{4\pi r}. \tag{3.16}$$

The magnetic field can now be found from Eq. (3.4). The operation curl is most conveniently performed in this case in terms of spherical co-ordinates (see Eqs. (A.3) and (A.10)). Only the terms in $\mathbf{i}_\phi$ are non-zero, giving

$$\begin{aligned}\mathbf{B} = \text{curl}\,\mathbf{A} &= \mathbf{i}_\phi \frac{1}{r}\left[\frac{\partial(rA_\theta)}{\partial r} - \frac{\partial A_r}{\partial \theta}\right] \\ &= \mathbf{i}_\phi \frac{\mu_0 l I_0 \sin\theta}{4\pi}\left\{-\frac{\omega \cos[\omega(t - r/c)]}{cr} - \frac{\sin[\omega(t - r/c)]}{r^2}\right\}.\end{aligned} \tag{3.17}$$

The second term on the right-hand side of Eq. (3.17) decreases more quickly with distance r than the first term. It is, apart from the presence of the retardation factor, the magnetic field that would be obtained by using Eq. (3.3). The more important term is the first, since it has the longer range and is the dominant term when $r \gg \lambda$. It is known as the *radiation field.* In terms of the dipole strength

$$p_0 = lq_0 = lI_0/\omega$$

it has the value

$$\mathbf{B}^{(\text{rad})} = -\mathbf{i}_\phi \frac{\mu_0 p_0 \omega^2 \sin\theta}{4\pi c r} \cos[\omega(t - r/c)]. \tag{3.18}$$

To find the electric component **E** of the radiation field we may assume that at large distances the outgoing radiation approximates to a plane wave over small areas. It follows that **E** is perpendicular to both **r** and **B**, and to be consistent with Eq. (2.38) it must have the form

$$\mathbf{E}^{(\text{rad})} = -\mathbf{i}_\theta \frac{\mu_0 p_0 \omega^2 \sin\theta}{4\pi r} \cos[\omega(t - r/c)]. \tag{3.19}$$

This can also be deduced by finding ϕ from Eq. (3.6) and then **E** from Eq. (3.8) (and then the shorter-range terms in r^{-2} and r^{-3} are also obtained). Note that the electromagnetic radiation is polarized in the plane of the electric dipole, as we might expect.

The dipole illustrated in Fig. 3.4 is not the only form of Hertzian dipole. Two charges with fixed magnitude $\pm q_0$ but with oscillating positions $\pm\mathbf{i}_z(l/2)\cos\omega t$ also give the oscillating electric dipole moment $\mathbf{i}_z p_0 \cos\omega t$ (with $p_0 = q_0 l$). The radiation field of this dipole is still given by Eqs. (3.18) and (3.19).

Figure 3.5 shows the form taken by **E** at a particular instant of time. Very near to the Hertzian dipole, in the region for which $kr \ll 1$, the instantaneous field is similar to that of a static dipole. This region is called the *static zone.* Further out the field lines start to form closed loops. The region in which this happens, for which $kr \sim 1$, is called the *intermediate zone.* As the loops move outwards, the radial components of the field diminish and the field becomes more and more transverse in character. Finally, in the *wave zone* or *radiation zone,* for which $kr \gg 1$, the radiation field given by Eqs. (3.18) and (3.19) is established. The radiation in this zone is known as *electric-dipole radiation.*

The field in the wave zone carries away energy from the Hertzian dipole. Using Eqs. (3.18) and (3.19) we see that Poynting's vector is

$$\mathbf{N} = \mathbf{E} \wedge \mathbf{H} = \mathbf{i}_r \frac{\mu_0 p_0^2 \omega^4 \sin^2\theta \cos^2[\omega(t - r/c)]}{16\pi^2 c r^2}.$$

This energy flow is directed radially outwards, and it has the $\sin^2\theta$ angular dependence shown in Fig. 3.6. Note that no radiation is emitted along the direction of oscillation of the dipole. The time-averaged energy flow in the

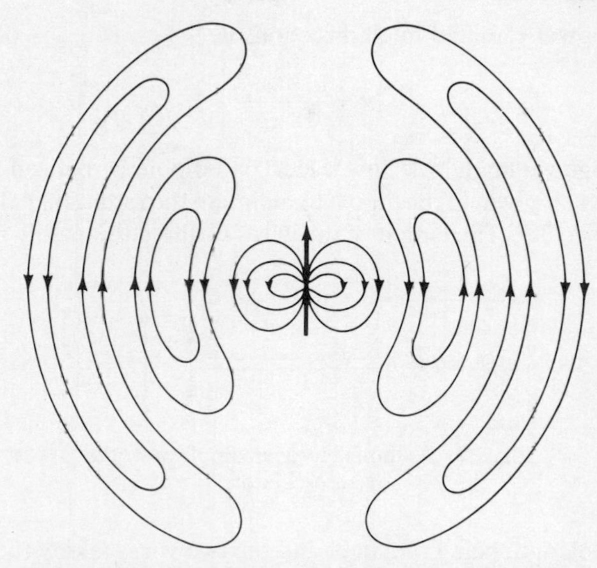

Fig. 3.5. The electric field of the Hertzian dipole, when $\omega t = 2n\pi$. The arrow at the origin shows the direction of the dipole moment. The field has cylindrical symmetry about the vertical axis through the dipole moment.

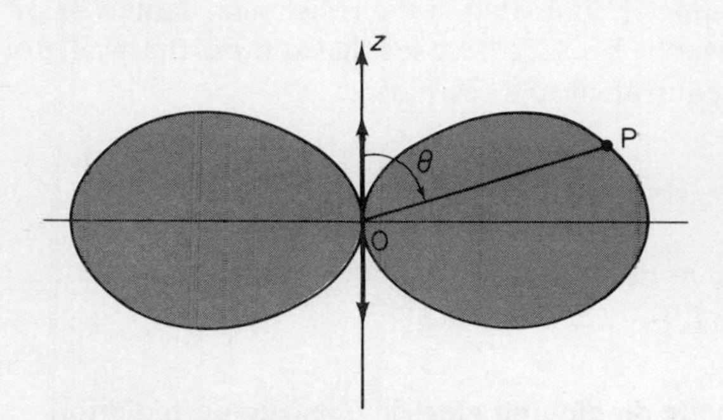

Fig. 3.6. Polar diagram showing the angular dependence of the energy flow for a Hertzian dipole. The power radiated in the direction θ is proportional to the length OP (and is independent of the azimuthal angle ϕ).

direction θ is

$$\bar{\mathbf{N}} = \mathbf{i}_r \frac{\mu_0 p_0^2 \omega^4 \sin^2 \theta}{32\pi^2 c r^2}. \tag{3.20}$$

By integrating this over the surface of a sphere of radius r we find that the total mean power radiated in all directions is

$$P = \frac{\mu_0 p_0^2 \omega^4}{12\pi c}. \tag{3.21}$$

At this stage we might ask how a Hertzian dipole is realized in practice, and how power is given to the dipole to maintain the radiation field. One way is shown in Fig. 3.7. The length of the wires forming the dipole is much less

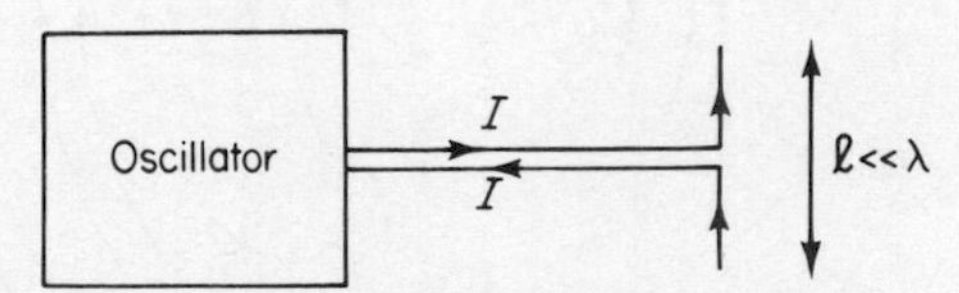

Fig. 3.7. A simple Hertzian dipole powered by an oscillator.

than the wavelength being radiated, and the two wires taking the current to and from the dipole are assumed to be close enough together that they do not add extra contributions to the radiation field. The power radiated by the dipole must of course be supplied by the oscillator circuit. If this were supplying the current I $(= -I_0 \sin \omega t)$ to a resistance R the power needed would be the power dissipated in the resistance, namely $\frac{1}{2}I_0^2 R$. By equating this to the power by Eq. (3.21) we see that as far as the oscillator is concerned the dipole is equivalent to a resistance

$$R = \frac{\mu_0 p_0^2 \omega^4}{6\pi c I_0^2} = \frac{2\pi}{3}\left(\frac{\mu_0}{\varepsilon_0}\right)^{1/2}\left(\frac{l}{\lambda}\right)^2. \tag{3.22}$$

This is known as the *radiation resistance* of the dipole. It has the numerical value $790(l/\lambda)^2\ \Omega$.

3.1.3 Magnetic dipole and electric quadrupole radiation

The radiation that we have studied in the previous section is the electric dipole radiation produced by an oscillating electric dipole. We shall see in Chapter 4 that it is associated with photons which carry away one unit ($\hbar$) of angular momentum, and therefore it is sometimes referred to as E1 radiation. Oscillating magnetic dipoles can also produce classical outgoing electromagnetic radiation, which is called *magnetic dipole radiation* (or M1 radiation). Suppose for example that the magnetic dipole has the form of a circular loop of area A, having its normal in the z direction and carrying an oscillating current $I_0 \cos \omega t$. The magnetic dipole moment is then

$$\mathbf{M} = \mathbf{i} M_0 \cos \omega t,$$

where

$$M_0 = AI_0. \tag{3.23}$$

Since the charge carriers are distributed evenly around the loop it does not have an electric dipole moment and so cannot emit E1 radiation. It can be shown that it produces the radiation field (see Jackson[6])

$$\mathbf{E}^{(\mathrm{M1})} = \mathbf{i}_\phi \frac{\mu_0 M_0 \omega^2 \sin\theta}{4\pi c r} \cos(\omega t - kr) \tag{3.24}$$

and

$$\mathbf{B}^{(\mathrm{M1})} = \mathbf{i}_\phi \frac{\mu_0 M_0 \omega^2 \sin\theta}{4\pi c^2 r} \cos(\omega t - kr). \tag{3.25}$$

The roles of **E** and **B** are thus interchanged with respect to the electric dipole field, and the direction of polarization of the radiation is normal to the direction of the magnetic dipole. The total mean power radiated is

$$P^{(\mathrm{M1})} = \frac{\mu_0 M_0^2 \omega^4}{24\pi c^3}. \tag{3.26}$$

If the current amplitude I_0 of the magnetic dipole oscillator is the same as that of the Hertzian dipole, the ratio of their powers becomes

$$\frac{P^{(\mathrm{M1})}}{P^{(\mathrm{E1})}} = 2\pi^2 \left(\frac{A}{\lambda l}\right)^2 \tag{3.27}$$

which is much less than unity if the dimensions of the dipoles are much less than λ. For this reason it is usually more efficient to transmit radio waves from an electric dipole aerial, rather than by means of a loop aerial. The same reasoning can also be applied qualitatively to the radiation emitted by excited atoms; because the emitted wavelengths are much longer than atomic dimensions, the radiation is usually electric dipole in character (see Chapter 4).

It is possible in principle to produce classical electromagnetic radiation which is neither E1 nor M1 radiation. For example the system of charges shown if Fig. 3.8 has a zero electric dipole moment and a zero magnetic

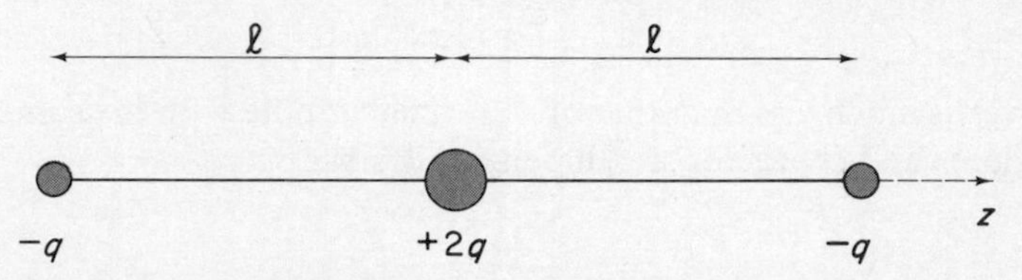

Fig. 3.8. A system of charges having a non-zero electric quadrupole moment. If the two charges $-q$ oscillate sinusoidally in the z direction, about the mean positions shown in the figure, the system emits electric quadrupole radiation.

dipole moment at all times. The system therefore cannot produce E1 or M1 radiation. The electric field produced by it is not zero however, even when the charges are stationary. The system can be regarded as two electric dipoles of equal but opposite moments placed end-to-end: because the dipole are not superimposed the resultant field is not zero. This type of system is said to have an *electric quadrupole moment.* Two dipoles placed side-by-side give a similar effect (see Problem 3.3). The magnitude of the electric quadrupole moment, for the charges shown in the figure, is

$$Q=\sum_i q_i(3z_i^2-r_i^2)=4ql^2$$

where $\mathbf{r}_i$ is the postion of the ith charge measured from the centre of the system (which in this case is the position of the fixed charge), and l is the distance between the fixed charge and either of the moving charges (for a definition of the electric quadrupole moment, see Jackson[6]). The oscillating system emits *electric quadrupole radiation.* Although the photons still have an intrinsic angular momentum of one unit ($\hbar$) they carry away extra angular momentum relative to the source, giving a total radiated angular momentum of up to 2 units per photon (see Chapter 4). Electric quadrupole radiation is therefore often called E2 radiation.

For a system having the oscillating quadrupole moment $Q_0 \cos \omega t$ (which can be achieved by a sinusoidal variation in the magnitude of the charges, rather than an oscillation of their positions), the radiation field is

$$\mathbf{E}^{(\mathrm{E2})}=-\mathbf{i}_\theta \frac{\mu_0\omega^3 Q_0 \sin\theta\cos\theta}{16\pi c r}\cos(\omega t-kr), \tag{3.28}$$

$$\mathbf{B}^{(\mathrm{E2})}=-\mathbf{i}_\phi \frac{\mu_0\omega^3 Q_0 \sin\theta\cos\theta}{16\pi c^2 r}\cos(\omega t-kr). \tag{3.29}$$

The power radiated at the angle θ to the z-axis is proportional to $\sin^2\theta\cos^2\theta$ (see Fig. 3.9), and the total mean power radiated is

$$P^{(\mathrm{E2})}=\frac{\mu_0\omega^6 Q_0^2}{960\pi c^3}. \tag{3.30}$$

This power is much less than that of a Hertzian dipole with the same charge amplitude q_0 and characteristic dimension l, since

$$\frac{P^{(\mathrm{E2})}}{P^{(\mathrm{E1})}}=\frac{\pi^2 Q_0^2}{20\lambda^2 p_0^2}=\frac{\pi^2}{5}\left(\frac{l}{\lambda}\right)^2. \tag{3.31}$$

Other more complicated forms of radiation also exist (E3, M2, M3, etc., see Section 4.5).

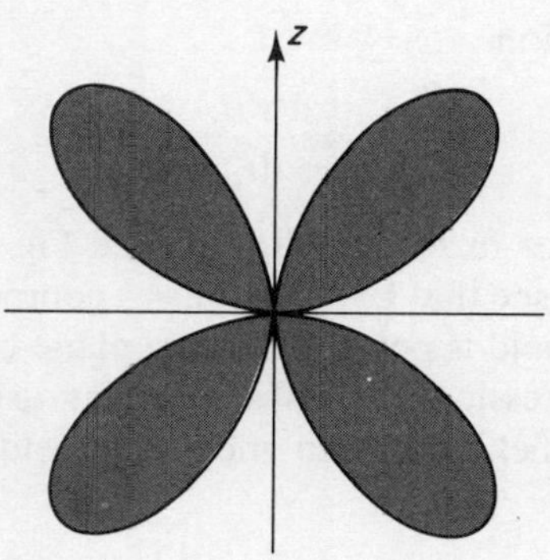

Fig. 3.9. Polar diagram of the power radiated by the oscillating electric quadrupole shown in Fig. 3.8. The power at the angle θ to the z-axis is proportional to $\sin^2\theta\cos^2\theta$ and is independent of the azimuthal direction ϕ.

3.1.4 The radiation from an accelerating charge

In Fig. 3.3 we saw that the electric field at a position **r** and time t is related to the earlier position $\mathbf{r}'(t-\tau)$ of the point charge producing the field, where

$$s=|\mathbf{r}-\mathbf{r}'(t-\tau)|=c\tau.$$

The 'retarded time' t_r at which the field at **r** and t is 'radiated' from the position $\mathbf{r}'$ is

$$t_r=t-\frac{s(t,t_r)}{c}. \tag{3.32}$$

Here we have written $s(t, t_r)$ for s to emphasize that it depends on t *and* t_r. As in the case of the field of a Hertzian dipole, the field radiated by an accelerating charge consists of short range fields (having the distance dependences s^{-2} and s^{-3}), together with a radiation field that varies with distance as s^{-1}. When the motion of the particle is non-relativistic (i.e., when its velocity is much less than c), the elastic and magnetic field strengths of this radiation field are given by (see Jackson[6])

$$\mathbf{E}=\frac{q}{4\pi\varepsilon_0c^2}\frac{\mathbf{i}_s\wedge(\mathbf{i}_s\wedge\mathbf{a})}{s} \tag{3.33}$$

and

$$\mathbf{B}=\frac{1}{c}\mathbf{i}_s\wedge\mathbf{E}. \tag{3.34}$$

Here **a** is the acceleration

$$\mathbf{a}=\frac{\mathrm{d}^2\mathbf{r}'(t_\mathrm{r})}{\mathrm{d}t_\mathrm{r}^2},$$

and $\mathbf{i}_s$ is the unit vector in the direction **s** (see Fig. 3.10). Since $(\mathbf{i}_s \wedge \mathbf{a})$ is perpendicular to **a**, we see that $\mathbf{i}_s \wedge (\mathbf{i}_s \wedge \mathbf{a})$ has a component parallel to **a** and therefore the electric field is polarized in the plane containing $\mathbf{i}_s$ and **a**.

An alternative expression that is sometimes useful (giving the same long-range radiation field but also including fields varying at s^{-2}, see Problem 3.4) is

$$\mathbf{E}=\frac{q}{4\pi\varepsilon_0 c^2}\frac{\mathrm{d}^2\mathbf{i}_s}{\mathrm{d}t^2} \tag{3.35}$$

(and **B** is still given by Eq. (2.38)).

Expressions (3.33) and (3.34) can be derived by first calculating the retarded potentials ϕ and **A** for the moving particle. These potentials are

$$\phi(\mathbf{r},t)=\frac{q}{4\pi\varepsilon_0 s}\frac{1}{1-\mathbf{i}_s\cdot\mathbf{v}/c} \tag{3.36}$$

and

$$\mathbf{A}(\mathbf{r},t)=\frac{\mu_0 q}{4\pi s}\frac{\mathbf{v}}{1-\mathbf{i}_s\cdot\mathbf{v}/c}, \tag{3.37}$$

where $\mathbf{v}$ $(=\mathrm{d}\mathbf{r}'(t_\mathrm{r})/\mathrm{d}t_\mathrm{r})$ is the velocity of the particle. These single particle retarded potentials are called the *Liénard–Wiechert potentials*. They apply to relativistic, as well as non-relativistic, values of v. The potential ϕ becomes the familiar Coulomb potential when v/c is very small.

Equation (3.35) is a convenient starting point for finding the power radiated by a non-relativistic accelerating charge of magnitude $\pm e$. If the angle between $\mathbf{i}_s$ and the acceleration **a** is θ, as in Fig. 3.10, the product $\mathbf{i}_s \wedge (\mathbf{i}_s \wedge \mathbf{a})$ has the magnitude

$$|\mathbf{i}_s \wedge (\mathbf{i}_s \wedge \mathbf{a})| = a \sin\theta.$$

Hence using expression (2.41) for Poynting's vector **N** we find that at the distance s,

$$N(\theta)=\varepsilon_0 c E^2=\frac{e^2 a^2 \sin^2\theta}{16\pi^2\varepsilon_0 c^3 s^2}.$$

This angular dependence is shown in the figure. The total power radiated is

$$P=\int N(\theta)s^2\sin\theta\,\mathrm{d}\theta\,\mathrm{d}\phi=\frac{e^2a^2}{6\pi\varepsilon_0 c^3}. \tag{3.38}$$

This is known as *Larmor's formula*.

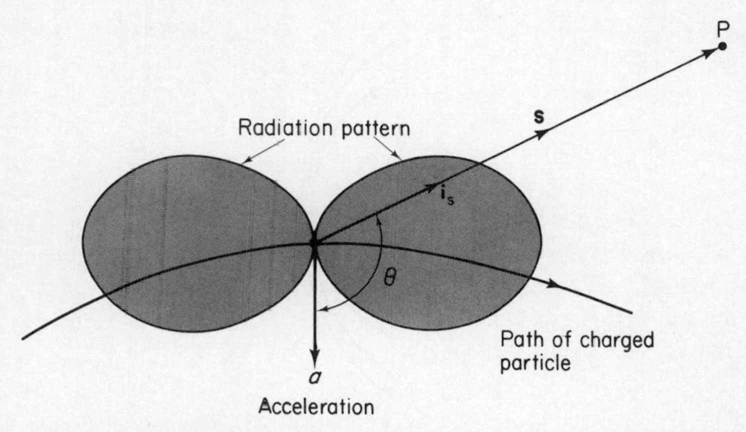

Fig. 3.10. The radiation pattern of a non-relativistic accelerating charge.

3.1.5 Synchrotron radiation

Another important example of radiation from an accelerating charge is the radiation emitted by a charged particle moving at a relativistic speed in a circular path. This type of motion occurs in the synchrotron accelerators used for producing or storing very energetic electrons. The particles are continually accelerated (by a magnetic field) towards the centre of their orbit, as shown in Fig. 3.11. Because the particles are travelling at nearly the velocity of the radiation that they generate, this radiation tends to form a shock-wave in the direction of motion of the particles. The observer therefore receives a brief pulse of radiation once in every revolution, when the particle (or bunch of particles) is moving directly towards him. Between these pulses the received radiation is much less intense. Other sources of synchrotron radiation are also known, for example when the trajectories of energetic cosmic ray particles are curved by inter-stellar magnetic fields.

We can obtain a qualitative understanding of the properties of synchrotron radiation by considering a specific case in which a particle of charge e moves at the velocity $0.999c$ (in the case of electrons, the energy is then 11.4 MeV), in an orbital radius R of 1 m and with a distance l of 30 m between the observer and the point b shown in the figure. For each angular position θ of the particle we know the time t_i ($= \theta R/v$, if the zero of time is taken when $\theta = 0$), and we can calculate the distance s between the particle and the observer. The time t_f at which the generated radiation reaches the observer (namely $t_f = t_i + s/c$) is therefore easily calculated, as is the direction of the unit vector $\mathbf{i}_s$ corresponding to this time. Table 3.1 shows some values of these quantities (only positive values are shown because t_i, t_f and $(\mathbf{i}_s)_z$ are all approximately symmetric about zero).

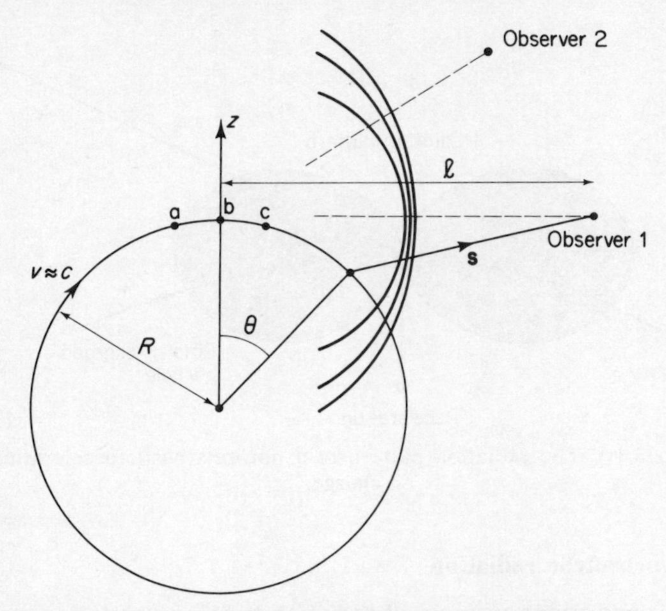

Fig. 3.11. Showing three wavefronts emitted when a relativistic charged particle is at the three positions marked a, b and c. The wavefronts arrive at observer 1 at nearly the same time, giving a short and intense pulse of radiation. Observer 2 does not see this 'pile-up' effect at this time. The figure also shows the quantities R, θ, s, and l used in Table 3.1.

Table 3.1 Values of the angular position θ of the particle shown in Fig. 3.11, the time t_i at which it has this position, the time t_f at which the radiated energy is received by an observer situated at a distance l from the point b of the orbit (see figure), and the z component of $\mathbf{i}_s$. The constant time l/c has been subtracted from t_f

θ	t_i	$t_f - \frac{l}{c}$	$(\mathbf{i}_s)_z$
0	0	0	0
1×10^{-2}	0.33×10^{-10}	0.34×10^{-13}	0.17×10^{-5}
2	0.67	0.71	0.67
3	1.00	1.15	1.50
4	1.33	1.69	2.67
5	1.67	2.36	4.17

We see from the table that the radiation is received in a time interval Δt_f that is much shorter than the interval Δt_i over which it was generated. This change in time scale is greatest at $t_i = 0$, when

$$\left(\frac{\mathrm{d}t_f}{\mathrm{d}t_i}\right)_{t_i=0} = 1 - \frac{v}{c} = 0.001.$$

The observer sees the direction of $\mathbf{i}_s$ change very quickly in his own time scale t_f. The acceleration $\mathrm{d}^2 i_{sz}/\mathrm{d}t_f^2$ is shown in Fig. 3.12. The radiation field is proportional to this acceleration (see Eq. (3.35)) and therefore the length τ of the pulse of radiation received by the observer is approximately 4×10^{-13} s. This is a much shorter time than the interval between the pulses, which is

$$T = \frac{2\pi R}{v} = 1.9 \times 10^{-8}\ \mathrm{s}.$$

The short pulses of radiation contain a wide range of frequencies and wavelengths. We can find the frequency distribution by Fourier analysing the pulse shown in Fig. 3.12. The amplitude of the component of frequency ν is proportional to

$$F(\nu) = \int_{-\infty}^{+\infty} f(t) \cos(2\pi\nu t)\, \mathrm{d}t, \tag{3.39}$$

where $f(t)$ is the time dependence shown in the figure (the cosine function is used because $f(t)$ is symmetric about $t = 0$). For frequencies that are much larger than $1/\tau$, the integrand of Eq. (3.39) goes through many oscillations in the time τ and $F(\nu)$ is small. On the other hand for small values of ν ($\ll 1/\tau$) the pulse is approximately equivalent to a delta function, the Fourier amplitudes of which are independent of ν. In the present example the frequency $1/\tau$ has a value of about 2.5×10^{12} Hz, corresponding to a wavelength of about 10^{-4} m. We deduce therefore that the synchrotron

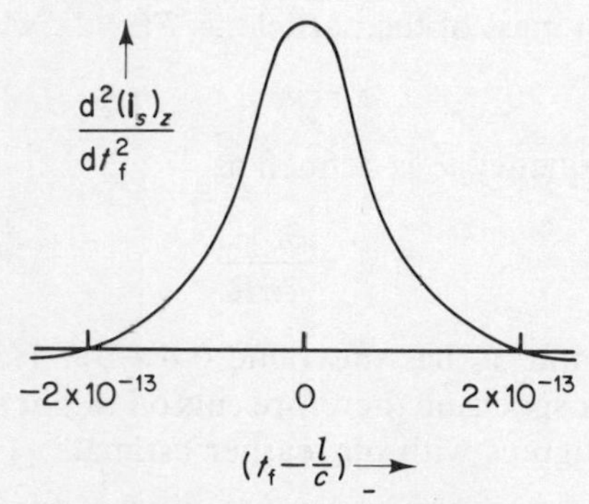

Fig. 3.12. The dependence of $\mathrm{d}^2 i_{sz}/\mathrm{d}t^2$ on t_f, for the example defined in Fig. 3.11.

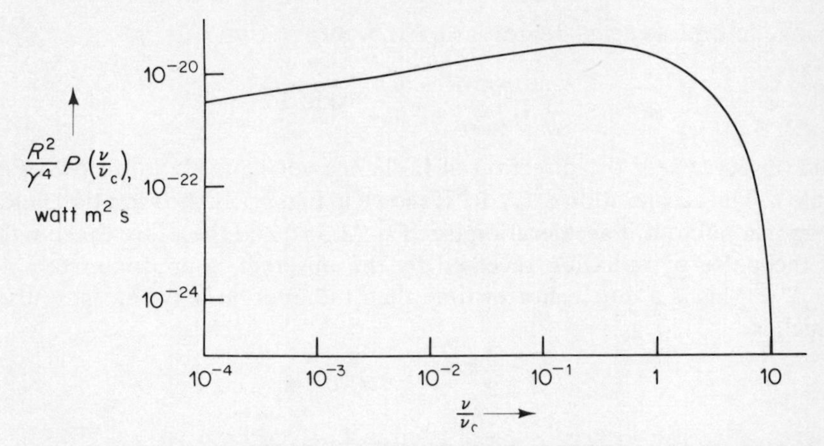

Fig. 3.13. $P(x)\,\mathrm{d}x$, where $x = \nu/\nu_c$, is the power radiated in the frequency interval from ν to $\nu + \mathrm{d}\nu$ by an electron orbiting with a radius R and an energy $\gamma m_0 c^2$. The critical frequency ν_c is $3c\gamma^3/(4\pi R)$. The power P has been multiplied by R^2/γ^4 to give a universal function. This curve has been derived from the work of D. H. Tomboulian and P. L. Hartman (*Physical Review*, **102** (1956), 1423), which was itself derived from the original work of J. Schwinger (*Physical Review*, **75** (1959), 1912).

radiation in this example is approximately 'white' (that is, has an approximately constant $F(\nu)$) for $\nu \lesssim 2.4 \times 10^{12}$ Hz ($\lambda \gtrsim 10^{-4}$ m) and is much less intense for higher frequencies (shorter wavelengths).

These conclusions are borne out by an exact calculation, which must take account of relativistic and quantal effects (see Jackson[6]), as shown in Fig. 3.13. The total power radiated in the frequency interval ν to $\nu + \mathrm{d}\nu$ by a single electron or proton of total energy

$$E = \gamma m_0 c^2,$$

where m_0 is the rest mass of the particle, is $P(x)\,\mathrm{d}x$ where

$$x = \nu/\nu_c,$$

and the 'critical' frequency ν_c is defined as

$$\nu_c = \frac{3c\gamma^3}{4\pi R}. \tag{3.40}$$

In the present example ν_c has the value 0.8×10^{12} Hz. We see from the figure that the power spectrum therefore cuts off rather sharply above about 3×10^{12} Hz, which agrees with our earlier estimate. The wavelength corresponding to ν_c is

$$\lambda_c = \frac{4\pi R}{3\gamma^3}. \tag{3.41}$$

The total power radiated over all frequencies is proportional to γ^4/R^2. In the limit of very large γ it is

$$P = \frac{e^2 c}{6\pi\varepsilon_0} \frac{\gamma^4}{R^2}, \tag{3.42}$$

which can be seen to be the Larmor power (Eq. (3.38), with $a = c^2/R$) multiplied by γ^4. This power can be rather large. For example the 2 GeV electron storage ring at Daresbury, England, has a bending radius of 5.56 m, and so each electron radiates the power 3.51×10^{-7} W, which causes it to lose an energy of 0.26 MeV per revolution if it is not continually re-accelerated. When the stored current is 1 A the total power that must be supplied to counteract the radiation loss is more than 0.25 MW. The value of λ_c for this storage ring is 0.39 nm, corresponding to $h\nu = 3.2$ keV; the emitted radiation therefore extends well into the hard X-ray region (see Fig. 1.1), which means that this and similar sources are extremely useful for a wide range of studies.

As mentioned before, synchrotron radiation is emitted when high energy electrons travel through interstellar magnetic fields. For an energy of 10^{10} eV and a field of 10^{-8} W m^{-2} for example, the orbital radius is approximately 3×10^{11} m, which on a cosmic scale is comparatively small. Equation (3.41) gives a critical wavelength of the order of 20 cm. Many such broad-band sources of microwaves and radio-waves have been found by radio-astronomers.

The qualitative reasoning based on Fig. 3.11 showed us that the radiation tends to be emitted in the forward direction, tangentially to the particle's orbit. More detailed calculations show that it is limited to a narrow cone of half-angle

$$\theta_{1/2} \sim \frac{m_0 c^2}{E}. \tag{3.43}$$

In the Daresbury storage ring for example this angle is 2.6×10^{-4} rad ($= 0.88'$). Another property of the radiation is that it is almost completely polarized in the plane of the orbit (see Eq. (3.33)).

3.1.6 Bremsstrahlung

A charged particle passing through a medium interacts in several different ways with the atoms and nuclei of the medium, one of which is that the particle is deflected by the Coulomb fields of the nuclei that it passes near to, as illustrated in Fig. 3.14. The acceleration of the particle causes radiation to be emitted, as in the case of synchrotron radiation. The emitted radiation is known as *bremsstrahlung* (from the German 'braking radiation').

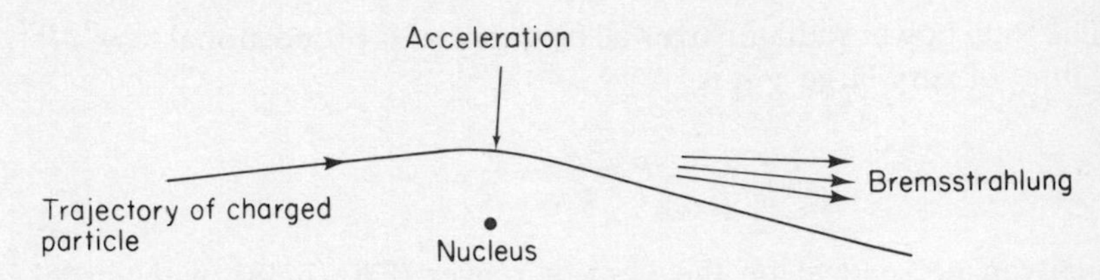

Fig. 3.14. Creation of bremsstrahlung during the acceleration of a charged particle (mass M, charge Z_1e) in the field of a nucleus (charge Z_2e).

The acceleration of a particle of mass M and charge Z_1e in the coulomb field of a nucleus of charge Z_2e is proportional to Z_1Z_2/M (assuming that M is much less than the mass of the nucleus), and therefore it follows from Eq. (3.35) that the radiated intensity is proportional to $Z_1^2(Z_1Z_2/M)^2 = Z_1^4Z_2^2/M^2$. The proportionality constant, and also the frequency range of the radiation, depend of course on the closeness of the encounter. The energy radiated by the particle results in a decrease in its kinetic energy, contributing to the general slowing-down of the charged particle as it passes through the medium. This energy loss process is most important when the particle mass is small (that is, for electrons) and when the medium contains atoms of high atomic number. It is for example an important source of power loss in plasmas containing ions of high atomic number and is the dominant mode of energy loss for electrons of high energy ($\gtrsim 1$ GeV) passing through matter.

A classical calculation using the methods described above would show that when the incident particle passes very close to a nucleus the acceleration can be large enough for the total energy radiated in a single deflection event to be greater than the total kinetic energy K of the particle. This is clearly impossible. It is possible however for a single photon to carry away nearly the whole of the available energy, in which case the photon would have the frequency K/h and the wavelength

$$\lambda_{\min} = \frac{hc}{K}. \tag{3.44}$$

This is therefore the minimum wavelength contained in the bremsstrahlung spectrum. When the particle in question is an electron accelerated in an X-ray tube it has the energy eV, where V is the potential difference across the tube, and then Eq. (3.44) becomes

$$\lambda_{\min} = \frac{hc}{eV}. \tag{3.45}$$

When this relationship was first discovered experimentally it was regarded not as a trivial consequence of the existence of photons, but as a new and unexplained law (see Chapter 1).

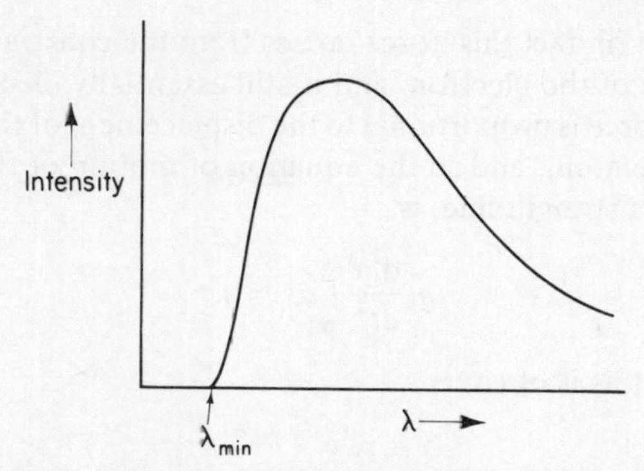

Fig. 3.15. The continuous bremsstrahlung spectrum.

Exact calculations (which must include quantal effects) show that the distribution of intensity over wavelength has the general form shown in Fig. 3.15. This continuous spectrum would usually have superimposed on it the 'characteristic' X-ray lines of atoms that have been excited and ionized by the photoelectric process (see Section 8.2.4).

3.1.7 The radiation from a 'classical' atom

After the discovery of the electron, but before the advent of the quantum theory of matter and radiation, it had not been possible to understand why atoms in their ground states are stable. It had been thought that the electrons in an atom must be moving, which implies, as we have seen, that they must be continually losing energy through radiation. This would cause the total energy of the atom to reduce to the point at which all the electrons have collapsed into the nucleus. The solution of this puzzle is now well known, namely that the wave nature of the electron requires that an atom can exist only at certain discrete energy levels, the lowest of which is stable because there is no way in which an atom in this level can lose further energy. Nevertheless it is instructive at this point to consider the atom in classical terms since this will introduce the concepts of lifetimes and natural widths of spectral lines, and also, perhaps surprisingly, it will give us reliable order-of-magnitude estimated of these quantities. The model will also prove useful in the discussion of refractive index in Section 3.2.2.

In its simplest form the 'classical' atom contains a single electron oscillating at a single frequency. The oscillation may be linear simple harmonic motion or uniform motion in a circular orbit (or more generally, planetary motion in an elliptical orbit). This type of model is sometimes called a *Lorentz atom*. The restoring force that acts on the electron is a combination of the Coulomb force from the rest of the atom and a 'non-electrical' force of undefined origin (in fact this 'force' arises from the constraints imposed by the wave nature of the electron, and is still essentially electrical in origin). The combined force is proportional to the displacement of the electron from the centre of the atom, and so the equation of motion of the electron, if it oscillates along a straight line, is

$$m\frac{d^2x}{dt^2}+ax=0.$$

The solution of this is of course

$$x=x_0\,e^{j(\omega_0 t+\phi)},$$

where $\omega_0=(a/m)^{1/2}$ is the natural angular frequency of the oscillation. But the loss of power by radiation disturbs this simple harmonic motion, causing it to be damped. We see from Larmor's formula (3.38) that the rate of loss of energy is

$$P=\frac{e^2a^2}{6\pi\varepsilon_0c^3}=\frac{e^2\omega_0^4x_0^2}{6\pi\varepsilon_0c^3}\,e^{2j(\omega_0t+\phi)}. \tag{3.46}$$

Now if this energy loss were caused instead by a damping force F acting on the electron, the rate of loss would be the rate at which work is done against the force, namely

$$P=F\frac{dx}{dt}=Fj\omega_0x_0\,e^{j(\omega_0t+\phi)}. \tag{3.47}$$

Comparing (3.46) and (3.47) we can now put

$$F=-\frac{je^2\omega_0^3x_0}{6\pi\varepsilon_0c^3}\,e^{j(\omega_0t+\phi)}=-\frac{e^2\omega_0^2}{6\pi\varepsilon_0c^3}\frac{dx}{dt}. \tag{3.48}$$

This force represents the effect of the radiation loss. It is known as the *radiative reaction*. Its presence causes the equation of motion of the electron to be modified to

$$m\frac{d^2x}{dt^2}+m\gamma\frac{dx}{dt}+m\omega_0^2x=0, \tag{3.49}$$

where the *decay constant* (or damping constant) γ is given by

$$\gamma=\frac{e^2\omega_0^2}{6\pi\varepsilon_0mc^3}. \tag{3.50}$$

Equation (3.49) is the familiar equation of a *damped harmonic oscillator*. Because γ is negligible compared with ω_0 the damped frequency is nearly the same as the undamped frequency, and so the solution of the equation is

$$x=x_0\,e^{-\gamma t/2}\,e^{j(\omega_0t+\phi)}. \tag{3.51}$$

The time dependence of the real part of x,

$$x_r = x_0\, e^{-\gamma t/2} \cos(\omega_0 t + \phi),$$

is illustrated in Fig. 3.16.

The total energy (the sum of the kinetic and potential energies) of the motion is

$$\mathscr{E} = \tfrac{1}{2}m\left(\frac{dx_r}{dt}\right)^2 + \tfrac{1}{2}ax_r^2 = \tfrac{1}{2}m\omega_0^2 x_0^2\, e^{-\gamma t}.$$

This can also be expressed as

$$\mathscr{E} = \mathscr{E}_0\, e^{-t/\tau}, \tag{3.52}$$

where

$$\tau = \gamma^{-1}. \tag{3.53}$$

We see that τ is the time in which the energy of the atom, and also the intensity of the radiation field, decreases by the factor e. In reality the energy of an excited atom does not decrease continuously, as supposed by this model, but changes suddenly, when a photon is emitted. The probability that it decays at time t is proportional to $\exp(-t/\tau)$, where τ is the *lifetime* of the excited state of the atom. We can therefore associate the τ given in Eq. (3.52) with the lifetime of the state that emits radiation of angular frequency

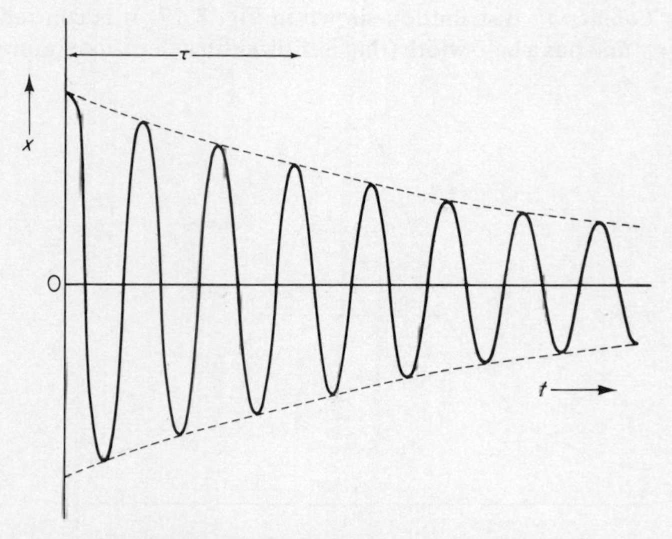

Fig. 3.16. Damped harmonic motion of the 'classical' atom. The amount of damping is shown greatly exaggerated. τ is the time in which the energy ($\propto x^2$) is reduced by the factor e.

ω_0 and wavelength λ_0 ($=2\pi c/\omega_0$). Expressing τ in terms of λ_0 we obtain

$$\tau = \lambda_0^2 \times 4.5 \times 10^4 \text{ s}. \tag{3.54}$$

This equation does in fact give the approximate wavelength dependence and the order-of-magnitude value of many of the observed lifetimes of excited states of atoms. For example, at $\lambda_0 = 500$ nm (yellow light) it gives $\tau = 1.1 \times 10^{-8}$ s, which is a typical lifetime for an optically allowed transition of this wavelength (see Chapter 4).

The electric field strength E of the emitted radiation has the same time dependence as the dipole moment ex (apart from a phase factor),

$$E = E_0 \, \mathrm{e}^{-\gamma t/2} \, \mathrm{e}^{\mathrm{j}(\omega_0 t + \phi')}.$$

As in the case of synchrotron radiation we can find the frequency content of this by performing a Fourier analysis:

$$F(\nu) \propto \int \mathrm{e}^{-\frac{1}{2}\gamma t + \mathrm{j}2\pi\nu_0 t} \, \mathrm{e}^{-\mathrm{j}2\pi\nu t} \, \mathrm{d}t = \frac{1}{2\pi \mathrm{j}(\nu - \nu_0) + \gamma/2}.$$

The intensity radiated per unit frequency interval at the frequency ν is therefore

$$P(\nu) \propto |F(\nu)|^2 \propto \frac{1}{(\nu - \nu_0)^2 + (\gamma/4\pi)^2}. \tag{3.55}$$

This is the *Lorentzian* distribution shown in Fig. 3.17. It is centred at the frequency ν_0, and has a half-width (that is, full width at half-maximum) given

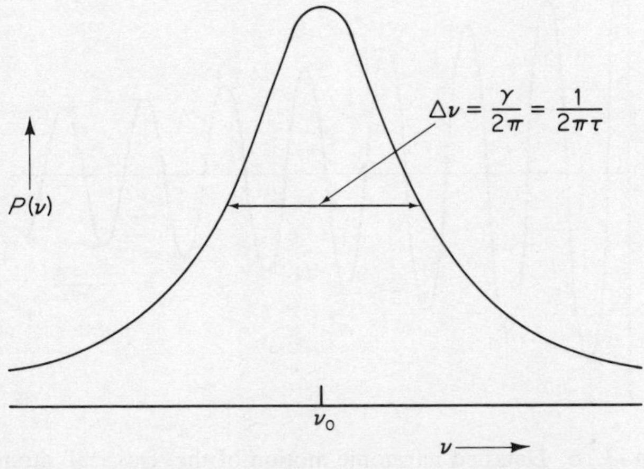

Fig. 3.17. The frequency distribution of the radiation emitted by the damped harmonic oscillator.

by

$$\Delta\nu = \gamma/2\pi = \frac{1}{2\pi\tau}. \tag{3.56}$$

This is referred to as the *natural width* of the line. It is interesting to note from Eq. (3.55) and Fig. 3.17 the long-range character of the wings of the Lorentzian distribution (the intensity in the wings decreases asymptotically as $(\nu-\nu_0)^{-2}$, which is far slower than the decrease for a Gaussian distribution); this fact is important in considering refractive indices (see Section 3.2) and various non-linear effects (see Section 4.5).

If the intensity is measured as a function of wavelength, rather than frequency, the natural width of the line is (cf. Eq. (2.31))

$$\Delta\lambda = \Delta\nu\frac{d\lambda}{d\nu} = \frac{\gamma}{2\pi}\frac{c}{\nu_0^2} = 1.18\times10^{-14}\ \text{m}.$$

Observed spectral lines do indeed have natural widths which are usually of this order of magnitude. The quantal description of these widths will be discussed in Chapter 4.

3.1.8 Cerenkov radiation

We have seen that a charged particle moving in free space can radiate only if it has a non-zero acceleration (which in practice is usually a sideways acceleration). This is not true when the particle is travelling through a dielectric medium. Cerenkov discovered in 1937 that if the velocity of the particle is greater than the phase velocity of light in the medium the particle can radiate even when travelling in a straight line with an essentially uniform velocity.

As in the case of synchrotron radiation, a qualitative understanding of the phenomenon can be obtained by considering the wavefronts emitted by the particle. Figure 3.18 shows two examples of the spherical wavefronts emitted by a particle at four equally spaced instants of time. The particle has the same velocity v in both examples. In Fig. 3.18(a) v is less than the velocity c' of the radiation in the medium, and the wavefronts tend to pile-up in the forward direction, as in Fig. 3.11. In Fig. 3.18(b) v is greater than c' and we see that a *shock-wave* is formed. This is analogous to the acoustic shock-wave that is created when an aeroplane travels through air at a speed greater, than that of sound, or the bow-wave that is formed when a boat travels through water at a speed greater than the phase velocity of water waves.

In terms of the refractive index n of the medium (see Section 2.4) the condition for the shock wave to be formed is

$$c' = \frac{c}{n} < v, \tag{3.57}$$

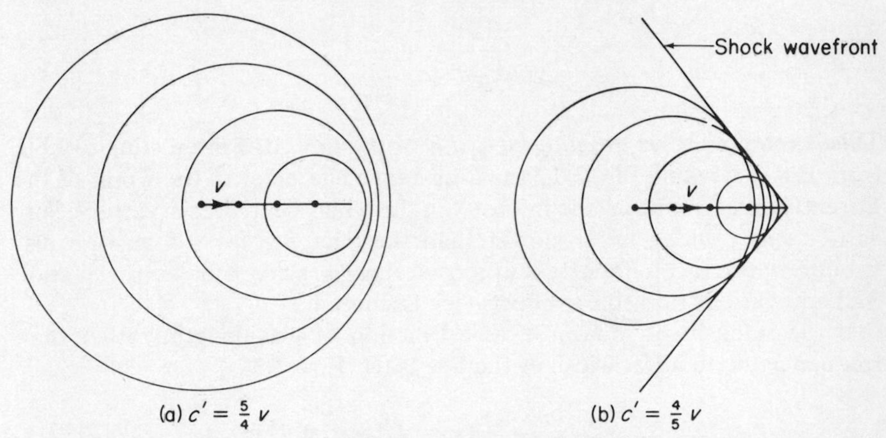

Fig. 3.18. Four wavefronts emitted by a particle which travels at the velocity v. The velocity of radiation in the medium is c'.

where c is the velocity of light in free space. The half-angle θ at the apex of the shock wavefront is given by

$$\sin\theta = \frac{c'}{v} = \frac{c}{nv}. \tag{3.58}$$

The light travels in a direction normal to the wavefront, at an angle $(\pi/2-\theta)$ to the particle direction.

A rigorous treatment shows that Eq. (3.58) still determines the angle of the emitted photons, and that the number emitted per unit length of travel of the particle in the frequency interval from ν to $\nu+\mathrm{d}\nu$ is (see Jackson[6])

$$\frac{\mathrm{d}N(\nu)}{\mathrm{d}x}\,\mathrm{d}\nu = \frac{2\pi\alpha Z^2}{c}\left(1-\frac{c^2}{n^2v^2}\right)\mathrm{d}\nu. \tag{3.59}$$

Here Ze is the charge of the particle, n is the refractive index of the medium at the frequency ν, and α is a fundamental dimensionless constant called the fine structure constant, having the value 1/137. For example a particle of $Z=1$ moving at nearly the velocity of light through crown glass emits approximately 30 photons in the visible region ($\lambda = 400$ to 750 nm) per mm of travel. An interesting example of the use of Eq. (3.59) is in explaining the origin of the light flashes that are seen by astronauts. These flashes are thought to be due to the Cerenkov radiation generated by cosmic particles passing through the vitreous humour ($n=1.34$, length $\simeq 1$ cm) of the eye. The eye therefore 'sees' approximately $80Z^2$ photons (in the sensitive range from 440 to 540 nm) for each cosmic particle of charge Ze and velocity c. Since the threshold sensitivity of the eye is approximately 3000 or 4000 photons under these conditions, the particles causing the observed flashes must have $Z \gtrsim 6$.

3.2 REFRACTION

The reader is now well aware that classical methods can provide considerable insight into the physics of the generation and propagation of electromagnetic radiation. These methods help also in understanding the processes of absorption and refraction of radiation, as we shall see in the present section. For example by treating the atoms or molecules of a gas as damped harmonic oscillators it is possible to understand the origin of the refractive index of the gas, and to explain the dependence of the refractive index on the frequency of the light. This can be achieved without any reference to the details of the internal structure of the atoms. (The quantum-mechanical treatment of these atomic processes is deferred until the next chapter.)

We have seen in Chapter 2 (Section 2.4) that the velocity of propagation of electromagnetic radiation in a dielectric medium is

$$v = \omega/k = c/n, \tag{3.60}$$

where

$$n = \varepsilon^{1/2} \tag{3.61}$$

is the refractive index of the medium (assuming that the relative permeability μ is unity). There are a few particular cases in which the values of ε and n can be simply related to the properties of the electrons, atoms, or molecules of which the medium is composed. We now deal with three of these cases.

3.2.1 Free electron gas

The presence of charged particles in the earth's upper atmosphere strongly affects the propagation of radio waves in this region. The charged particles are electrons and positive ions which are formed when the ultraviolet light from the sun is absorbed by the atoms and molecules of the upper atmosphere. The particles tend to be trapped by the magnetic field of the earth and thus stay in the upper regions, forming the *ionosphere*. Because the electrons are much lighter than the positive ions they are accelerated more by the electric fields of radio waves passing through the ionosphere, and so are more effective in modifying the propagation of the radio waves. The ionosphere may therefore be regarded as a tenuous gas of free electrons, that is, as a *free electron gas*. Other examples of free electron gases are the sun's corona and laboratory-generated plasmas. Even the conduction electrons of a metal sometimes behave as a free electron gas when scattering electromagnetic radiation (see Section 3.2.2).

A free electron (of charge $-e$) in the electric field

$$\mathbf{E} = \mathbf{i}_x E_{0x}\, \mathrm{e}^{\mathrm{j}\omega t}$$

of an electromagnetic plane wave has the equation of motion

$$m\frac{d^2x}{dt^2} = -eE_{0x}\,e^{j\omega t}.$$

Its displacement is therefore

$$x = \left(\frac{eE_{0x}}{m\omega^2}\right)e^{j\omega t},$$

and it gives rise to an electric dipole moment

$$\mathbf{p} = -\frac{e^2}{m\omega^2}\mathbf{E}.$$

Each of the electrons of the electron gas experiences the externally applied field **E** together with an internal field caused by the induced dipole moments of the other electrons. If the electron density N is sufficiently small (as in the ionosphere) this second contribution can be neglected and the volume polarization becomes

$$\mathbf{P} = -\frac{Ne^2}{m\omega^2}\mathbf{E}.$$

Hence the relative permittivity ε of the electron gas is given by (see Eq. (2.11))

$$\varepsilon = 1 - \frac{Ne^2}{\varepsilon_0 m\omega^2}.$$

By putting

$$\nu_p = \frac{1}{2\pi}\left(\frac{Ne^2}{\varepsilon_0 m}\right)^{1/2}, \tag{3.62}$$

we can write

$$\varepsilon = 1 - \left(\frac{\nu_p}{\nu}\right)^2 \quad \text{and} \quad n = \left(1 - \frac{\nu_p^2}{\nu^2}\right)^{1/2}. \tag{3.63}$$

Clearly something strange happens when ν is equal to ν_p: ε and n become zero! When ν is less than ν_p, ε becomes negative and n becomes imaginary!

A zero value of ε implies that the electric displacement **D** is zero even when the electric field **E** is not. Now we know that **D** depends on the free charges only (because $\operatorname{div}\mathbf{D} = \rho_f$), while **E** depends on all the charges, including the induced polarization charges. Therefore the vanishing of ε implies that it may be possible for an oscillating **E** to exist in the electron gas even in the absence of the free charges which give rise to the external fields, which means the absence of external radiation falling on the gas. An oscillation at the frequency ν_p is therefore self-sustaining (we are neglecting dissipative effects) once it has been initiated by some means. This type of oscillation is known to exist, and is called a plasma oscillation. The frequency defined by Eq. (3.62) is known as the *plasma frequency*. It has the numerical value

$$\nu_p = 8.98N^{1/2}\ \text{Hz}. \tag{3.64}$$

In the case of the ionosphere the maximum value of N occurs during the day-time (when the flux from the sun is greatest) and in the higher regions (approximately 300 km above the earth's surface). Here N reaches approximately $4\times10^{11}\ \text{m}^{-3}$, corresponding to $\nu_p \simeq 6$ MHz.

Let us consider now what happens to radio waves travelling through the ionosphere.

(i) n is real but less than unity ($\nu > \nu_p$)

According to Snell's law (2.61) a radio-wave would be refracted *away* from the normal when it strikes the ionosphere. The angle of refraction becomes 90° when

$$\sin\theta_i = n,$$

and for angles of incidence greater than this the wave is totally externally reflected. In fact the edge of the ionosphere is not sharp (and also there is some refraction in the lower, un-ionized part of the atmosphere), but reflection must occur for all angles of incidence given by

$$\sin\theta_i \geqslant n_{min},$$

where n_{min} is the minimum refractive index of the ionosphere, occurring at the height of maximum electron density (see Eqs. (3.62) and (3.63)). These effects are illustrated in Fig. 3.19(*a*) and (*b*).

The fact that n is less than unity does not mean that information can be transmitted at a speed faster than c. The velocity given by Eq. (3.60) is the *phase velocity* v_p while the velocity at which information is transmitted by the radio waves is the *group velocity*,

$$v_g = \frac{d\omega}{dk}. \tag{3.65}$$

In the present case

$$k = \frac{\omega}{v_p} = \frac{\omega}{c}\left(1 - \frac{4\pi^2\nu_p^2}{\omega^2}\right)^{1/2},$$

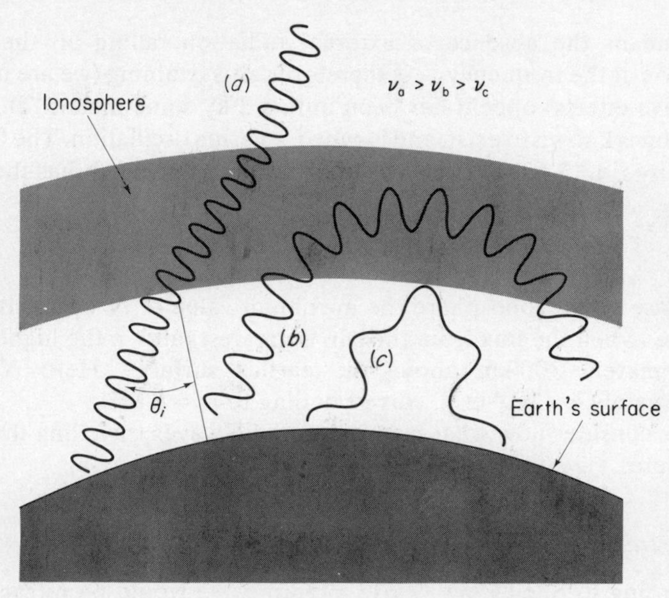

Fig. 3.19. Refraction and reflection of radio waves by the ionosphere. (*a*) For the highest frequency waves the refractive index n of the ionosphere is approximately unity, and the waves are slightly refracted away from the normal. (*b*) At lower frequencies n is smaller and the waves are refracted back towards the earth; the part of the ionosphere at which they travel horizontally has $n = \sin \theta_i$. (*c*) At still lower frequencies n is less than $\sin \theta_i$ at the bottom of the ionosphere, and the wave is totally externally reflected.

and therefore

$$v_g = c\left(1 - \frac{\nu_p^2}{\nu^2}\right)^{1/2}, \tag{3.66}$$

which is less than c.

(ii) n is imaginary ($\nu \leqslant \nu_p$)

This occurs when $\varepsilon \leqslant 0$. It is then convenient to express the refractive index as

$$n = \pm j\kappa$$

where κ is real and positive. The expression for the electric field in the wave then becomes

$$\mathbf{E} = \mathbf{i}_x E_{0x}\, e^{j(\omega t - kz)} = \mathbf{i}_x E_{0x}\, e^{j\omega(t - nz/c)} = \mathbf{i}_x E_{0x}\, e^{j\omega t}\, e^{\pm\omega\kappa z/c}.$$

The positive exponent gives an exponentially increasing field, which is clearly an unphysical result. We must therefore take the negative exponent, giving a field which decreases exponentially with distance. There is no net flow of energy because the magnitude of Poynting's vector is (see Eqs. (2.7) and (2.40))

$$N = \mathrm{Re}\,(E) \times \mathrm{Re}\,(H) \propto \mathrm{Re}\,(\mathrm{e}^{\mathrm{j}\omega t}) \times \mathrm{Re}\,(\mathrm{j}\,\mathrm{e}^{\mathrm{j}\omega t}) = -\cos\omega t \sin\omega t,$$

which has a time-averaged value of zero. This is also clear from the fact that the group velocity (Eq. (3.66)) is zero or imaginary. We see also that no energy is absorbed by the medium because the rate at which work is done on the electrons is proportional to

$$E_x \frac{\mathrm{d}x}{\mathrm{d}t} \propto \cos\omega t \sin\omega t,$$

which also has a time-averaged value of zero.

Because there is no energy flow and no energy absorption we must conclude that the wave is completely reflected by the medium, for *any* angle of incidence. This can also be seen in the case of normal incidence by using Eq. (2.59) for the reflection coefficient R. With a slight modification of this equation to allow for the imaginary refractive index, we find that

$$R = \left|\frac{n_2 - n_1}{n_2 + n_1}\right|^2 = \left|\frac{-\mathrm{j}\kappa - 1}{-\mathrm{j}\kappa + 1}\right|^2 = 1,$$

so that the ionic medium behaves as a perfect mirror at these frequencies.

The electron density of the highest region of the ionosphere is always greater than $5 \times 10^{10}\ \mathrm{m}^{-3}$, corresponding to $\nu_\mathrm{p} \simeq 2$ MHz ($\lambda_\mathrm{p} \simeq 150$ m). All frequencies less than this are therefore always totally reflected by the ionosphere, even at night-time, when the electron densities are least. The processes of refraction and reflection of radio waves by the ionosphere provide an important means of long range communication. Multiple reflections between the ionosphere and the earth can also occur (with an attenuation in amplitude), enabling communication to take place over many thousands of kilometres.

3.2.2 A gas of 'classical' atoms

Our second example is that of a gas of atoms treated as a system of damped harmonic oscillators. We have seen in Section 3.1.7 that the equation of motion of an electron in such an atom is

$$\frac{\mathrm{d}^2 x}{\mathrm{d}t^2} + \frac{1}{\tau}\frac{\mathrm{d}x}{\mathrm{d}t} + \omega_0^2 x = 0,$$

where τ is the classical lifetime and ω_0 is the natural frequency of oscillation. In the presence of an electric field of amplitude E_0 and frequency ω the equation of motion becomes

$$\frac{\mathrm{d}^2x}{\mathrm{d}t^2}+\frac{1}{\tau}\frac{\mathrm{d}x}{\mathrm{d}t}+\omega_0^2x=-\frac{eE_{0x}}{m}\mathrm{e}^{\mathrm{j}\omega t}, \tag{3.67}$$

which has the steady state solution

$$x=-\frac{eE_{0x}}{m}\frac{\mathrm{e}^{\mathrm{j}\omega t}}{\omega_0^2-\omega^2+\mathrm{j}\omega/\tau}. \tag{3.68}$$

The dipole moment of the atom is therefore

$$p=\frac{e^2E_{0x}\,\mathrm{e}^{\mathrm{j}\omega t}}{m(\omega_0^2-\omega^2+\mathrm{j}\omega/\tau)}. \tag{3.69}$$

If there are N electrons per unit volume (where N is assumed to be small) the overall dielectric constant of the medium is (see Eq. (2.11))

$$\varepsilon=1+\frac{Ne^2}{m\varepsilon_0(\omega_0^2-\omega^2+\mathrm{j}\omega/\tau)}. \tag{3.70}$$

This is a complex quantity, as is its square root, the *complex refractive index*

$$n_\mathrm{c}=\left[1+\frac{Ne^2}{m\varepsilon_0(\omega_0^2-\omega^2+\mathrm{j}\omega/\tau)}\right]^{1/2}. \tag{3.71}$$

It is usual to write n_c in the form

$$n_\mathrm{c}=n-\mathrm{j}\kappa, \tag{3.72}$$

where the real part n is still known as the refractive index, and κ is called the *extinction coefficient*, or sometimes the *absorption index*. Simple expressions can be obtained for n and κ when the density of the medium is low (so that $\varepsilon-1\ll 1$) and the forced motion is nearly resonant (that is, when $\omega\simeq\omega_0$). Then we obtain

$$\boxed{n=1-\frac{Ne^2}{4m\varepsilon_0\omega_0}\frac{\omega-\omega_0}{(\omega-\omega_0)^2+1/(2\tau)^2}} \tag{3.73}$$

and

$$\boxed{\kappa=\frac{Ne^2}{8m\varepsilon_0\tau\omega_0}\frac{1}{(\omega-\omega_0)^2+1/(2\tau)^2}} \tag{3.74}$$

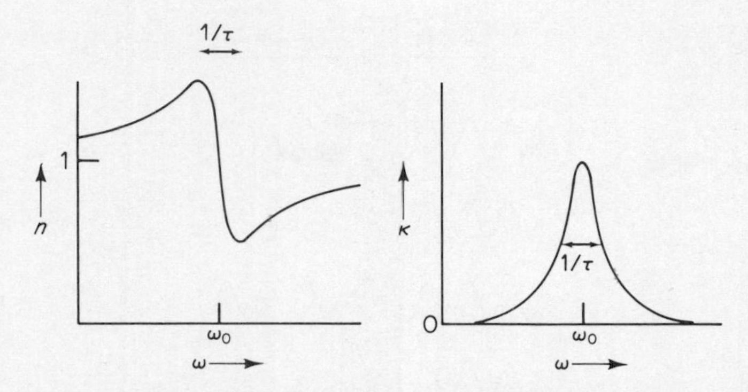

Fig. 3.20. The real refractive index n and extinction coefficient κ.

We have assumed that the damping is light ($\tau^{-1} \ll \omega_0$, see Section 3.1.7). The frequency dependences of n and κ are shown in Fig. 3.20.

The extinction coefficient has the familiar *Lorentzian shape*, with a width (full width at half maximum) equal to $1/\tau$. This shape is precisely that shown in Fig. 3.17 for the frequency distribution of the light emitted by the excited atom (the intimate connection between the absorption and emission processes will be dealt with more rigorously in Chapter 4). On the other hand the refractive index n has what is known as a *dispersion shape*. Both these shapes apply to a gas of stationary atoms. In practice the thermal motion of the atoms causes an increase in the width of the features (see Section 6.3).

The significance of a complex refractive index is seen by considering the electric field strength of a plane wave travelling through the medium. This is

$$\mathbf{E}_x = \mathbf{i}_x E_{0x}\, e^{j(\omega t - kz)} = \mathbf{i}_x E_{0x}\, e^{j\omega(t - n_c z/c)} = \mathbf{i}_x E_{0x}\, e^{j\omega(t - nz/c)}\, e^{-\omega\kappa z/c}.$$

The wave therefore propagates with a phase velocity c/n, but with an amplitude that decreases exponentially with distance. The intensity of the wave at the position z is

$$I(z) \propto |\mathbf{E}|^2 \propto e^{-\mu z},$$

where the absorption coefficient μ is given by

$$\mu = \frac{2\omega\kappa}{c} = \frac{Ne^2}{4\tau m\varepsilon_0 c[\omega - \omega_0)^2 + 1/(2\tau)^2]}. \tag{3.75}$$

We can find a different way of describing the absorption process by considering the absorption which takes place in a thin slab of the gas, of thickness dz, as shown in Fig. 3.21. The fraction f of the intensity that is absorbed in this slab is

$$f = \frac{I_0(1 - e^{-\mu\, dz})}{I_0} = \mu\, dz.$$

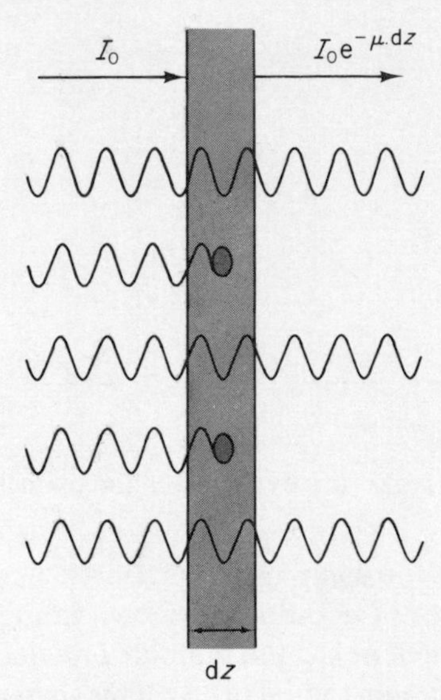

Fig. 3.21. The attenuation of a beam of electromagnetic radiation through absorption by the atoms or molecules contained in a thin slab of gas, each of the atoms having an absorption cross-section σ.

If we now associate an effective cross-sectional area σ with each of the atoms in the slab, of which there are N dz per unit area, the fractional area covered by these atoms is

$$f = \sigma N \, dz.$$

By equating these two fractions we find that the effective *cross-section for absorption* is

$$\sigma = \frac{\mu}{N} = \frac{e^2}{4\tau m \varepsilon_0 c[(\omega - \omega_0)^2 + 1/(2\tau)^2]} \tag{3.76}$$

per atom.

Real atoms have many excited states and hence many frequencies ω_i at which they can absorb. We cannot suppose that each electron in an atom possesses its own characteristic frequency ω_i, because the number of excited states is larger than the number of electrons in the atom. Instead we must suppose that there is a probability f_i of an electron taking part at the absorption frequency ω_i, where the sum of the probabilities is the total number of electrons in the atom,

$$\sum_i f_i = Z. \tag{3.77}$$

The probabilities f_i are known as the *oscillator strengths* of the atom. The absorption coefficient then becomes (using the frequency ν_i rather than angular frequency ω_i)

$$\mu = \sum_i \frac{Ne^2 f_i}{16\pi^2 \tau_i m \varepsilon_0 c[(\nu - \nu_i)^2 + 1/(4\pi\tau_i)^2]} \tag{3.78}$$

where the summation is over all the excited states of the atom, and N is now the number of *atoms* per unit volume. The analogous generalization of the real part of the refractive index is (from Eqs. (3.71) and (3.72))

$$n = 1 - \sum_i \frac{Ne^2 f_i(\nu^2 - \nu_i^2)}{8\pi^2 m\varepsilon_0[(\nu^2 - \nu_i^2)^2 + (\nu/4\pi\tau_i)^2]} \tag{3.79}$$

The dependence on ν of n and μ is illustrated in Fig. 3.22.

We see from the figure that when μ is negligible the refractive index increases as ν increases. This is known as *normal dispersion*. A transparent refracting object, such as a glass prism, therefore always causes high frequency light to deviate through larger angles than light of lower frequency.

Another feature of Fig. 3.22 is that n is less than unity at frequencies higher than the highest characteristic frequency ν_i (which is the frequency required to remove the most tightly bound electrons). At very high frequencies the refractive index becomes (Eqs. (3.77) and (3.79))

$$n \to 1 - \frac{(NZ)e^2}{8\pi^2 \varepsilon_0 m\nu^2} \qquad (\nu \gg \text{all } \nu_i)$$

This is also the limiting value of the refractive index for free electrons (Eq. (3.63)), since at these high frequencies the atomic electrons can be considered as being essentially free. The difference $(n-1)$ is usually very small even if the density N is that of a liquid or solid, and so in these circumstances the present derivation is still valid. Thus X-rays and γ-rays suffer very little refraction or change in velocity when passing through liquids or solids (although they can be attenuated by the scattering and absorption processes that we shall consider in Chapter 8).

3.2.3 Refraction in liquids and solids

In the case of dense media we cannot assume that ε is nearly unity, and so cannot obtain the simple expressions for n and κ given by Eqs. (3.73) and

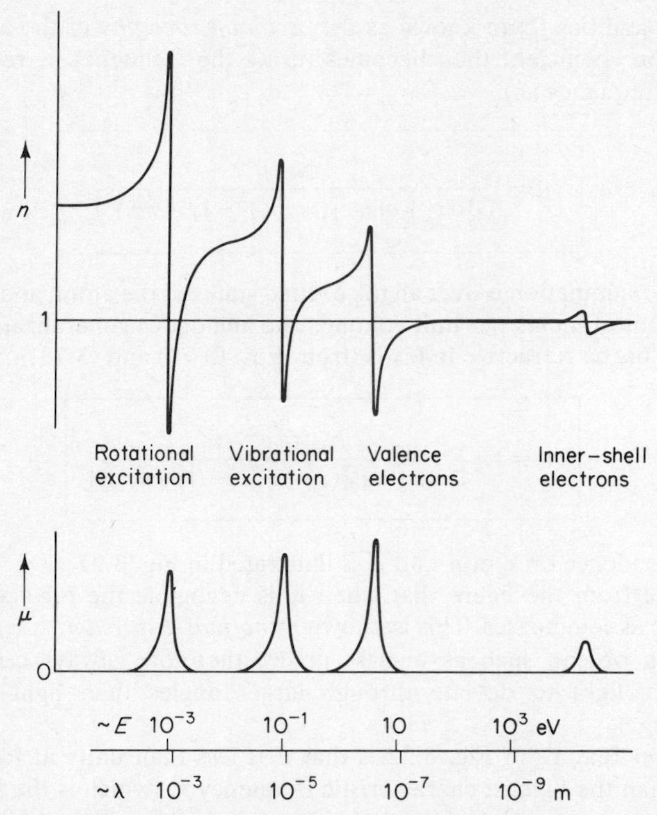

Fig. 3.22. Simplified form of the dependence on wavelength and energy of the refractive index n and absorption coefficient μ of a molecular gas. Only one excited state of each type is shown. The widths of the peaks have been exaggerated for clarity.

(3.74). Instead we must return to Eq. (3.71). Representing the real refractive index by N_r and the extinction coefficient by K, we find

$$(N_r - jK)^2 = 1 + \sum_i \frac{Ne^2 f_i}{m\varepsilon_0(\omega_i^2 - \omega^2 + j\omega/\tau_i)}. \tag{3.80}$$

The frequency dependence of N_r and K near a resonance then loses the symmetry possessed by the dispersion and Lorentzian forms of a tenuous gas (see Fig. 3.23). This point is also illustrated in Problem 3.10.

Another approximation made in the previous section was to take the local electric field $\mathbf{E}^{\text{local}}$ acting at any point in the gas to be equal to the externally applied field $\mathbf{E}$. In other words the extra field caused by the polarization $\mathbf{P}$ of the gas was assumed to be negligible compared with $\mathbf{E}$. This is usually not

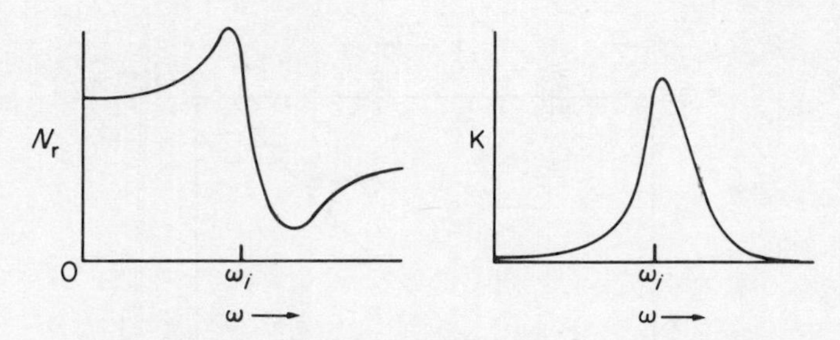

Fig. 3.23. The refractive index and extinction coefficient for a dense medium, in the region of a resonance frequency ω_i.

true in the case of liquids and solids, for which it can be shown that

$$\mathbf{E}^{\text{local}} = \mathbf{E} \Big/ \left(1 - \frac{N\alpha}{3}\right),$$

where α is the polarizability of the atoms or molecules of the medium. The relationship between ε and α is then given by the Clausius–Mossotti formula

$$\frac{\varepsilon - 1}{\varepsilon + 2} = \frac{N\alpha}{3}. \tag{3.81}$$

In the limit in which $N\alpha \ll 1$, and $(\varepsilon - 1) \ll 1$, this becomes

$$\varepsilon = 1 + N\alpha,$$

which is the relationship implied by Eq. (3.70). On the other hand when $N\alpha$ is not small Eqs. (3.80) and (3.81) give

$$\frac{(N_r - jK)^2 - 1}{(N_r - jK)^2 + 2} = \frac{Ne^2}{3m\varepsilon_0} \sum_i \frac{f_i}{(\omega_i^2 - \omega^2 + j\omega/\tau_i)}. \tag{3.82}$$

This equation cannot be simplified in the way that was possible in the case of a low pressure gas, but the general behaviour of N_r and K is still similar to that shown in Figs. 3.22 and 3.23. For example the dispersion is still normal in the regions of transparency. Figure 3.24 shows this for ordinary crown glass in the visible region.

3.2.4 Propagation in anisotropic media

There are some media for which the relative permittivity ε and refractive index n depend on the directions of propagation and polarization of the light. For example a crystal composed of molecules which are more easily polarized in one direction than another is like this. Figure 3.25 shows a model of a medium of this sort. It is composed of linear molecules which are assumed to be more easily polarized along their length than transversely.

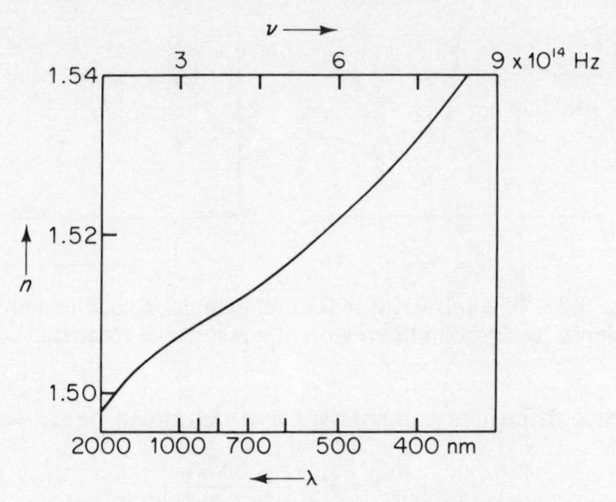

Fig. 3.24. The dependence on wavelength of the refractive index of ordinary crown glass in the visible region.

The molecules are all aligned in the same direction, which we take to be the z-direction (and we take the transverse direction in the figure to be the x-direction). When the electric vector **E** of a plane wave is in the z-direction the polarization is largest, giving the largest values of ε and n. This occurs for the wave labelled (a) in the figure. At the other extreme, when **E** is in the x (or y) direction, the polarization is smallest, as are ε and n. This occurs whenever the direction of polarization is perpendicular to the plane of the figure, as is the case for wave (c). It also occurs for the wave (d), for *both* directions of polarization (i.e. parallel to x or y). An intermediate case is that of the wave (b). Wave (b) also differs in another respect from waves (a), (c), and (d): the induced polarization for this wave is not parallel to **E**, and so the electric displacement **D** is also not parallel to **E**.

The direction of wave (d) is known as the *optic axis*: for this direction of propagation the refractive index is independent of the plane of polarization. Because there is only one such axis the medium is called *uniaxial*. Many crystalline dielectrics are of this type. Waves polarized perpendicular to the optic axis, such as wave (c), are known as *ordinary waves*: for these the refractive index is constant and independent of the direction of propagation. The wavevector for these waves is

$$k = n_0 k_0, \tag{3.83}$$

where k_0 is the magnitude of the wavevector in free space and n_0 is the *ordinary refractive index*. Waves polarized in a plane containing the optic axis, such as waves (a) and (b) are known as *extraordinary waves*: for these

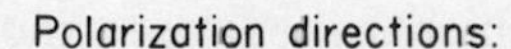
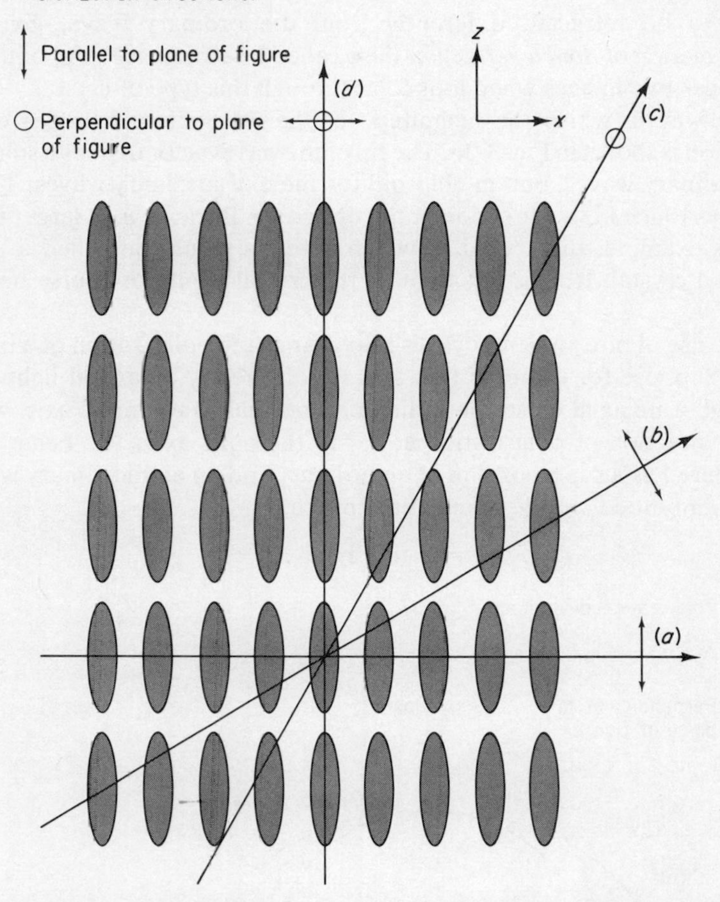

Fig. 3.25. Schematic illustration of a form of optical anisotropy. The molecules of this medium are more easily polarized along their length than transversely, and they are all aligned in the same direction (z). The amount by which they are polarized, and hence the refractive index, depends on the directions of propagation and polarization of the light.

the refractive index depends on the direction of propagation. The wave-vector for these waves is given by

$$\frac{k_z^2}{n_0^2}+\frac{k_\perp^2}{n_e^2}=k_0^2, \tag{3.84}$$

where n_e is the *extraordinary refractive index* and the subscript $\perp$ denotes any direction perpendicular to the optic axis. In general $\mathbf{E}$ is not transverse to

k, and Poynting's vector is not parallel to **k**, which causes the extraordinary wave to be refracted differently from the ordinary wave, giving the phenomenon of *double refraction* (also called birefringence). A point object produces two images when looked at through this type of crystal.

The way in which the magnitude of the wavevector **k** varies with its direction is shown in Fig. 3.26. The tip of the wavevector defines a sphere for the ordinary waves, but an ellipsoid for the extraordinary waves. The two surfaces touch in the direction of the optic axis. Because n_e is larger than n_0 in this example, the crystal to which it refers would be called a *positive* uniaxial crystal. If n_e is less than n_0 the crystal would of course be called *negative*.

One use of birefringent crystals is to change the polarization of a beam of light. Suppose for example that a beam of linearly polarized light is sent through a uniaxial crystal in a direction normal to the optic axis, with an initial direction of polarization at 45° to the optic axis. The beam can be considered as a superposition of an ordinary and an extraordinary wave, of equal amplitude and the same initial phase,

$$\mathbf{E} = E_0(\mathbf{i}_z + \mathbf{i}_x)\, \mathrm{e}^{\mathrm{j}(\omega t - k_0 y)} \tag{3.85}$$

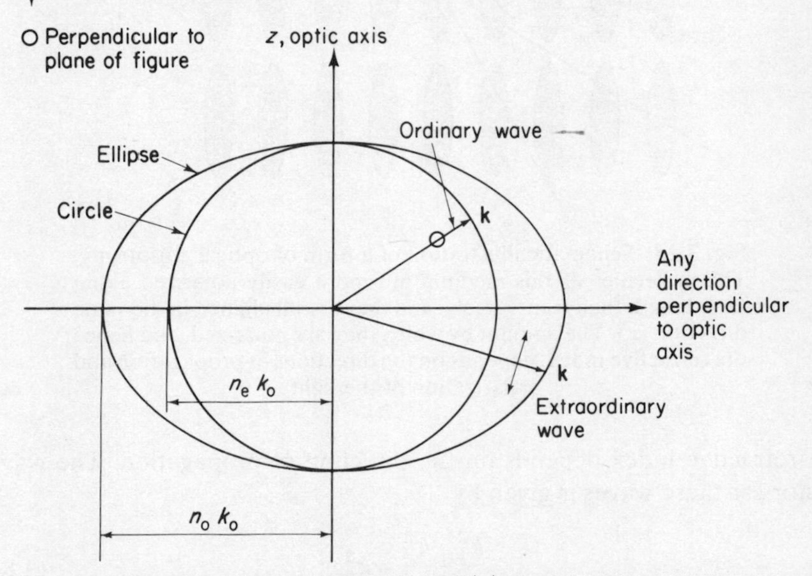

Fig. 3.26. The variation with direction of $|\mathbf{k}|$ for the ordinary and extraordinary waves in a uniaxial crystal. The extraordinary wave is polarized in the plane containing the optic axis (the plane of the figure), while the ordinary wave is polarized perpendicular to this plane.

(using the axes shown in Fig. 3.25). Inside the crystal the two components travel with different velocities, and the combined wave is

$$\mathbf{E}=E\,\mathrm{e}^{\mathrm{j}\omega t}(\mathbf{i}_z\,\mathrm{e}^{-\mathrm{j}n_o k_o y}+\mathbf{i}_x\,\mathrm{e}^{-\mathrm{j}n_e k_o y}).$$

After traversing a length l of the crystal it emerges as the wave

$$\mathbf{E}=E_0(\mathbf{i}_z+\mathbf{i}_x\,\mathrm{e}^{\mathrm{j}\Delta\phi})\,\mathrm{e}^{\mathrm{j}(\omega t-k_o y+\phi)},$$

where the phase difference $\Delta\phi$ between the two components is given by

$$\frac{\Delta\phi}{2\pi}=\frac{(n_0-n_e)k_0 l}{2\pi}=\frac{(n_0-n_e)l}{\lambda_0}. \tag{3.86}$$

For example if $\Delta\phi/2\pi=\frac{1}{4}$ the wave emerges as a left-hand circularly polarized wave (see Eq. (2.36)). The resulting device is called a *quarter-wave plate.* It gives a very convenient means of producing circularly polarized light from linearly polarized light, or *vice versa.* An equally useful device is the *half-wave plate,* for which $\Delta\phi/2\pi=\frac{1}{2}$. It converts the wave (3.85) having a linear polarization in the direction $(\mathbf{i}_z+\mathbf{i}_x)$ to a wave having linear polarization in the direction $(\mathbf{i}_z-\mathbf{i}_x)$; in other words the direction of linear polarization is *rotated* by 90°.

A dielectric medium can sometimes be made optically anisotropic by applying a static electric field to it. This happens for example in the case of a liquid composed of molecules which have a permanent electric dipole moment, because they become partially aligned in the direction of the electric field. The refractive index of the liquid in the direction of the field is then different from its refractive index at right angles to this direction, and so the liquid is uniaxial, with the optic axis in the field direction. This is known as the *Kerr effect.* The value of (n_0-n_e) depends on E^2 rather than E, since the polarizability in the alignment direction is necessarily a positive quantity and is not changed by a reversal of the field direction. Equation (3.86) is therefore replaced by

$$\Delta\phi\propto E^2 l, \tag{3.87}$$

where l is the length of travel through the liquid. If on the other hand the medium is a crystal which has an anisotropic refractive index even when $E=0$, the application of a non-zero E causes the difference (n_0-n_e) to change by an amount proportional to E. This is the *Pockels effect.* Figure 3.27 shows how these effects can be used to control the intensity of a beam of light.

Other more complicated forms of optical anisotropy exist also, but are beyond the scope of the present treatment. Before leaving the subject however, one important associated effect deserves mention, namely that of *anisotropic absorption.* Different directions of polarization of a transmitted beam can be associated with different absorption coefficients as well as with

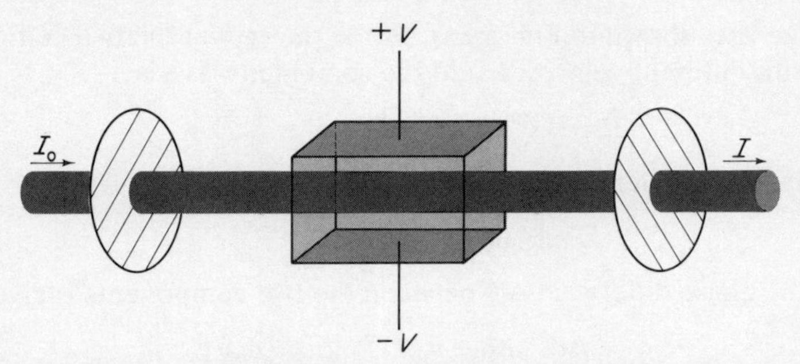

Fig. 3.27. Schematic example of how the Kerr or Pockels effects can be used to control the intensity of a beam of light. The medium (liquid or crystal) is placed in a cell between two crossed polarizers. The voltage V (which can also be applied longitudinally in the case of the Pockels effect) is switched between two values. For the first the cell acts as a half-wave plate, causing the plane of polarization of the light to be rotated by 90°, and the intensity I to be half the intensity I_0 of the incident beam (assumed unpolarized). For the other value of V the plane of polarization is unaltered, and no light is passed through the second linear polarizer. The cell itself is known as a Kerr cell or Pockels cell.

different refractive indices. This phenomenon exists in anisotropic crystals, but is present to a much more marked degree in certain man-made materials such as polaroid. In these one direction of linear polarization is absorbed almost completely, leaving only the other direction of polarization in the transmitted beam.

3.2.5 Non-linear effects

Some interesting and useful effects occur when the electric field strength in an intense beam (this invariably means a laser beam) approaches the internal electric field strength that exists in a medium. To estimate the magnitude of a typical internal field, let us consider the field acting on an electron at the site of a positive ion in a crystal. If the field is effectively that of a point charge at a distance r_0 of approximately 0.1 to 0.2 nm from the electron, we find that

$$E^{\text{internal}} \sim \frac{e^2}{4\pi\varepsilon_0 r_0^2} \sim 10^{11}\ \text{V/m}.$$

If E^{beam} is very much smaller than this it will act on the electron in the manner described in the previous sections. On the other hand if $E^{\text{beam}} \gtrsim E^{\text{internal}}$, the medium will almost certainly break down. The non-linear effects that we want to discuss occur at intermediate fields, from $\sim 10^7$ V/m to $\sim 10^9$ V/m, for which the polarization P can be expressed as

$$P = \varepsilon_0(\chi_1 E + \chi_2 E^2 + \chi_3 E^3 + \cdots),$$

where χ_1 is the electric susceptibility of the dielectric and χ_2, χ_3, etc., are higher order coefficients. In general **P** and **E** are not in the same direction for anisotropic (for example crystalline) dielectrics, so that the χ_n depend on the directions of **P** and **E**, but this is not important for the present qualitative treatment. Also some of the χ_n may vanish for symmetry reasons, but we confine our attention to those media for which χ_2 is non-zero.

When the incident beam contains the single angular frequency ω_1,

$$E = E_0 \cos(\omega_1 t - k_1 z),$$

the time dependence of the polarization is given by

$$P/\varepsilon_0 = \chi_1 E_0 \cos(\omega_1 t - k_1 z) + \tfrac{1}{2}\chi_2 E_0^2\{1 + \cos[2(\omega_1 t - k_1 z)]\} + \tfrac{1}{4}\chi_3 E_0^3\{3\cos(\omega_1 t - k_1 z) + \cos[3(\omega_1 t - k_1 z)]\} + \cdots$$

The fluctuating polarization P itself generates electromagnetic radiation, and we see that this contains the harmonics $2\omega_1$, $3\omega_1$, etc., as well as the fundamental frequency ω_1. The frequency doubled term usually has a much higher intensity than the higher order harmonics. The process is called *frequency doubling*.

Figure 3.28 illustrates how the frequency doubled radiation is produced and how it propagates through the medium. The positions of the anti-nodes of the polarization term $\cos[2(\omega_1 t - k_1 z)]$ are represented by the shaded

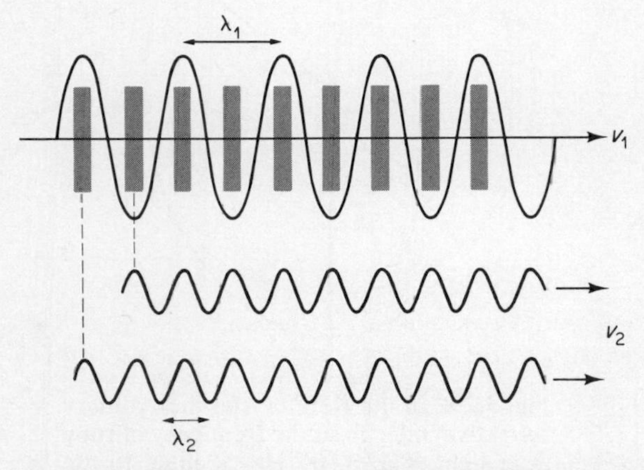

Fig. 3.28. The incident beam of wavelength λ_1 travels through the medium with velocity v_1. The shaded regions represent the maxima of the polarization term which generates the frequency doubled radiation of wavelength λ_2 and velocity v_2.

regions on the figure. We can regard these anti-nodes as the 'generators' of the frequency doubled radiation. The wavelets produced by these generators all have the same phase in the forward direction (as illustrated in the figure for two of the wavelets) if the phase velocity (v_2) of the frequency doubled radiation is the same as the phase velocity (v_1) of the generators. This *phase-matching* condition can also be written as

$$n(\omega_2) = n(2\omega_1) = n(\omega_1), \tag{3.88}$$

where $n(\omega)$ is the refractive index of the medium at the angular frequency ω. When this is satisfied the wavelets add coherently in the forward direction, while in other directions their sum averages to zero. The medium must of course be transparent at the frequency $2\omega_1$ as well as at ω_1.

In a region of normal dispersion, $n(2\omega)$ is greater than $n(\omega)$, and therefore condition (3.88) cannot be satisfied. A way out of this difficulty comes from the existence of birefringent crystals (see previous section). For example, the KDP (potassium dihydrogen phosphate, KH_2PO_4) crystal, a negative uniaxial crystal, can be used for frequency doubling. The ordinary and extraordinary refractive indices $n_0(\omega)$ and $n_e(\omega)$ both increase with ω, but at a

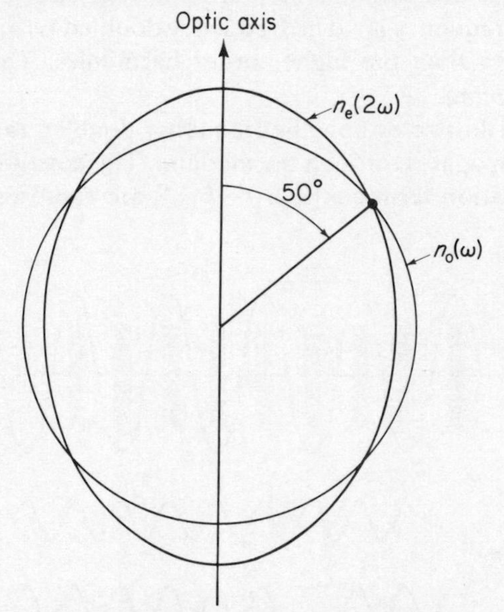

Fig. 3.29. In the KDP crystal the ordinary refractive index n_0 at the frequency of ruby laser light (4.32×10^{14} Hz) is equal to the extraordinary refractive index n_e at twice this frequency, when the direction of propagation of the light is at 50° to the optic axis of the crystal. The difference between n_0 and n_e is exaggerated for clarity.

given value of ω $n_e(\omega)$ is always less than $n_O(\omega)$ for this crystal. This makes it possible to find frequencies and directions of propagation for which

$$n_e(2\omega) = n_O(\omega). \tag{3.89}$$

This technique is known as *index-matching*. Figure 3.29 shows the values of $n_e(2\omega)$ and $n_O(\omega)$ for the KDP crystal, at the frequency corresponding to the 694 nm line of the ruby laser. Index matching occurs when the beam is at 50° to the optic axis, and then the efficiency for producing the frequency doubled light of wavelength 347 nm has its maximum value of approximately 20 to 30%.

PROBLEMS: CHAPTER 3

(Answers to selected problems are given in Appendix E)

3.1 An infinitely large, flat conducting sheet lies in the x–y plane and has a vanishingly small thickness in the z direction. A surface current of density $\mathbf{I} = \mathbf{i}_x I_0 \exp(j\omega t)$ ampère per metre flows in the sheet. Show that Maxwell's equations are satisfied if the surrounding electromagnetic field is

$$\mathbf{E} = \mathbf{i}_x E_0 \exp(j\omega t - kz), \qquad \mathbf{B} = \mathbf{i}_y E_0/c$$

at $z > 0$, and

$$\mathbf{E} = \mathbf{i}_x E_0 \exp(j\omega t + kz), \qquad \mathbf{B} = -\mathbf{i}_y E_0/c$$

at $z < 0$, where $E_0 = -\mu_0 c I_0/2$. Show also that the power carried away by the radiation field is equal to the power expended in maintaining the current flow in the presence of the electric field.

3.2 A point charge q oscillates sinusoidally along a straight line, with an amplitude l_0 and a maximum velocity v_0. Show that if the charge acts as a Hertzian dipole and satisfies the condition $l_0 \ll \lambda$, then its velocity is non-relativistic.

3.3 Electric quadrupole radiation is produced when two Hertzian dipoles of equal strength but oscillating 180° out of phase are placed side by side with a small separation between them. Show that the strength of the radiation field is proportional to the cube of the frequency.

3.4 Show that Eqs. (3.33) and (3.35) for the radiation field of an accelerating particle are equivalent (neglecting the shorter range terms varying as $1/s^2$).

3.5 Cerenkov detectors are detectors of charged particles in which the emitted radiation is used to establish the charge and energy of the particles. If the medium of such a detector has a dielectric constant of 2.0, and if the angle between the Cerenkov radiation and the direction of a proton travelling through the detector is measured to be $55 \pm 1°$, what is the energy of the particle?

3.6 When the same pulsar is observed at 300 MHz and 900 MHz it is found that the low frequency signal is delayed by 0.1 s with respect to the high frequency signal. If the delay is caused by the presence of electrons in the intervening space, and the group velocity is given by Eq. (3.66), show that there are 7.5×10^{22} electrons per m^2 column in the line of sight.

3.7 Show that Eq. (3.65) for the group velocity can be written also in the form

$$v_g = v_p\left(1 - \frac{\lambda}{n}\frac{dn}{d\lambda}\right)^{-1}.$$

How much longer would a 1 ns pulse of light of mean wavelength 550 nm take to travel 10 km in air than in a vacuum if the refractive index of the air is

$$n = 1.00027 + 1.5 \times 10^{-18}/\lambda^2$$

(where λ is in metres)?

3.8 When a spacecraft re-enters the earth's outer atmosphere its velocity and temperature are sufficient to ionize the atoms in its vicinity. If the density of the plasma formed in this way is $10^{14}\ \mathrm{m}^{-3}$, what is the effect on radio communications between the spacecraft and earth?

3.9 Estimate the range of angles of incidence for which a polished graphite surface can be used to totally externally reflect a beam of 2.5 keV X-rays in vacuum.

3.10 Show that when the function

$$\varepsilon = 1 + \frac{a}{(\omega_0^2 - \omega^2) + jb}$$

is plotted on an Argand diagram it lies on a circle of radius $a/2b$, centred at $1 - ja/2b$. Draw this circle for $a = 1$, $b = 0.2$, $\omega_0 = 1$, marking a few points for representative values of ω. Use this Argand diagram to estimate $\varepsilon^{1/2}$, and plot $\mathrm{Re}\,(\varepsilon^{1/2})$ and $-\mathrm{Im}\,(\varepsilon^{1/2})$ as functions of ω. Note the asymmetries in these functions. Compare them with those shown in Fig. 3.23. Give a mathematical reason for the fact that n cannot be negative.

3.11 A uniaxial crystal is placed between crossed linear polarizers having their axes at 45° to the optic axis of the crystal. It is found that when the thickness of crystal is 0.5025 mm the attenuation of this system is least at a wavelength of 500 nm. It is known also that at this wavelength $n_o - n_e = 0.1$, and that this difference changes by 0.1% for a 1% change in the wavelength. What range of wavelengths is passed when the system is illuminated with broad-band light in the neighbourhood of 500 nm?

CHAPTER 4

The introduction of quantum ideas

The propagation and generation of radiation have been treated so far in almost purely classical terms. In this chapter we shall initially continue in this vein in discussing the 'thermal' radiation emitted by hot bodies, but shall discover that the use of classical concepts alone leads to a physically impossible situation. This is the so-called '*ultraviolet catastrophe*', the fly in the ointment that spoiled an otherwise almost unblemished description of electromagnetic radiation for the 19th century physicists. The way in which this problem was resolved, through the introduction of the idea of quantized energies, is the subject of the first section of the chapter: we shall see how the properties of thermal radiation are explained by the existence of the *photon*, the quantum of the radiation field. In the second section the quantum idea is applied to the processes of absorption and emission of radiation by atoms, enabling the transition probabilities to be related to the oscillator strengths discussed in purely classical terms in Chapter 3.

The next important step, that of relating the oscillator strengths and transition probabilities to the internal structure of the atomic states, is discussed in the third section. The approach here is *semi-classical*, in the sense that the atomic states are treated quantum-mechanically while the radiation field and its interaction with the atoms is treated classically. We shall see that the transition probabilities of absorption and emission are completely determined by the wavefunctions of the participating atomic states. In the fourth section we look at the transition probabilities from a different point of view, in terms of the *quantum-numbers* of the atomic states. These quantum numbers are often known even when there is very little knowledge of the spatial dependence of the atomic wavefunctions. We shall see that there are *selection rules* which determine whether a given transition is likely to be strong or weak. Finally we shall consider the weak transitions, and discuss the relevant selection rules and transition probabilities.

4.1 RADIATION IN THERMAL EQUILIBRIUM

We want to find the energy of a classical radiation field inside a closed cavity, when the radiation and cavity walls are in thermal equilibrium at temperature T. The easiest type of cavity to consider is a cubic cavity with perfectly reflecting and conducting internal walls, enclosing a vacuum.

The electric field $\mathbf{E}$ inside the cavity must satisfy three conditions:

(i) It must satisfy the boundary condition (2.55) for $E_{\parallel}$, which implies that $E_{\parallel}$ is zero very near to the cavity walls because it is zero inside the material of the walls.

(ii) It must satisfy Maxwell's equation (2.1)

$$\operatorname{div} \mathbf{E} = 0$$

everywhere inside the cavity.

(iii) It must satisfy the wave equation (2.18)

$$\nabla^2 \mathbf{E} = \frac{1}{c^2}\frac{\partial^2 \mathbf{E}}{\partial t^2}$$

everywhere inside the cavity.

If these three conditions are satisfied then all the remaining boundary conditions and the remaining Maxwell's equations are also satisfied, for $\mathbf{B}$ as well as $\mathbf{E}$.

A particular electric field, representing a *standing wave* inside the cubic cavity, is the field

$$\begin{aligned}\mathbf{E} = \Big[&\mathbf{i}_x E_{0x} \cos\frac{l\pi x}{a} \sin\frac{m\pi y}{a} \sin\frac{n\pi z}{a} \\ &+ \mathbf{i}_y E_{0y} \sin\frac{l\pi x}{a} \cos\frac{m\pi y}{a} \sin\frac{n\pi z}{a} \\ &+ \mathbf{i}_z E_{0z} \sin\frac{l\pi x}{a} \sin\frac{m\pi y}{a} \cos\frac{n\pi z}{a}\Big] \cos(\omega t + \phi), \end{aligned} \tag{4.1}$$

where the cube occupies the space $0 \leqslant x \leqslant a$, $0 \leqslant y \leqslant a$, $0 \leqslant z \leqslant a$, and where l, m, and n are arbitrary integers (positive or zero) which determine the number of nodes and antinodes in the standing wave pattern. We see that the first of the three conditions is satisfied for this field. For example at the two faces $x = 0$ and a, the parallel components of the field (that is the y and z components) are zero because $\sin(l\pi x/a)$ is zero at both these faces. Applying the second condition, we find

$$lE_{0x} + mE_{0y} + nE_{0z} = 0.$$

This can also be expressed as

$$\mathbf{k} \cdot \mathbf{E}_0 = 0, \tag{4.2}$$

where

$$\mathbf{k} = \frac{\pi}{a}(\mathbf{i}_x l + \mathbf{i}_y m + \mathbf{i}_z n), \tag{4.3}$$

and so we see that the electric field is transverse to the wavevector $\mathbf{k}$, as in the case of a freely propagating wave. Finally, the third condition gives

$$(l^2 + m^2 + n^2)\left(\frac{\pi}{a}\right)^2 = \left(\frac{\omega}{c}\right)^2.$$

The frequency ν is therefore given by

$$\nu = \frac{c}{2a}(l^2 + m^2 + n^2)^{1/2}. \tag{4.4}$$

Each field of the type (4.1) is known as a *normal mode* of the cavity. The wavevector and frequency are completely defined by the integers l, m, and n. There are two independent directions of $\mathbf{E}_0$ (that is, two directions of polarization) that satisfy condition (4.2), and thus two independent normal modes are associated with each set of values (l, m, n). The energy in a mode is proportional to $|\mathbf{E}_0|^2$.

The next step to finding the total energy in the cavity is to find the *number* of normal modes in the frequency interval from ν to $\nu + \mathrm{d}\nu$. Figure 4.1 will help us to do this. Here we have taken two dimensions only (x and y), and have represented each mode (l, m) by a point having the co-ordinates $(lc/2a, mc/2a)$. We see from Eq. (4.4) that the frequency of the mode is then the distance from the origin to the point. To find the number of modes $N(\nu)\,\mathrm{d}\nu$ between ν and $\nu + \mathrm{d}\nu$ we need only count the number of points between the two arcs shown in the figure. This number is approximately equal to the area enclosed by the arcs $(\frac{1}{2}\pi\nu\,\mathrm{d}\nu)$ divided by the area occupied by each point $((c/2a)^2)$. The extension to three dimensions is straightforward. The arcs become octants of spherical surfaces, and we find

$$N(\nu)\,\mathrm{d}\nu = \frac{\frac{1}{8}(4\pi\nu^2\,\mathrm{d}\nu)}{(c/2a)^3} \times 2.$$

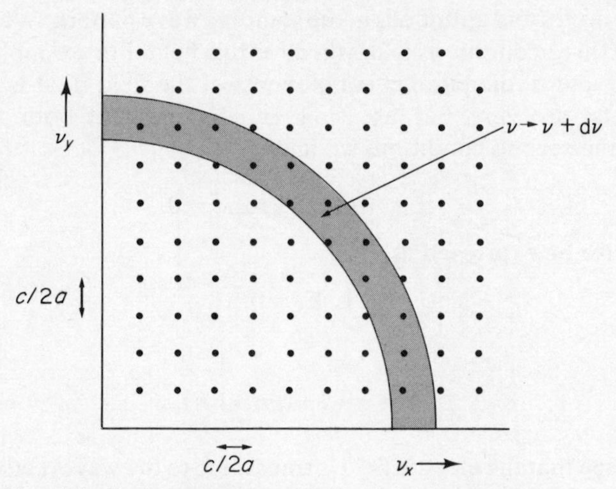

Fig. 4.1. Each normal model (l, m) is represented by a point having the co-ordinates $(lc/2a, mc/2a)$. The distance from the origin to a point is the frequency of the corresponding normal mode. The two arcs have radii ν and $\nu+d\nu$. The number of points enclosed between them is the number of normal modes having frequencies between ν and $\nu+d\nu$.

The factor 2 arises from the two directions of polarization associated with each point. $N(\nu)$ is proportional to the volume of the cavity, and so the number of modes per unit volume and per unit frequency interval, called the *mode density*, is

$$\boxed{\rho(\nu)=\frac{8\pi\nu^2}{c^3}} \tag{4.5}$$

It can be shown that this expression holds for a cavity of any shape.

Because the walls of the cavity and the thermal radiation contained within it are in thermal equilibrium at temperature T, classical theory requires that the probability $P(E)\,dE$ that any one of the modes has an energy between E and $E+dE$ is given by

$$P(E)=(1/kT)\,e^{-E/kT} \tag{4.6}$$

where the exponential factor is known as *Boltzmann's factor*, and k is *Boltzmann's constant.* The normalization of $P(E)$ ensures that the total probability is unity,

$$\int_0^\infty P(E)\,dE=1.$$

The mean energy of a mode is given by

$$\bar{E}_{\mathrm{cl}}=\int E\,P(E)\,\mathrm{d}E=kT. \tag{4.7}$$

Therefore all modes possess the *same* mean energy, regardless of their frequency. The energy contained per unit volume and per unit frequency interval is therefore predicted to be

$$u(T,\nu)=\frac{8\pi\nu^2kT}{c^3}. \tag{4.8}$$

This is known as the *Rayleigh–Jeans formula*. It fits experimentally observed energy densities at low frequencies (we shall see below how low these frequencies must be), but it fails completely at high frequencies. Also it has the obvious weakness of giving an infinite total energy U per unit volume, since

$$U(T)=\int_0^\infty u(T,\nu)\,\mathrm{d}\nu=\frac{8\pi kT}{c^3}\int_0^\infty \nu^2\,\mathrm{d}\nu=\infty.$$

This is the *ultraviolet catastrophe* referred to earlier, so called because the high frequency (u.v.) end of the spectrum gives rise to the infinity.

This obvious absurdity was resolved by the introduction by Planck, in 1901, of a radically new concept—a concept that eventually led to the quantum theory of matter and radiation. Planck took a model in which the walls of the cavity are assumed to contain localized resonators (for example Hertzian dipoles) that can emit and absorb radiation. His work implied that the resonators do not possess a continuous range of energies but exist instead only at the discrete energies $nh\nu$, where n is any integer, ν is the frequency of oscillation and h is a universal constant now known as *Planck's constant.* In other words, the energy of the oscillators is *quantized.* Four years later Einstein went further and suggested that the radiation field is itself quantized, the quantum of energy being $h\nu$ at the frequency ν. These radiation quanta are now known as *photons.*

Let us see how the existence of photons affects the radiation energy density. Each normal mode must contain a whole number of photons. The total energy of a mode can therefore take only the *discrete* values

$$E_n=nh\nu, \qquad n=0,1,2,\ldots, \tag{4.9}$$

as opposed to the *continuum* of energies that is assumed to exist in the classical approach*. If the Boltzmann factor is retained, the probability $P(n)$ that a mode contains n photons is given by

$$P(n)=\alpha\,\mathrm{e}^{-nh\nu/kT}. \tag{4.10}$$

* Each mode has also an invariable zero-point energy $\frac{1}{2}h\nu$, which does not affect the present treatment (see Appendix C).

The proportionality constant α can be found by requiring that the total probability be unity, giving

$$\alpha = 1 - e^{-h\nu/kT}.$$

The mean energy in the mode is therefore

$$\bar{E}_{qu} = \sum_{n=0}^{\infty} E_n P(n) = (1 - e^{-h\nu/kT}) \sum_{n=0}^{\infty} nh\nu\, e^{-nh\nu/kT},$$

which can be simplified to

$$\bar{E}_{qu} = \frac{h\nu}{e^{h\nu/kT} - 1}. \tag{4.11}$$

This new value of the mean energy is the same as the classical result in the limit $h\nu \ll kT$. From Eqs. (4.7) and (4.11) we have

$$\frac{\bar{E}_{cl}}{\bar{E}_{qu}} = \frac{e^{h\nu/kT} - 1}{h\nu/kT} = 1 + \frac{1}{2!}\left(\frac{h\nu}{kT}\right) + \frac{1}{3!}\left(\frac{h\nu}{kT}\right)^2 + \cdots.$$

This is why the classical Rayleigh–Jeans formula agrees with the experimental results at low frequencies ($h\nu \ll kT$). On the other hand $\bar{E}_{qu}$ is much smaller than $\bar{E}_{cl}$ at large ν (when $h\nu \gg kT$). This therefore removes the ultraviolet catastrophe.

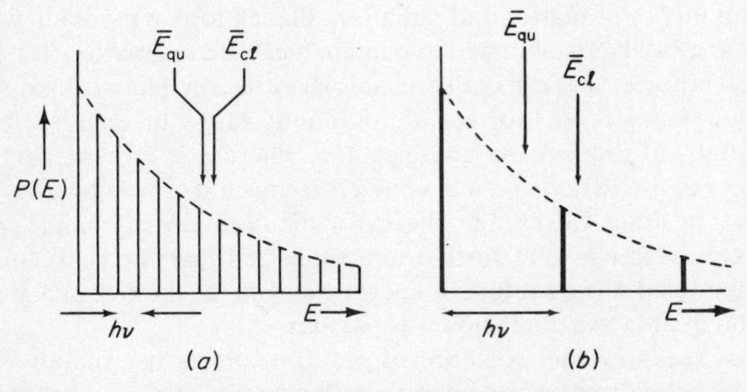

Fig. 4.2. Relative probabilities of a mode of frequency ν having an energy E at temperature T when (a) $h\nu \ll kT$, and (b) $h\nu \sim kT$. The broken curves show the Boltzmann factor $e^{-E/kT}$.

These points are illustrated in Fig. 4.2, which shows the quantal probability distribution for E (Eq. (4.10)) for two cases, $h\nu \ll kT$ and $h\nu \sim kT$. In the second case the high probability of the zero energy mode causes the mean energy to be significantly less than kT.

The radiation energy per unit volume and unit frequency interval at the temperature T is obtained by multiplying the mode density (Eq. (4.5)) by the mean energy per mode (Eq. (4.11)), giving

$$u(T, \nu) = \frac{8\pi\nu^2}{c^3} \frac{h\nu}{e^{h\nu/kT} - 1} \tag{4.12}$$

This is the celebrated *Planck radiation law*. It has been found to accurately represent observed energy densities at all the frequencies and temperatures that have been investigated.

It is often more convenient for experimental comparisons to express the radiant energy density in terms of wavelength rather than frequency. Using the relation $c = \nu\lambda$ we see that the wavelength range $d\lambda$ spanned by the frequency interval $d\nu$ is given by

$$|d\lambda| = \frac{c}{\nu^2}|d\nu|,$$

and therefore the energy density in the interval from λ to $\lambda + d\lambda$ is

$$u(T, \lambda)\, d\lambda = \frac{8\pi ch}{\lambda^5(e^{hc/\lambda kT} - 1)}\, d\lambda. \tag{4.13}$$

In reality the walls of the cavity are not perfectly reflecting, as we assumed in the derivation above, but are surfaces which absorb part of the radiation falling on them. They also emit thermal radiation. When the walls are in thermal equilibrium with the cavity radiation the rates of absorption and emission of radiation energy must be the same, or else the temperature of the walls, and of the radiation in the cavity, would change. Furthermore this equality of absorption and emission rates must apply at all frequencies. The radiation energy density is therefore independent of the nature of the walls, and Planck's law applies whenever there is thermal equilibrium.

The energy density $u(T, \lambda)$ of radiation in space is measured less easily than the power $w(T, \lambda)$ emitted per unit area and unit wavelength interval by a hot surface. We must therefore establish the connection between these two quantities. A particular idealized form of surface for which to do this is the surface which absorbs *all* the radiation falling onto it. It is called a *black-body surface.* Because it absorbs more than any other surface it also emits more. To find how much it emits we must first consider the angular distribution of the emitted radiation. The form of this can be deduced from the fact that the sun looks like a uniformly illuminated flat disc. This apparent uniformity applies as well to any other body which is hot enough for it to radiate much more light than it reflects. We see from Fig. 4.3 that the radiation emitted from an area A, at an angle θ to the normal, appears to an observer to have come from a projected area $A \cos\theta$. Because the received

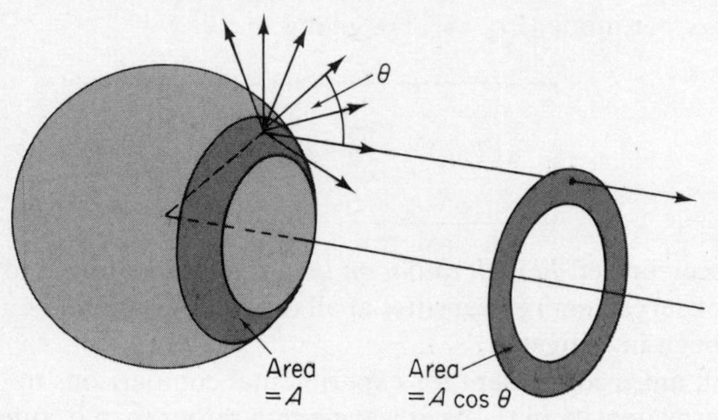

Fig. 4.3. The radiation emitted from an area A at an angle θ to the surface normal, appears to have come from a projected area $A \cos \theta$.

radiation power per unit projected area is constant, the emitted radiation power must be proportional to $\cos \theta$. This proportionality is known as *Lambert's cosine law*. Since the total power emitted per unit area and wavelength interval is $w(T, \lambda)$, the part emitted into a solid angle $d\Omega$ in the direction θ must be

$$\pi^{-1} w(T, \lambda) \cos \theta \, d\Omega = b(T, \lambda) \cos \theta \, d\Omega, \quad \text{say.} \tag{4.14}$$

Now we know the angular distribution we can find the connection between $u(T, \lambda)$ and $w(T, \lambda)$ by considering the energy contained in a small region of arbitrary shape situated at the centre of a spherical cavity of radius R (Fig. 4.4). The part of the region shown shaded in the figure has a length l and a projected area ΔA, and is traversed by radiation emitted normally from the element of area dS, in the direction of the cone of solid angle $\Delta A/R^2$ (where $l \ll R$). The radiant power emitted per unit wavelength interval into this cone is $b(T, \lambda) \, dS(\Delta A/R^2)$, and since the radiation travels a distance c in one second the energy contained within the shaded volume is

$$dE = b(T, \lambda) \, dS\left(\frac{\Delta A}{R^2}\right)\frac{l}{c}.$$

By considering other cones emanating from dS and traversing the central region, we find that the total energy per unit wavelength interval in this region, caused by the element dS, is

$$\Delta E = \frac{b(T, \lambda)}{R^2 c} \Sigma(\Delta A \, l) = \frac{b(T, \lambda)}{R^2 c} V,$$

where V is the volume of the region. Therefore the energy density within V

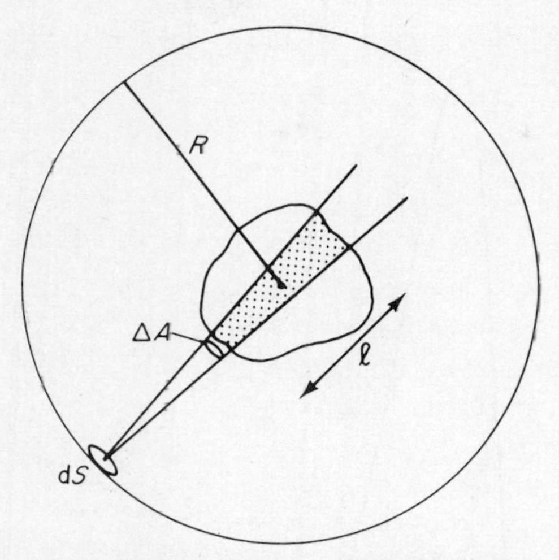

Fig. 4.4. The shaded volume contains energy emanating from the area dS on the inner surface of a spherical cavity at temperature T.

caused by the whole of the area $4\pi R^2$ of the cavity walls is

$$u(T, \lambda) = \frac{b(T, \lambda)4\pi}{c} = \frac{4w(T, \lambda)}{c}.$$

The *emissive power* of a black-body surface at temperature T is therefore

$$w(T, \lambda) = \frac{2\pi c^2 h}{\lambda^5(e^{hc/\lambda kT} - 1)}. \tag{4.15}$$

This form of the radiation law is plotted in Fig. 4.5, for three different temperatures.

We can see from Fig. 4.5 that the wavelength λ_m at which $w(T, \lambda)$ is a maximum decreases as T increases. The exact dependence of λ_m on T can be deduced from Eq. (4.15). It is

$$\lambda_m T = \text{constant}. \tag{4.16}$$

This is known as *Wien's displacement law*. The value of the constant is 2.896×10^{-3} m K.

The total power radiated by unit area of a black-body at temperature T is obtained by integrating Eq. (4.15). The result is

$$W = \sigma T^4, \tag{4.17}$$

where

$$\sigma = \frac{2\pi^5 k^4}{15c^2 h^3} = 5.67 \times 10^{-8}\ \text{J K}^{-4}\ \text{m}^{-2}\ \text{s}^{-1}$$

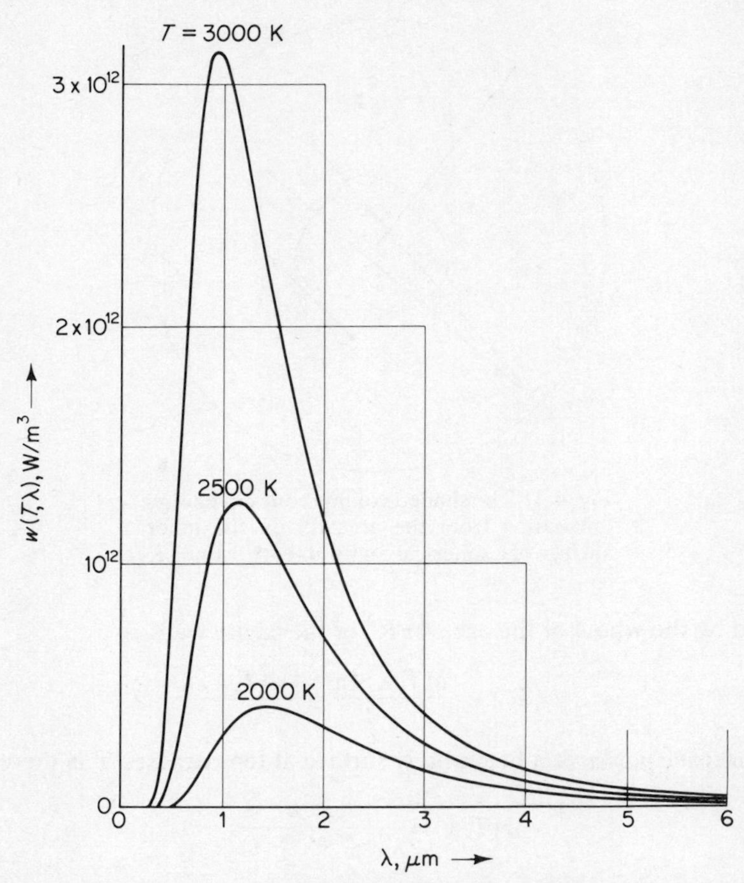

Fig. 4.5. The power $w(T, \lambda)$ radiated per unit area and unit frequency interval by a black body at temperature T.

is *Stefan's constant*, and Eq. (4.17) is known as the *Stefan–Boltzmann law*.

It is interesting to note that Planck's own derivation of his law (see Appendix B) was based on the assumption that the radiation field has a continuum of energies and that only the resonators of the wall are quantized. Einstein's hypothesis that the radiation field is itself quantized has proved to be far more valuable, but was not generally accepted until almost 20 years after it was first proposed (see Chapter 1).

4.2 EINSTEIN'S *A* AND *B* COEFFICIENTS

An atom in a radiation field can absorb energy at certain characteristic frequencies, and in doing so is raised to an excited state. If it is already in an excited state it can spontaneously emit radiation, falling to a lower state. The existence of both these processes is self-evident, but in 1916 Einstein showed theoretically that a less obvious process must exist also, namely that in which a radiation field stimulates an already excited atom to emit radiation. This is called *stimulated emission.*

Absorption can also be described as the process of *annihilation* of a photon, and spontaneous and stimulated emission as the processes of spontaneous and stimulated *creation* of a photon. In these terms the three processes can be expressed as

$$\text{absorption:} \qquad h\nu + A \rightarrow A^*, \tag{4.18}$$

$$\text{spontaneous emission:} \qquad A^* \rightarrow A + h\nu, \tag{4.19}$$

$$\text{stimulated emission:} \qquad h\nu + A^* \rightarrow A + 2h\nu, \tag{4.20}$$

where A represents an atom in a state of energy $\mathscr{E}_1$, A^* represents an atom in a state of higher energy $\mathscr{E}_2$, and $h\nu$ represents a single photon of energy

$$h\nu = \mathscr{E}_2 - \mathscr{E}_1.$$

The three processes are illustrated in Fig. 4.6.

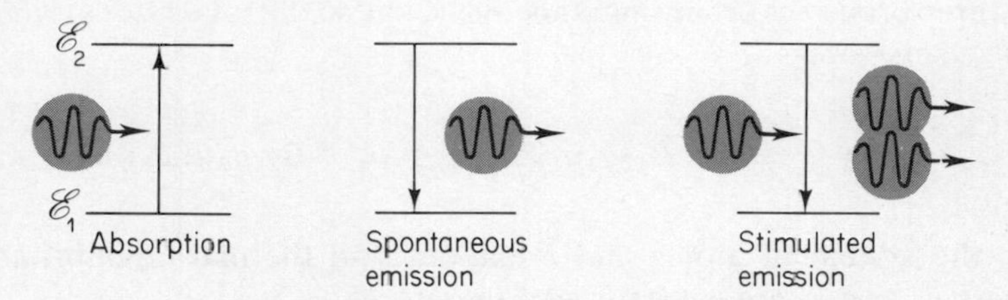

Fig. 4.6. The three fundamental processes of absorption, spontaneous emission and stimulated emission.

We can investigate the connections between the three processes (and at the same time show the existence of stimulated emission) by considering a low density gas composed of atoms (or molecules) that have only the two levels of energy $\mathscr{E}_1$ and $\mathscr{E}_2$ (real atoms have more than two levels of course, but this model suffices for the present purpose). The atoms are in thermal equilibrium with a radiation field at temperature T, and the energy density of the radiation is $u(T, \nu)$ per unit frequency interval. The number of photons absorbed in time $\mathrm{d}t$ by atoms in level 1 is proportional to the number n_1 of atoms in this level, and it is proportional also to $u(T, \nu)$, where

$$\nu = (\mathscr{E}_2 - \mathscr{E}_1)/h.$$

Therefore the rate of decrease of n_1 can be expressed as

$$-\frac{dn_1}{dt} = B_{12} n_1 u(T, \nu) = \frac{dn_2}{dt}, \tag{4.21}$$

where B_{12} is called the *Einstein coefficient for absorption*. The last part of this equation follows because the rate of decrease of n_1 must equal the rate of increase of the number n_2 in the upper level. In the process of spontaneous emission the rate of decrease of n_2 is proportional to n_2 but is independent of $u(T, \nu)$, and so we can write

$$-\frac{dn_2}{dt} = \frac{dn_1}{dt} = A_{21} n_2, \tag{4.22}$$

where A_{21} is called the *Einstein coefficient for spontaneous emission*. Finally, for the process of stimulated emission the rate of decrease of n_2 is proportional to n_2 and is proportional also to $u(T, \nu)$, and so

$$-\frac{dn_2}{dt} = \frac{dn_1}{dt} = B_{21} n_2 u(T, \nu), \tag{4.23}$$

where B_{21} is called the *Einstein coefficient for stimulated emission*.

All three processes occur simultaneously, and so the total rates at which n_1 and n_2 change are

$$\frac{dn_1}{dt} = -\frac{dn_2}{dt} = -B_{12} n_1 u(T, \nu) + A_{21} n_2 + B_{21} n_2 u(T, \nu).$$

When the system of atoms and radiation is in thermal equilibrium the numbers n_1 and n_2 are constant, and therefore

$$B_{12} n_1 u(T, \nu) = A_{21} n_2 + B_{21} n_2 u(T, \nu).$$

We know also that in thermal equilibrium the number of atoms in a level of energy $\mathscr{E}$ is proportional to the Boltzmann factor $\exp(-\mathscr{E}/kT)$, which implies that

$$\frac{n_2}{n_1} = e^{-(\mathscr{E}_2 - \mathscr{E}_1)/kT}.$$

Combining these two relationships gives

$$B_{12} u(T, \nu) = [B_{21} u(T, \nu) + A_{21}]\, e^{-(\mathscr{E}_2 - \mathscr{E}_1)/kT}. \tag{4.24}$$

Since this must apply for *all* values of ν and T, including arbitrarily high values of T for which $u(T, \nu)$ tends to infinity and the exponential factor tends to 1, it follows that

$$\boxed{B_{12} = B_{21}} \tag{4.25}$$

We see that there is a basic symmetry between the processes of annihilation and stimulated creation of photons. Substitution of Eq. (4.22) into Eq. (4.21) now gives

$$u(T, \nu) = \frac{A_{21}/B_{21}}{e^{(\mathscr{E}_2 - \mathscr{E}_1)/kT} - 1}, \tag{4.26}$$

which is clearly a form of Planck's law for thermal radiation (Eq. (4.12)). We see now why the process of stimulated emission must exist: only then are Planck's law and the Boltzmann factor consistent with the co-existence of atoms and radiation in thermal equilibrium. By comparing with Eq. (4.12) we see that

$$\boxed{A_{21} = \frac{8\pi h\nu^3}{c^3} B_{21}} \tag{4.27}$$

which gives a second relationship connecting the three Einstein coefficients.

Relative importance of stimulated and spontaneous emission

But what is the relative importance of the stimulated and spontaneous emission processes? The ratio of their transition rate is

$$\frac{B_{21} u(T, \nu)}{A_{21}} = \frac{u(T, \nu) c^3}{8\pi h\nu^3} = \frac{u(T, \nu)/h\nu}{\rho(\nu)}, \tag{4.28}$$

which is the number of *photons* per unit volume and unit frequency interval, divided by the number of *modes* per unit volume and unit frequency interval. In other words the ratio of rates is the average number of *photons per mode*. This is true even when the system of atoms and radiation is not in thermal equilibrium. For example we shall see in Chapter 6 that the average number of photons in certain modes of a laser cavity can be extremely large, and then stimulated emission occurs much more frequently than spontaneous emission. For thermal radiation, on the other hand, the average number of photons in a mode of frequency ν at temperature T is

$$m = \frac{1}{e^{h\nu/kT} - 1}$$

which is much less than unity when $h\nu \gg kT$.

Degenerate levels

The discussion has so far been restricted to a two-state model. Real atoms and molecules have many different states and many different energy levels. The states are specified by various quantum numbers that we need not concern ourselves with at the moment, and sometimes more than one state can have the same energy $\mathscr{E}_i$ and hence belong to the energy level i. The number of different states belonging to a level i is the *degeneracy* g_i of the level.

In generalizing the equations that we have obtained for the two-state non-degenerate model, we could decide that the populations n and the coefficients A and B now refer to the individual states and the transitions between them. It is more conventional, though, to work in terms of the total populations of the degenerate levels. Consider for example absorption from a level i of degeneracy g_i to a level j of degeneracy g_j. If the constituent states of the level i are labelled by m_i ($m_i = 1$ to g_i) the Einstein coefficient for absorption from the state m_i to the state m_j of the higher level is $B_{m_i m_j}$, and the absorption rate is proportional to $n_{m_i} B_{m_i m_j}$. The absorption rate for transitions from the state m_i to any of the states m_j is proportional to

$$n_{m_i} \sum_{m_j=1}^{g_j} B_{m_i m_j}.$$

Summing this over the states m_i, and dividing by g_i to give an average rate from *one* of the lower states to *any* of the upper states, we find that this average rate is proportional to

$$\frac{n_i}{g_i} \sum_{m_i=1}^{g_i} \sum_{m_j=1}^{g_j} B_{m_i m_j},$$

where n_i is the total population of the lower energy level. Therefore the Einstein coefficient which is appropriate to the total population is

$$B_{ij} = \frac{1}{g_i} \sum_{m_i=1}^{g_i} \sum_{m_j=1}^{g_j} B_{m_i m_j}.$$

The Einstein coefficients for spontaneous and stimulated emission are given similarly by

$$A_{ji} = \frac{1}{g_j} \sum_{m_i=1}^{g_i} \sum_{m_j=1}^{g_j} A_{m_j m_i},$$

$$B_{ji} = \frac{1}{g_j} \sum_{m_i=1}^{g_i} \sum_{m_j=1}^{g_j} B_{m_j m_i}.$$

The two equations (4.25) and (4.27) connecting the coefficients now become

$$\boxed{g_i B_{ij} = g_j B_{ji}} \tag{4.29}$$

and

$$\boxed{A_{ji} = \frac{8\pi h\nu^3}{c^3} B_{ji}} \tag{4.30}$$

Lifetimes and absorption cross-sections

Spontaneous emission is the *only* process that can occur for a level j if the radiation density $u(T, \nu)$ is negligible at all the frequencies corresponding to transitions from j to other levels i. The level j decays spontaneously to levels i of lower energy, and the total rate of change of its population is therefore

$$\frac{dn_j}{dt} = -n_j \sum_i A_{ji}$$

(if we disregard any re-population of level j by spontaneous decays from higher levels). The solution of this equation is

$$n_j = n_{j0}\, e^{-t/\tau_j},$$

where

$$\tau_j = \left(\sum_i A_{ji}\right)^{-1}, \tag{4.31}$$

and the summation is over all the levels of energy less than $\mathscr{E}_j$. The time τ_j is the *radiative lifetime* (see Section 3.1.7) of the level j. This connection between lifetimes and the absorption coefficient A_{ji} is sometimes useful in establishing the absolute magnitudes of the coefficients A_{ji} (and hence the absolute magnitudes of B_{ji} and B_{ij}) from experimental measurements of atomic lifetimes.

Another way of finding the absolute magnitudes of the Einstein coefficients is by making use of the expressions obtained in Chapter 3 for the cross-section for absorption from a level i to a level j. Near the resonant frequency ν_{ij}, given by

$$h\nu_{ij} = \mathscr{E}_j - \mathscr{E}_i, \tag{4.32}$$

the absorption cross-section per atom is (Eq. (3.76))

$$\sigma(\nu)=\frac{e^2 f_{ij}}{16\pi^2 m\varepsilon_0 c\tau_j[(\nu-\nu_{ij})^2+1/(4\pi\tau_j)^2]}. \tag{4.33}$$

Here τ_j is the lifetime of level j, and f_{ij} is the oscillator strength for the transition. Suppose that a beam having the radiation energy density $u(\nu)$ per unit volume and unit frequency interval passes through a region containing atoms in the level i. The amount of radiation energy in the interval ν to $\nu+\mathrm{d}\nu$ which passes through an area $\sigma(\nu)$ in one second is

$$\mathrm{d}\mathscr{E}=c\sigma(\nu)u(\nu)\,\mathrm{d}\nu.$$

The number of photons absorbed by each atom in one second is therefore

$$\mathrm{d}n=c\sigma(\nu)u(\nu)\,\mathrm{d}\nu/h\nu.$$

Integrating this over all frequencies (and assuming that $\tau_j^{-1} \ll \nu_{ij}$, and that $u(\nu)$ is uniform over the frequency region of absorption), we find that the average number of photons absorbed by each atom in one second is

$$\bar{n}=\frac{e^2 f_{ij} u(\nu_{ij})}{4m\varepsilon_0 h\nu_{ij}}.$$

But this number is also equal to $B_{ij}u(\nu_{ij})$ (Eq. (4.21)) and therefore the Einstein B coefficient is given by

$$\boxed{B_{ij}=\frac{e^2 f_{ij}}{4m\varepsilon_0 h\nu_{ij}}} \tag{4.34}$$

This connection between B_{ij} and f_{ij} does not depend on the value of τ_j, which is fortunate, since the value given by the purely classical theory (Section 3.1.7) is no more than an indication of the magnitude of actual lifetimes. The result (4.34) will be obtained again in the next section, by a quantum mechanical treatment. Two other results follow immediately by combining Eq. (4.34) with Eqs. (4.29), (4.30), and (4.31),

$$A_{ji}=\frac{2\pi e^2}{m\varepsilon_0 c^3}\frac{g_i\nu_{ij}^2 f_{ij}}{g_j}, \tag{4.35}$$

and

$$\tau_j=\frac{m\varepsilon_0 c^3}{2\pi e^2}g_j\left(\sum_i g_i\nu_{ij}^2 f_{ij}\right)^{-1}, \tag{4.36}$$

where, as before, the summation is taken only over the levels i having $\mathscr{E}_i<\mathscr{E}_j$.

4.3 THE CALCULATION OF TRANSITION PROBABILITIES

So far the only way in which the intrinsic properties of the atoms and molecules have entered our treatment is through the energies $\mathscr{E}_i$ of the excited states and the oscillator strengths f_{ij} connecting different states. In this section we relate the oscillator strengths to the wavefunctions of the atomic and molecular states.

Under the influence of an external perturbation an atom initially in a state m can be induced to undergo transitions to other states n. Our starting point is the result (obtained from first order perturbation theory*) that for small perturbations the probability that the atom is in a state n after a time t is $|a_n(t)|^2$, where $a_n(t)$ is given by

$$j\hbar\frac{da_n(t)}{dt}=e^{j\omega_{mn}t}\int \psi_n^* H' \psi_m \, d\tau. \tag{4.37}$$

Here H' is the interaction energy of the perturbation, ψ_m and ψ_n are the wavefunctions of the states m and n in the absence of the perturbation, the symbol $\hbar$ represents $h/2\pi$, and ω_{mn} is given by the energy difference of the two states,

$$\omega_{mn}=(\mathscr{E}_n-\mathscr{E}_m)/\hbar.$$

Of interest to us is the perturbation caused by a radiation field acting on the atom. The largest contribution to H' is then the interaction between the electric field $\mathbf{E}$ of the radiation and the electric dipole moment $\mathbf{p}$ of the atom. Ignoring the other contributions (which will be discussed in the next section), H' is given by

$$H'=-\mathbf{E}\cdot\mathbf{p}. \tag{4.38}$$

Consider now the effect of a particular radiation field, that of the linearly polarized plane wave

$$\mathbf{E}=\mathbf{i}_x E_0 \cos(\omega t-kz).$$

The energy density is

$$u=\tfrac{1}{2}\varepsilon_0 E_0^2,$$

and so the electric field acting on the atom can be written as

$$\mathbf{E}=\mathbf{i}_x\left(\frac{2u}{\varepsilon_0}\right)^{1/2}\tfrac{1}{2}[e^{j(\omega t-kz)}+e^{-j(\omega t-kz)}]. \tag{4.39}$$

Because the wavelength of visible light is much larger than atomic dimensions, the quantity kz $(=2\pi z/\lambda)$ varies very little over the volume of the atom, which allows us to put

$$e^{jkz}\simeq 1.$$

* See any book on quantum mechanics.

The perturbation energy is therefore

$$H' = -\left(\frac{u}{2\varepsilon_0}\right)^{1/2}(\mathrm{e}^{\mathrm{j}\omega t}+\mathrm{e}^{-\mathrm{j}\omega t})p_x. \tag{4.40}$$

The dipole moment of the atom is

$$\mathbf{p} = -\sum_{k=1}^{N} e\mathbf{r}_k, \tag{4.41}$$

where the $\mathbf{r}_k$ are the position vectors of the N atomic electrons, and the origin of co-ordinates is taken to be at the centre of the atom.

To calculate the amplitude $a_n(t)$ we substitute (4.40) into (4.37) and integrate with respect to time. The result is

$$a_n(t) = \left(\frac{u}{2\varepsilon_0\hbar^2}\right)^{1/2}(\mathbf{D}_{nm})\left[\frac{\mathrm{e}^{\mathrm{j}(\omega_{mn}+\omega)t}-1}{\omega_{mn}+\omega}+\frac{\mathrm{e}^{\mathrm{j}(\omega_{mn}-\omega)t}-1}{\omega_{mn}-\omega}\right], \tag{4.42}$$

where

$$\mathbf{D}_{nm} = \int \psi_n^*\left(-e\sum_{k=1}^{N}\mathbf{r}_k\right)\psi_m\,\mathrm{d}\tau. \tag{4.43}$$

The integration in $\mathbf{D}_{nm}$ is taken over the position co-ordinates of all the atomic electrons. The vector $\mathbf{D}_{nm}$ is called the *transition electric dipole moment* for the transition from m to n.

The first term in Eq. (4.42), with $\omega_{mn}+\omega$ in the denominator, is large if $\omega \simeq -\omega_{mn}$, which requires that a state n exists for which

$$\mathscr{E}_n = \mathscr{E}_m - \hbar\omega.$$

This term therefore applies to the process of stimulated emission, in which the perturbation of frequency ω stimulates the initial state to emit a photon. Similarly, the second term of the equation refers to absorption, and it is large if there exists a state n of energy

$$\mathscr{E}_n = \mathscr{E}_m + \hbar\omega.$$

These two types of transition are illustrated in Fig. 4.7. The states n and n' can emit or absorb further photons, but we restrict ourselves to time intervals that are sufficiently short for these secondary processes to be negligible.

Let us suppose that the frequency of the radiation is near to an absorption frequency, and that only absorption takes place. Then we need retain only the second term in Eq. (4.42), and so obtain

$$|a_n(t)|^2 = \frac{2u|\mathbf{D}_{nm}|^2}{3\varepsilon_0\hbar^2}\,\frac{\sin^2[\frac{1}{2}(\omega_{mn}-\omega)t]}{(\omega_{mn}-\omega)^2}.$$

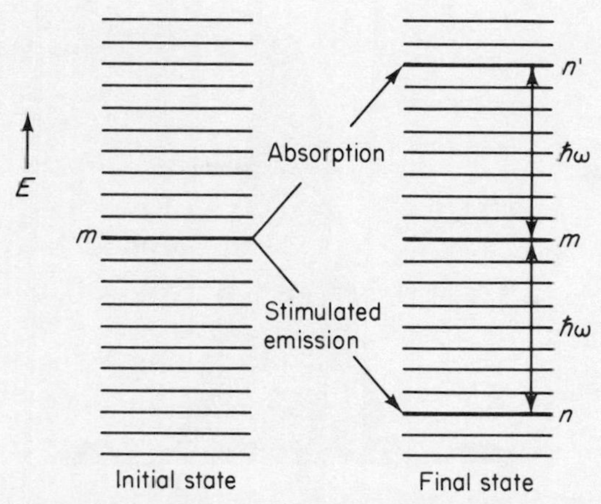

Fig. 4.7. A radiation field of angular frequency ω can cause either absorption or stimulated emission, if states of the appropriate energies exist. (The uniform level spacing is unrealistic, but is used for convenience).

Here we have put $|\mathbf{D}_{nm}|^2/3$ in place of $|(\mathbf{D}_{nm})_x|^2$, because the x, y and z components of $\mathbf{D}_{nm}$ are of equal magnitude. The probability $|a_n(t)|^2$ is shown in Fig. 4.8. The maximum probability occurs on resonance, when $\omega = \omega_{mn}$, and since

$$\lim_{x\to 0} \frac{\sin(xt)}{x} = t,$$

we see that this maximum value is proportional to t^2. The value of x at which $\sin^2(xt)/x^2$ has half its maximum value (t^2) is proportional to t^{-1}, and so the frequency width $\Delta\omega$ of the central peak of Fig. 4.8 is also proportional to t^{-1}.

In practice the radiation field cannot have a *single* frequency ω (see Chapter 5). The energy density u must be spread over a *range* of angular frequencies, giving an energy density $u(\omega)\,\mathrm{d}\omega$ in the interval from ω to $\omega + \mathrm{d}\omega$. Now we may assume that $u(\omega)$ is uniform over the interval for which $|a_n(t)|^2$ is significant: we are able to do this because the time in which absorption occurs can be made arbitrarily large by reducing the beam intensity, which makes $\Delta\omega$ arbitrarily small. The probability of absorption in the time t is then obtained by integrating over frequency

$$|a_n(t)|^2 = \frac{2|\mathbf{D}_{nm}|^2 u(\omega_{mn})}{3\varepsilon_0\hbar^2} \int \frac{\sin^2[\frac{1}{2}(\omega_{mn}-\omega)t]}{(\omega_{mn}-\omega)^2}\,\mathrm{d}\omega$$

$$= \frac{2\pi|\mathbf{D}_{nm}|^2 u(\omega_{mn})t}{6\varepsilon_0\hbar^2}. \tag{4.44}$$

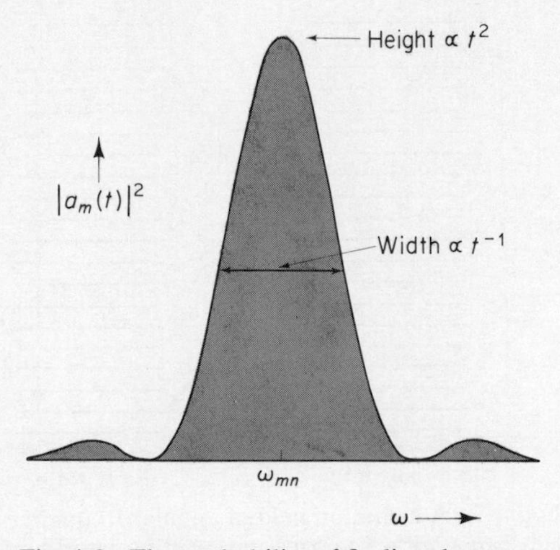

Fig. 4.8. The probability of finding the atom, initially in the state m, in the state n after irradiation at angular frequency ω for a time t. In practice the radiation field would usually have a uniform distribution of intensity in the region of ω_{mn}, and then the probability of being in state n is proportional to the shaded area, which is proportional to t.

We see that the probability is proportional to t. This proportionality also follows directly from Fig. 4.8, because the integral in Eq. (4.44) is proportional to the shaded area in the figure.

The *rate* of absorption is the probability per unit time, $|a_n(t)|^2/t$. To compare this with the absorption rate discussed in Section 4.2 it is convenient to express the energy density in terms of the frequency ν rather than the angular frequency ω. If a given interval corresponds to a range $d\nu$ in frequency and a range $d\omega$ in angular frequency, then clearly

$$d\nu = d\omega/2\pi.$$

If the energy in this interval can be expressed either as $u(\omega)\,d\omega$ or $u(\nu)\,d\nu$ then it follows that

$$u(\omega) = u(\nu)/2\pi.$$

We see therefore that the rate of absorption is given by

$$R = \frac{|\mathbf{D}_{nm}|^2 u(\nu_{mn})}{6\varepsilon_0\hbar^2}.$$

Comparing this with Eq. (4.18), we obtain a connection between the Einstein coefficient for absorption and the transition electric dipole moment,

$$B_{nm} = \frac{1}{6\varepsilon_0 \hbar^2} |\mathbf{D}_{nm}|^2. \tag{4.45}$$

To take account of possible degeneracies of the levels we follow the treatment used at the end of the previous section and redefine $|\mathbf{D}|^2$ by summing over the sub-levels of the final state and averaging over those of the initial state. For absorption from a degenerate level i to a degenerate level j,

$$|\mathbf{D}_{ij}|^2 = \frac{1}{g_i} \sum_{m_i=1}^{g_i} \sum_{m_j=1}^{g_j} |\mathbf{D}_{m_i m_j}|^2.$$

The form of Eq. (4.45) then remains unchanged,

$$\boxed{B_{ij} = \frac{1}{6\varepsilon_0 \hbar^2} |\mathbf{D}_{ij}|^2.} \tag{4.46}$$

Because

$$|\mathbf{D}_{m_i m_j}| = |\mathbf{D}_{m_j m_i}|,$$

we also have the relation

$$|\mathbf{D}_{ji}|^2 = \frac{g_i}{g_j} |\mathbf{D}_{ij}|^2.$$

This implies that

$$B_{ji} = \frac{g_i}{g_j} B_{ij},$$

as obtained previously (Eq. (4.29)). We can now use Eqs. (4.34) and (4.35) to obtain the Einstein coefficient for spontaneous emission, and the oscillator strength, in terms of $\mathbf{D}_{ij}$:

$$\boxed{A_{ij} = \frac{8\pi^2 \nu_{ij}^3}{3\varepsilon_0 \hbar c^3} |\mathbf{D}_{ij}|^2,} \tag{4.47}$$

$$\boxed{f_{ij} = \frac{4\pi m \nu_{ij}}{3\hbar e^2} |\mathbf{D}_{ij}|^2.} \tag{4.48}$$

Equations (4.46), (4.47), and (4.48), together with the equation for **D** (4.43), allow the experimentally useful quantities A, B and f to be calculated, if the wavefunctions of the participating states are known. There also exist sum rules for the f_{ij}, such as the sum rule (3.77) mentioned in Section 3.2.2, which are useful in checking calculational and experimental results.

We finish this section on a less rigorous note, with a qualitative classical interpretation of the process of spontaneous emission. We see from Eq. (4.47) that the spontaneous decay rate is proportional to $|\mathbf{D}_{ij}|^2$, and from Eq. (4.43) that $\mathbf{D}_{ij}$ refers to a charge distribution having the density $-e\psi_j^*\psi_i$. We can picture the atom as oscillating between the states i and j during an emission event, so that the charge density of the mixed, or transition, state acts as a Hertzian dipole of strength $\mathbf{D}_{ij}$ and angular frequency ω_{ij}. An electromagnetic field of frequency ω_{ij} is created. When this radiation field manifests itself as a single photon the atom is stabilized in the state of lower energy. The mean time for this to happen is proportional to $\hbar\omega_{ij}/P$, where P $(\propto|\mathbf{D}_{ij}|^2\omega_{ij}^4)$ is the mean power radiated (Eq. (3.21)). The transition rate is therefore proportional to $\omega_{ij}^3|\mathbf{D}_{ij}|^2$, as given by Eq. (4.47). This model can be taken further (see Corney[7]) to give the numerical factor in Eq. (4.47).

4.4 SELECTION RULES

The only restriction that we have met so far on whether or not a transition can take place is that energy must be conserved. Even when this condition is met there are other factors which can cause the transition probability to be very small or, in extreme cases, zero. These factors arise from the need to conserve linear momentum, angular momentum and parity. Conservation of linear momentum has a comparatively trivial effect (see the discussion of the Doppler effect in Chapter 9), but conservation of angular momentum and parity places certain restrictions, known as *selection rules*, on the allowed changes in the state of the atom, molecule or nucleus.

We start with a brief reminder of the meaning of the angular momentum and parity of an atom, and of the rules governing the addition of angular momenta and the combination of parities.

4.4.1 Angular momentum and parity of atomic states

The electrons of an atom in a state i can have a total angular momentum of magnitude $\sqrt{J_i(J_i+1)}\hbar$, where J_i is a positive integer (including zero) if there is an even number of electrons in the atom, or is an integer plus one half if there is an odd number of electrons. The nucleus of the atom also has an angular momentum, and this contributes to the total angular momentum of the atom, but this contribution can be ignored for the present discussion. We shall therefore refer to $\sqrt{J_i(J_i+1)}\hbar$ as the angular momentum of the atom. The component of the angular momentum along one fixed direction, called the direction of quantization, can have the values $M_i\hbar$, where

$$M_i = -J_i, -J_i+1, \ldots, J_i-1, J_i. \tag{4.49}$$

J_i and M_i are the angular momentum *quantum numbers* of the atom in state *i*. The angular momentum of the atom is often represented as $\mathbf{J}_i\hbar$, where the vector $\mathbf{J}_i$ has the length $\sqrt{J_i(J_i+1)}$, and the component M_i in the quantization direction. Other systems in which one or more particles move about a fixed point all have angular momenta which are quantized in this way.

When two separate systems having angular momenta $\mathbf{J}_1\hbar$ and $\mathbf{J}_2\hbar$ are combined to give a total angular momentum $\mathbf{J}_3\hbar$,

$$\mathbf{J}_1\hbar + \mathbf{J}_2\hbar = \mathbf{J}_3\hbar,$$

the magnitude of $\mathbf{J}_3$ is restricted to the set of values

$$J_3 = |J_1 - J_2|, |J_1 - J_2|+1, \ldots, J_1+J_2-1, J_1+J_2. \tag{4.50}$$

This can be expressed more succinctly as the *triangulation rule*

$$\Delta(J_1J_2J_3) \neq 0, \tag{4.51}$$

where Δ is a function which is non-zero only if J_3 is one of the set given by Eq. (4.50). The concept of a triangle enters because condition (4.50) is also the condition under which three lines of length J_1, J_2, and J_3 can be joined to form a triangle (see Fig. 4.9), if we accept that straight lines (and even points) are special forms of a triangle. Less obviously, the triangulation rule is also the condition that three lines of length $\sqrt{J_1(J_1+1)}$, $\sqrt{J_2(J_2+1)}$, and $\sqrt{J_3(J_3+1)}$ can form a triangle, and so we see that the rule simply expresses the fact that an angular momentum $\mathbf{J}\hbar$ behaves as a vector and obeys the normal laws of vector addition, with the proviso that its length is $\sqrt{J(J+1)}\hbar$. One other condition that must be satisfied when $\mathbf{J}_1$ and $\mathbf{J}_2$

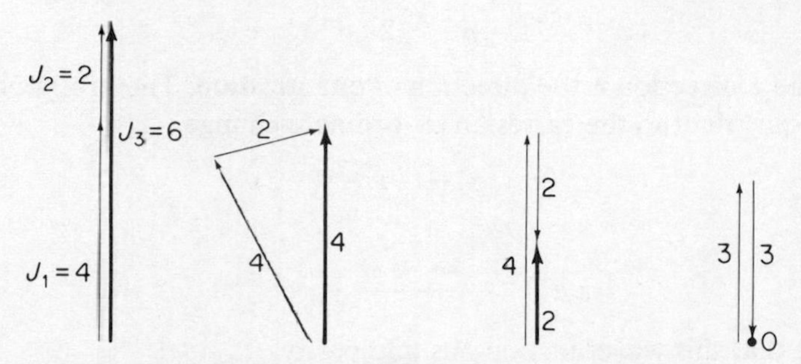

Fig. 4.9. The angular momenta $\mathbf{J}_1\hbar$ and $\mathbf{J}_2\hbar$ combine to form the angular momentum $\mathbf{J}_3\hbar$. These examples all satisfy the triangulation rule.

combine to form $\mathbf{J}_3$ is that the components in the direction of quantization must add correctly,

$$M_1+M_2=M_3. \tag{4.52}$$

The parity of a state describes how the wavefunction of the state behaves when the co-ordinate system is inverted about the origin,

$$\mathbf{r}\to-\mathbf{r}. \tag{4.53}$$

It can be shown that only two types of inversion behaviour exist for atomic states.* Taking the nucleus as the origin of co-ordinates, either the wavefunction remains unchanged under the inversion,

$$\psi(-\mathbf{r}_1,-\mathbf{r}_2,\ldots,-\mathbf{r}_N)=\psi(\mathbf{r}_1,\mathbf{r}_2,\ldots,\mathbf{r}_N), \tag{4.54}$$

in which case it is said to have *even parity*, or the wavefunction is multiplied by -1,

$$\psi(-\mathbf{r}_1,-\mathbf{r}_2,\ldots,-\mathbf{r}_N)=-\psi(\mathbf{r}_1,\mathbf{r}_2,\ldots,\mathbf{r}_N), \tag{4.55}$$

in which case it has *odd parity*. In writing these equations we have assumed that the wavefunctions depend only on the vector positions $\mathbf{r}_k$ of the N electrons in the atom, but the statement that only these two types of behaviour exist is still true when the wavefunctions depend on other internal parameters of the atom, such as the intrinsic spins of the electrons.

An example of a state of even parity is the ground state of the hydrogen atom, which has the wavefunction

$$\psi^{(1s)}=\pi^{-1/2}a_0^{-3/2}\,\mathrm{e}^{-r/a_0}, \tag{4.56}$$

where a_0 is the Bohr radius. Under the inversion operation (4.53) the direction of $\mathbf{r}$ changes but not its magnitude r, and so

$$\mathrm{e}^{-r/a_0}\xrightarrow{\text{inversion}}\mathrm{e}^{-r/a_0}.$$

On the other hand the $2p$ excited state with $M=0$ has the wavefunction

$$\psi^{(2p,M=0)}=\pi^{-1/2}(2a_0)^{-5/2}z\,\mathrm{e}^{-r/2a_0}, \tag{4.57}$$

where the z direction is the direction of quantization. The inversion operation is equivalent to the cartesian co-ordinate change

$$(x,y,z)\to(-x,-y,-z),$$

and so

$$z\,\mathrm{e}^{-r/2a_0}\xrightarrow{\text{inversion}}-z\,\mathrm{e}^{-r/2a_0},$$

showing that this wavefunction has odd parity.

* Provided that we ignore certain minor effects which are in any case extremely difficult to detect experimentally.

Expressing Eqs. (4.54) and (4.55) as the single equation

$$\psi(-\mathbf{r}_1, -\mathbf{r}_2, \ldots, -\mathbf{r}_N) = \Pi \times \psi(\mathbf{r}_1, \mathbf{r}_2, \ldots, \mathbf{r}_N), \tag{4.58}$$

the value of Π is $+1$ for even parity and -1 for odd parity. When a wavefunction ψ_3 can be considered as the product of the two separate wavefunctions ψ_1 and ψ_2,

$$\psi_3 = \psi_1 \times \psi_2,$$

its parity is the product of the parities of the separate parts,

$$\Pi_3 = \Pi_1 \times \Pi_2. \tag{4.59}$$

This equation represents the law of combination of parities. The extension to products of more than two parts is obvious.

4.4.2 Angular momentum and parity associated with a photon in an electric dipole transition

Before we can apply the laws of combination of angular momentum and parity to the processes of emission and absorption of photons, we must find the angular momentum and parity associated with the photon. Let us start by assuming that these quantities have the definite values $\sqrt{L_p(L_p+1)}\hbar$ and Π_p. Conservation of angular momentum then requires, for the emission process

$$A_i \rightarrow A_f + h\nu,$$

that

$$\mathbf{J}_i = \mathbf{J}_f + \mathbf{L}_p,$$

which implies the conditions

$$\Delta(J_i J_f L_p) \neq 0 \tag{4.60}$$

and

$$M_i = M_f + M_p. \tag{4.61}$$

Conservation of parity requires that

$$\Pi_i = \Pi_f \times \Pi_p. \tag{4.62}$$

Analogous relationships apply to the absorption process

$$A_i + h\nu \rightarrow A_f.$$

The absorption and emission transition probabilities are proportional to $|\mathbf{D}_{fi}|^2$, and therefore the conditions (4.60), (4.61), and (4.62) must represent the conditions under which $\mathbf{D}_{fi}$ is non-zero. This implies that we can find the values of L_p, M_p, and Π_p by considering the form of $\mathbf{D}_{fi}$.

We see from Eqs. (4.41) and (4.43) that $\mathbf{D}_{fi}$ is given by

$$\mathbf{D}_{fi} = \int \psi_f^* \mathbf{p} \psi_i \, d\tau,$$

where $\mathbf{p}$ is the operator representing the electric dipole moment of the atom,

$$\mathbf{p} = \sum_{k=1}^{N} (-e\mathbf{r}_k).$$

In the one-electron case the z-component of this is

$$p_z = -ez.$$

This has the same angular dependence ($\cos\theta$, in spherical polar coordinates) as the wavefunction (4.57) of the $2p$ state of the hydrogen atom, with $M=0$. Since the $2p$ state corresponds to an angular momentum quantum number $L=1$, we see that the operator p_z is associated with the angular momentum quantum numbers $L=1, M=0$. This is true also for the multi-electron case. Therefore each of the three factors in the product $\psi_f^* p_z \psi_i$ which appears in $(\mathbf{D}_{fi})_z$ corresponds to a definite angular momentum. Clearly we can now say that the angular momentum quantum numbers associated with the photon are those given by the operator p_z. These are $L_p=1$, $M_p=0$. We see also that p_z has odd parity, giving $\Pi_p=-1$.

To see if these values of L_p, M_p, and Π_p apply to the x and y components of $\mathbf{p}$, we can express $\mathbf{p}$ in terms of the spherical harmonics $Y_{l,m}$, which are functions describing the angular dependence of states having the angular momentum $\sqrt{l(l+1)}\hbar$ and z-component $m\hbar$. We find that only the functions $Y_{1,-1}$, $Y_{1,0}$, and $Y_{1,1}$ are needed,* and that

$$\mathbf{p} \propto \mathbf{i}_x(Y_{1,1}+Y_{1,-1}) + \mathbf{i}_y(Y_{1,1}-Y_{1,-1}) + \mathbf{i}_z\sqrt{2}\,Y_{1,0}, \tag{4.63}$$

and so again conclude that $L_p=1$. Because the photon carries away one unit of angular momentum in electric dipole transitions, this type of transition is often referred to as an *E1 transition*. We see also from Eq. (4.63) that the value of M_p is unique (and equal to 0) only when $\mathbf{p}$ is in the z-direction: otherwise a mixture of the three values -1, 0 and $+1$ exists. The fact that $Y_{l,m}$ has the parity $\Pi=(-1)^l$ shows that $\Pi_p=-1$ in the present case.

Specific values, or combinations of values, of M_p are associated with specific states of polarization of the light. For example, if all the photons resulting from the decay of atoms in a light source have $M_p=1$ the light travelling in the z-direction is right-hand circularly polarized (see Section 2.2). On the other hand when there is a coherent mixture of the values of

* These three spherical harmonics are

$$Y_{1,0} = \sqrt{\frac{3}{4\pi}}\cos\theta, \qquad Y_{1,\pm1} = \mp\sqrt{\frac{3}{8\pi}}\sin\theta\, e^{\pm i\phi}.$$

$M_p = +1$ and -1 (as happens for example when $(\mathbf{D}_{fi})_x$ or $(\mathbf{D}_{fi})_y$ are non-zero, see Eq. (4.63)), the light travelling in the x–y plane is linearly polarized in this plane. A random mixture of the three M values gives randomly polarized light.

A completely different way of finding L_p and M_p is to consider the angular momentum contained in the radiation field of a classical Hertzian dipole. If the dipole radiates a total angular momentum $\mathfrak{L}$ per second, and if it is surrounded by a sphere which absorbs all the radiation falling onto it, the net torque exerted on the sphere is equal to $\mathfrak{L}$. This torque can be calculated if the value of Poynting's vector $\mathbf{N}$ is known at the surface of the sphere, since the absorbed radiation exerts a force $\mathbf{N}/c$ per unit area (see Section 2.6), and therefore

$$\mathfrak{L} = \frac{1}{c} \int (\mathbf{r} \wedge \mathbf{N}) \, dS,$$

where the integration is over the surface of the sphere (see Fig. 4.10). The angular momentum is small because the largest components of $\mathbf{E}$ and $\mathbf{B}$ (i.e.

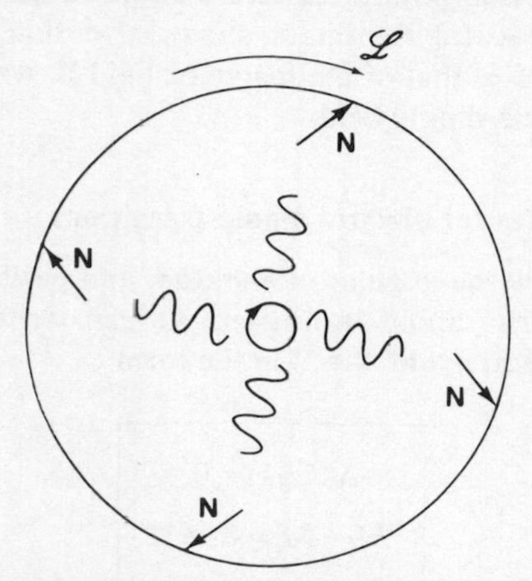

Fig. 4.10. The radiation from a classical dipole can have non-zero components of the momentum $\mathbf{N}$ in the transverse direction (indicated here by drawing non-radial arrows to represent $\mathbf{N}$). This leads to a non-zero total angular momentum for the whole field. The dipole shown here consists of a charge moving in a circular orbit. The outer circle represents a section of a spherical surface which absorbs the radiation, and to which the angular momentum of the radiation is imparted.

the radiation terms given by Eqs. (3.18) and (3.19)) give an **N** which is directed radially outwards, thus contributing nothing to the angular momentum of the field. But **E** and **B** have smaller components (see the discussion of Section 3.1.2) which give a non-radial contribution to **N**, and hence in some circumstances, a non-zero value of $\mathfrak{L}$. For example, it can be shown (see Corney[7] or Jackson[6]) that the z-component of $\mathfrak{L}$ for a charge moving in a circular orbit in the xy plane is related to the radiated power P by

$$\frac{\mathfrak{L}_z}{P}=\frac{1}{\omega}=\frac{\hbar}{\hbar\omega}.$$

This shows that each photon (of energy $\hbar\omega$) contributes an angular momentum component of magnitude $\hbar$, so that $M_\mathrm{p}=+1$. Other forms of the Hertzian dipole give

$$\frac{-\hbar}{\hbar\omega}\leqslant\frac{\mathfrak{L}_z}{P}\leqslant\frac{+\hbar}{\hbar\omega}. \tag{4.64}$$

This implies that M_p is in general a mixture of the values -1, 0, and $+1$, and that $L_\mathrm{p}=1$. The classical treatment shows also that the parity of the interaction energy H' is that of the magnetic field **B**, which is odd (see Eq. (3.18)) for the electric-dipole field.

4.4.3 Selection rules for electric dipole transitions

Now that we know the angular momentum and parity associated with a photon in an electric dipole transition, we can write the conservation conditions (4.60), (4.61), and (4.62) in the form

$$\Delta(J_\mathrm{i}J_\mathrm{f}1)\neq 0, \tag{4.65}$$

$$M_\mathrm{f}-M_\mathrm{i}=0, \pm 1, \tag{4.66}$$

$$\Pi_\mathrm{i}=-\Pi_\mathrm{f}. \tag{4.67}$$

These are the *selection rules* for E1 transitions. They apply to emission and absorption processes. The first rule is often expressed in the alternative form

$$\Delta J=J_\mathrm{f}-J_\mathrm{i}=0, \pm 1, \qquad \text{but} \quad J_\mathrm{i}=0\leftrightarrow J_\mathrm{f}=0 \text{ not allowed}. \tag{4.68}$$

When all the selection rules are obeyed the transition is said to be *optically allowed*. Otherwise it is described as *optically forbidden*: this does not mean, however, that the transition probability is exactly zero, because other forms of interaction between the atom and radiation field (to be discussed in the next section) usually result in a small but finite transition probability.

Even when a transition is optically allowed, there are sometimes other rules which, if not satisfied, cause the transition probability to be reduced. We may call these *approximate selection rules*. In the case of atoms they are connected with the *approximate quantum numbers* L and S. These arise from the way in which the individual angular momenta of the electrons in the atom couple together to give the total angular momentum $\mathbf{J}\hbar$. The individual angular momenta are of two sorts, the angular momenta $\mathbf{l}_k\hbar$ due to the orbital motion of the electrons, and the intrinsic spin angular momenta $\mathbf{s}_k\hbar$ (where $s_k = \frac{1}{2}$) of the electrons. The angular momenta $\mathbf{L}\hbar$ and $\mathbf{S}\hbar$ are defined by

$$\mathbf{L} = \sum_{k=1}^{N} \mathbf{l}_k,$$

$$\mathbf{S} = \sum_{k=1}^{N} \mathbf{s}_k.$$

In the heavier atoms the spin–orbit coupling between the individual $\mathbf{l}_k$ and $\mathbf{s}_k$ is strong, causing $\mathbf{L}$ and $\mathbf{S}$ to have indefinite values. In most other atoms, however, the spin–orbit interactions are sufficiently weak that $\mathbf{L}$ and $\mathbf{S}$ are approximately constant for a given state of the atom. The electric dipole interaction energy (Eq. (4.39)) involves the electron positions and not their intrinsic spins, which implies that

$$\mathbf{L}_{\mathrm{i}} = \mathbf{L}_{\mathrm{f}} + \mathbf{L}_{\mathrm{p}}$$

and

$$\mathbf{S}_{\mathrm{i}} = \mathbf{S}_{\mathrm{f}}$$

in the emission process $A_{\mathrm{i}} \rightarrow A_{\mathrm{f}} + h\nu$ (and with analogous formulas for the absorption process). In terms of the approximate quantum numbers L and S, this gives the approximate selection rules (for emission or absorption)

$$\Delta(L_{\mathrm{i}}L_{\mathrm{f}}1) \neq 0, \tag{4.69}$$

$$S_{\mathrm{i}} = S_{\mathrm{f}}. \tag{4.70}$$

The effect of the rigid and approximate selection rules for E1 transitions is illustrated in Fig. 4.11(a). In the spectroscopic notation used to label the levels of the C atom, the left superscript gives the value of $2S+1$, the letter denotes the value of L(S, P and D denoting $L = 0$, 1 and 2 respectively), the right subscript gives the values of J, and the right superscript, where it exists, denotes odd parity (the other states having even parity). All the transitions shown satisfy all the selection rules (4.65), (4.66), and (4.67), as well as the approximate rules (4.69) and (4.70).

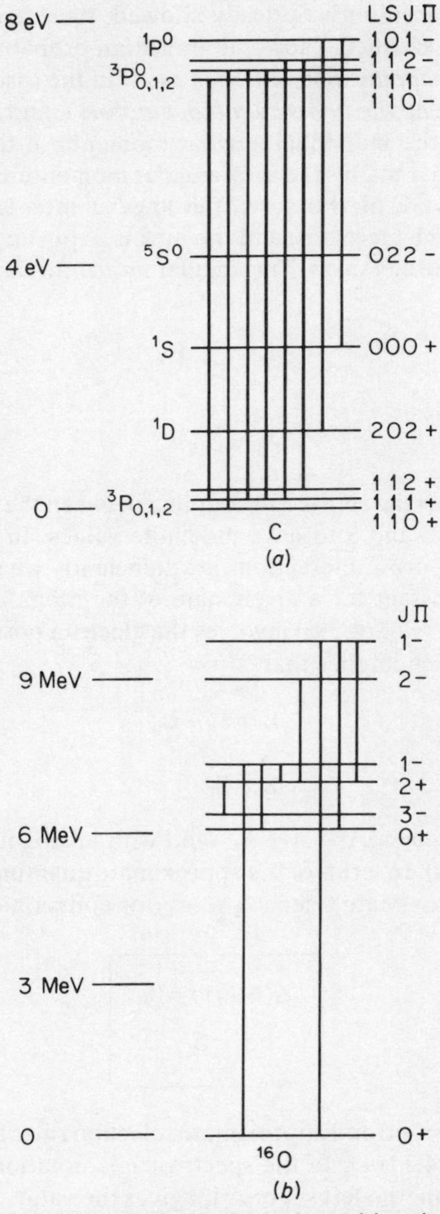

Fig. 4.11. Optically allowed transitions in the carbon atom and the ^{16}O nucleus. The transitions shown satisfy the selection rules (4.65), (4.66), and (4.67), and in the case of the carbon atom only, the selection rules (4.69) and (4.70) as well.

Radiative transitions can take place also between the excited states of an atomic nucleus. These states are characterized by their energy E, angular momentum quantum number J and parity Π, but the quantum numbers L and S are usually inappropriate. The emitted or absorbed radiation is of course in the form of γ-rays (with energies typically in the range from about 0.1 to 10 MeV). The selection rules (4.65), (4.66), and (4.67) apply for nuclear E1 transitions, as in the case of atoms. Figure 4.11(b) shows examples of E1 transitions between the lowest seven levels of the ^{16}O nucleus.

4.5 OPTICALLY FORBIDDEN TRANSITIONS

The violation of one or more of the electric dipole selection rules implies that the contribution $-\mathbf{E}\cdot\mathbf{p}$ (Eq. (4.38)) to the perturbation H' is unable to cause the radiative transition. In this section we investigate the effectiveness of the smaller contributions to H'.

Magnetic dipole transitions

The first of these arises from the interaction between the radiation field and the magnetic dipole moment of the atom. Because of their orbital motion and intrinsic magnetic moment the electrons of an atom give it the magnetic moment

$$\boldsymbol{\mu} = g_J \mu_B \mathbf{J}.$$

Here μ_B is the *Bohr magneton,*

$$\mu_B = \frac{e\hbar}{2m} = 9.274 \times 10^{-24}\ \mathrm{A\,m^2}, \tag{4.71}$$

$\mathbf{J}\hbar$ is the total angular momentum of the electrons of the atom, and g_J (the gyromagnetic ratio) is a constant of the order of unity (its exact value depends on the detailed structure of the atom). The magnitude of $\boldsymbol{\mu}$ is $g_J \mu_B \sqrt{J(J+1)}$, and its component in the direction of quantization is $g_J \mu_B M$. It interacts with the magnetic field $\mathbf{B}$ of the radiation field, giving the perturbation energy

$$H' = -\boldsymbol{\mu}\cdot\mathbf{B}. \tag{4.72}$$

The emission and absorption transitions which are induced by this perturbation are called *magnetic dipole* (M1) transitions.

The magnetic moment $\boldsymbol{\mu}$ is a vector, as is the electric dipole moment $\mathbf{p}$, and therefore the photons in M1 transitions also have $L_p = 1$ and carry away one unit of angular momentum. The selection rule in J is therefore still

$$\Delta(J_i J_f 1) \neq 0\ (\mathrm{M1}). \tag{4.73}$$

The vector $\boldsymbol{\mu}$ differs from $\mathbf{p}$ in one respect however, because the angular momentum of a system has the form of a cross-product of two vectors, $m\mathbf{r}\wedge\mathbf{v}$, where m is a mass and $\mathbf{v}$ a velocity: both $\mathbf{r}$ and $\mathbf{v}$ have odd parity, and so $\boldsymbol{\mu}$ has even parity. The parity selection rule is therefore

$$\Pi_i = \Pi_f \quad \text{(M1)}. \tag{4.74}$$

Electric quadrupole transitions

To see the origin of another contribution to H', we consider the potential energy of a system of particles with charges q_k and position vectors $\mathbf{r}_k$, situated in a static electric potential $\phi(\mathbf{r})$,

$$U = \sum_k q_k \phi(\mathbf{r}_k).$$

Taking a suitable origin of co-ordinates (i.e. at or near the centre of the system), ϕ can be expanded in a Taylor series, to give

$$\begin{aligned}\phi(\mathbf{r}_k) &= \phi(0) + \sum_i x_{ik}\frac{\partial\phi}{\partial x_i} + \tfrac{1}{2}\sum_{i,j} x_{ik}x_{jk}\frac{\partial^2\phi}{\partial x_i\,\partial x_j} + \cdots \\ &= \phi(0) - \sum_i x_{ik}E_i - \tfrac{1}{2}\sum_{i,j} x_{ik}x_{jk}\frac{\partial E_i}{\partial x_j} + \cdots,\end{aligned}$$

where x_{1k}, x_{2k}, and x_{3k} are now used instead of x_k, y_k, and z_k, and E_1, E_2, and E_3 denote E_x, E_y, and E_z. Then U can be expressed as

$$U = \left(\sum_k q_k\right)\phi(0) - \sum_k q_k(\mathbf{r}_k\cdot\mathbf{E}) - \tfrac{1}{2}\sum_k q_k\left(\sum_{i,j} x_{ik}x_{jk}\frac{\partial E_i}{\partial x_j}\right) + \cdots.$$

The first term gives the potential energy of the total charge, and has no effect on radiative transitions. The second set of terms corresponds to the electric dipole interaction considered in the previous section: it gives rise to the optically allowed transitions. The third set of terms is of interest to us at present. It corresponds to the interaction between the field and the system's various electric quadrupole moments, which are defined by

$$Q_{xx} = \sum_k q_k x_k^2, \qquad Q_{xy} = \sum_k q_k x_k y_k, \qquad \text{etc.}$$

(where we now revert to the notation $\mathbf{r}_k = (x_k, y_k, z_k)$). Figure 3.8 shows an example of a system for which Q_{zz} is non-zero. The part of the electric field (4.39) which is responsible for absorption is

$$\mathbf{E} = \mathbf{i}_x\frac{E_0}{2}\mathrm{e}^{-\mathrm{j}(\omega t - kz)} = \mathbf{i}_x\frac{E_0}{2}\mathrm{e}^{-\mathrm{j}\omega t}(1 + \mathrm{j}kz + \cdots),$$

and the only non-zero first derivative of this is

$$\frac{\partial E_x}{\partial z} = \frac{\mathrm{j}kE_0}{2} \mathrm{e}^{-\mathrm{j}\omega t}.$$

This field gradient interacts with the electric quadrupole moment Q_{xz}, giving the interaction energy

$$-\tfrac{1}{2}Q_{xz}\frac{\partial E_x}{\partial z} = \frac{\mathrm{j}keE_o}{4}\left(\sum_k x_k z_k\right) \mathrm{e}^{-\mathrm{j}\omega t}. \tag{4.75}$$

Transitions induced by this perturbation are called *electric quadrupole* (E2) transitions.

The electric quadrupole moments can be expressed in terms of the spherical harmonics $Y_{2,m}$, which implies that $L_\mathrm{p} = 2$ and that the photons carry away two units of angular momentum. A photon observed in a beam can of course have only one unit of angular momentum, but we see that in an E2 emission transition the photon has an angular momentum with respect to the emitting atom, and that it takes a total of two units from the atom. The angular momentum selection rule is therefore

$$\Delta(J_\mathrm{i}J_\mathrm{f}2) \neq 0 \quad \text{(E2)}. \tag{4.76}$$

The electric quadrupole moments have even parity, giving the parity selection rule

$$\Pi_\mathrm{i} = \Pi_\mathrm{f} \quad \text{(E2)}. \tag{4.77}$$

Relative probabilities of E1, M1 and E2 transitions

The relative magnitudes of the E1, M1 and E2 transition probabilities can be estimated by comparing the interaction energies. We see from Eq. (4.38) that the E1 perturbation energy has the magnitude

$$U^{\mathrm{E1}} \sim ea_0E_0,$$

where the Bohr radius

$$a_0 = \frac{4\pi\varepsilon_0\hbar^2}{me^2} = 5.29\times10^{-11}\ \mathrm{m}$$

is used as a typical atomic dimension, and E_0 is the amplitude of the electric field; from Eq. (4.72) we find

$$U^{\mathrm{M1}} \sim \mu_\mathrm{B}B_0 \sim \frac{e\hbar}{mc}E_0,$$

and from Eq. (4.75) we find

$$U^{\mathrm{E2}} \sim \frac{keE_0a_0^2}{4} \sim \frac{eE_0a_0^2}{\lambda}.$$

The transition probabilities are proportional to the oscillator strengths f, which are in turn approximately proportional to $|U|^2$ (cf. Eqs. (4.43), (4.48)). Therefore

$$f^{\mathrm{E1}}:f^{\mathrm{M1}}:f^{\mathrm{E2}} \sim 1:\alpha^2:(a_0/\lambda)^2, \tag{4.78}$$

where

$$\alpha = \frac{e^2}{4\pi\varepsilon_0\hbar c} = \frac{1}{137}$$

is the fine structure constant. The ratio a_0/λ is about 10^{-4} for visible light, and therefore

$$f^{\mathrm{E1}}:f^{\mathrm{M1}}:f^{\mathrm{E2}} \text{ (atomic)} \sim 1:10^{-4}:10^{-8}.$$

We see from this that M1 and E2 transitions are aptly described as *optically forbidden* transitions, since they are so much weaker than optically allowed E1 transitions of approximately the same transition frequency. An atomic excited state which can decay only *via* a forbidden transition would have a lifetime several orders of magnitude larger than the typical lifetime for an allowed transition. Such states are known as *metastable states.*

The ratios given by Eq. (4.78) can also be obtained from the classical expressions (3.21, 3.26, and 3.31) for the power radiated by oscillating electric dipoles, magnetic dipoles, and electric quadrupoles respectively. This emphasizes the fact that the ratio $f^{\mathrm{M1}}/f^{\mathrm{E1}}$ is small for atoms because of the comparative inefficiency of magnetic dipoles in producing radiation, and that the ratio $f^{\mathrm{E2}}/f^{\mathrm{E1}}$ is small because the dimensions of the atom are much smaller than the wavelength being radiated.

For radiative transitions in nuclei the appropriate dimension to use in place of a_0 is the nuclear radius R ($\sim 3\times 10^{-15}$ m), and the appropriate mass m is that of the proton, m_{p}. Typical γ-ray wavelengths are of the order of 10^{-12} m, giving

$$f^{\mathrm{E1}}:f^{\mathrm{M1}}:f^{\mathrm{E2}} \text{ (nuclear)} \sim 1:\left(\frac{\hbar}{m_{\mathrm{p}}cR}\right)^2:\left(\frac{R}{\lambda}\right)^2 \sim 1:10^{-2}:10^{-3}. \tag{4.79}$$

These differences are not as great as in the atomic case, and M1 and E2 transitions are often observed in nuclear radiative decays.

The E1 and E2 transitions are the first two members of a series (E1, E2, E3, etc.). Some of the higher members are present in nuclear decays (but not in atomic transitions). The selection rules for all the electric transitions, called *electric multipole* transitions, are

$$\Delta(J_{\mathrm{i}}J_{\mathrm{f}}L) \neq 0, \qquad \Pi_{\mathrm{i}} = (-1)^L\Pi_{\mathrm{f}} \quad \text{(EL)}, \tag{4.80}$$

Similarly, the selection rules for the *magnetic multipole* transitions (M1, M2, M3, etc.) are

$$\Delta(J_i J_f L) \neq 0, \qquad \Pi_i = (-1)^{L+1}\Pi_f \quad \text{(ML)}. \tag{4.81}$$

These selection rules often allow more than one type of transition between two states (see Problem 4.7), but usually one of the multipoles dominates. Examples of the higher order multipole transitions are shown in Fig. 4.12.

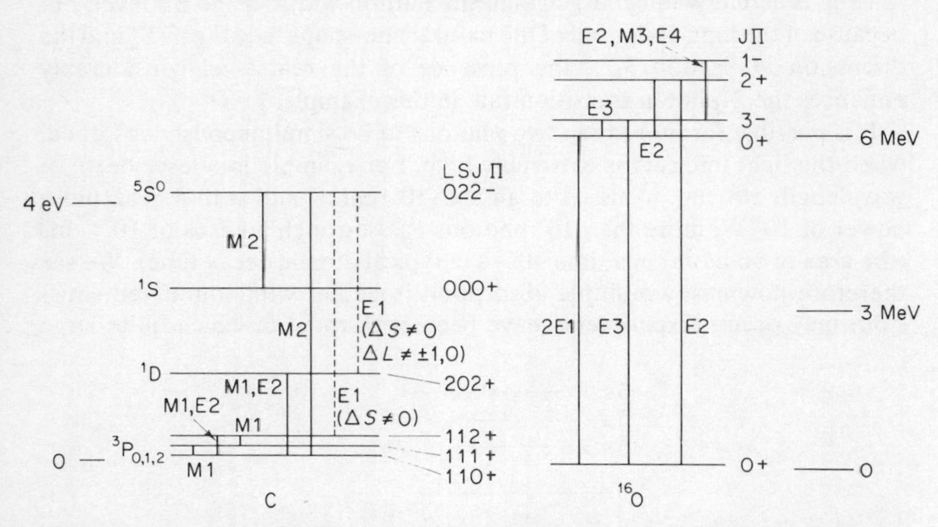

Fig. 4.12. Examples of optically forbidden transitions in the carbon atom and the ^{16}O nucleus. The dashed lines represent transitions which are optically forbidden because the selection rules in ΔS and ΔL are not obeyed.

Multi-photon transitions

A different type of transition is induced by the non-linear perturbation that exists when the electric field strength is very large, as in an intense laser beam (see also Section 3.2.5). The time dependence of this perturbation contains the angular frequencies 2ω and -2ω. It induces transitions between states separated by the energy $2\hbar\omega$, although the photons of the field have an energy of only $\hbar\omega$. This term is therefore responsible for transitions in which two photons are absorbed or emitted at the same time. If the radiation field contains more than one frequency the two photons may have different frequencies, but the sum of their energies must equal the energy differences of the atomic states. These transitions are called *two-photon transitions.* To find the selection rules we may regard the transitions as two separate one-photon transitions proceeding *via* a virtual (i.e. non-existent) intermediate state, at the appropriate intermediate energy. In Fig. 4.13 this virtual state is indicated by a broken line. Each of the separate transitions is usually an E1 transition (other electric or magnetic multipolarities would give a much smaller transition rate), and so the selection rules are

$$J_i - J_f = 0, \pm 1, \pm 2, \qquad \Pi_i = \Pi_f \quad (2E1). \tag{4.82}$$

Transitions of the type $J_i = 0$ to $J_f = 0$ are now possible, which is not the case for any of the single-photon multipole transitions. In the example shown in Fig. 4.13 the virtual state is near to a real level (the $3p$ level) of the atom. The spacing is actually much larger than the natural width of the $3p$ level, but because of the long-range tail of the natural line-shape (see Fig. 3.17 and the discussion in Section 3.7), the presence of the real level considerably enhances the 2-photon transition rate in this example.

It is possible for more than two photons to be simultaneously absorbed, when the light intensity is extremely high. For example in a laser beam of wavelength 500 nm, focussed to an area $(0.1 \text{ mm})^2$ and with a peak pulse power of 10^6 W, more than 10^4 photons pass through an area of 10^{-20} m^2 (the area of an atom) in a time 10^{-8} s (a typical atomic decay time). We see therefore how easily multiple absorption (together with stimulated emission) may occur. Experiments have been performed in which quite large

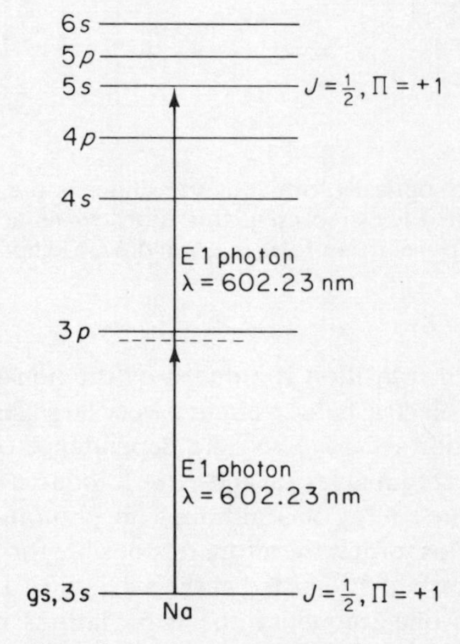

Fig. 4.13. An example of two-photon absorption in sodium. The photons are provided by a pulsed dye laser of high power. The broken line indicates the position of a virtual (i.e. non-existent) intermediate state.

numbers ($\geqslant 10$) of low energy photons have been used to excite and ionize atoms and molecules. The theoretical description of such multi-photon processes has produced some interesting problems (perturbation theory is of little use, and so new techniques must be developed). These processes may also lead soon to interesting practical applications.

Non-radiative transitions

Other mechanisms for transitions also exist. For example, some atomic excited states can decay by ejecting an electron, some molecular states can dissociate into atoms, and some nuclear states can decay by ejecting α or β particles. In the case of large molecules, internal re-distributions of the excitation energy are also possible. These transitions are called *non-radiative.* In general they exist in competition with radiative transitions, so that a state which can decay non-radiatively can usually also decay radiatively. The transition probability for the non-radiative decay route is however usually very much larger than that of the radiative decay. These states therefore have very short lifetimes, and only a small proportion of the decays are radiative.

PROBLEMS: CHAPTER 4

(Answers to selected problems are given in Appendix E)

4.1 A simple harmonic oscillator having the energy levels $\hbar\omega(n+\frac{1}{2})$ (for example, a vibrating diatomic molecule) is in thermal equilibrium at the temperature T. Calculate the two values of T for which the probability that the $n=1$ level is excited is 10^{-4}, if $\hbar\omega = 10^{-21}$ J.

4.2 A hollow vessel of heat capacity 10^4 J/K and internal volume 1 m^3 is heated from 290 K to 2000 K. Show (neglecting any heat losses) that only the fraction 7×10^{-10} of the heat that must be supplied is used to increase the radiant energy in the enclosure.

4.3 A 100 W tungsten filament lamp has a filament of diameter 40 μm and length 1 m. The filament is tightly coiled in such a way that only 20% of its true surface can be seen. Estimate the temperature fluctuation of the filament when the lamp is connected to a 50 Hz AC supply (the heat capacity of tungsten is approximately 3.5×10^6 J m^{-3} K^{-1}, and the loss of heat by conduction can be neglected).

4.4 What is the transition probability for spontaneous emission of the first excited state of an atom if its lifetime for radiative decay is 100 ns, the excitation energy is 2 eV, and the excited and ground states are both non-degenerate?

4.5 A molecule is in its first excited rotational state of energy 1 meV, and is situated in a thermal radiation field of temperature 300 K. What is the ratio of the stimulated to spontaneous emission rates? What is this ratio for an atom in its first excited state of energy 4 eV?

4.6 The oscillator strengths for the transition from the first excited state ($3p$) to the ground state ($3s$) of the sodium atom (see Fig. 4.13) is 0.94. The wavelength of the transition is 589 nm. Calculate the radiative lifetime of the excited state.

4.7 Give examples of the angular momentum and parity quantum numbers (J and Π) possessed by two states if the possible radiative transitions between them are (i) E1 only, (ii) M1 only, (iii) E2 only, (iv) M1 or E2, (v) E1 or E2, (vi) M2 or E3, and (vii) E1 or E3 only.

4.8 The figure shows the lowest 8 levels of the sodium atom. State which transitions between the levels are optically allowed.

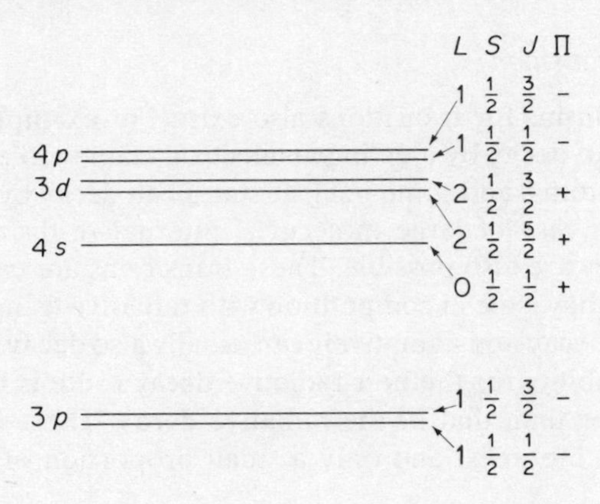

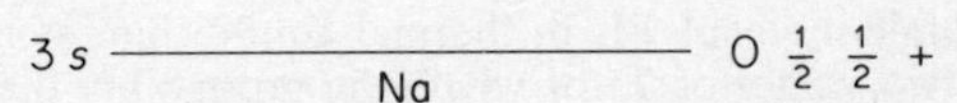

Fig. for Problem 4.8. Energy level diagram of the lower excited states of the sodium atom. The splittings of the p and d configurations are shown greatly exaggerated.

CHAPTER 5

Coherence

In this chapter we take a new direction, and discuss the types of radiation field that exist in practice, and the coherences that exist between different parts of the field. This is done partly to prepare the ground for the discussion of lasers in Chapter 6, and the discussion of the uses of coherent light (e.g. for holography) in Chapter 7. The subject of coherence is also of great interest in its own right.

We start by treating the radiation field as a classical electromagnetic field that can be described by giving the electric and magnetic field strengths as a function of position and time. The definition of coherence is discussed, and then we go on to investigate simple ways in which the degree of coherence can be measured, and how this degree of coherence depends on the properties of the radiation source. We shall then find that it is helpful, and in some cases indispensible, to consider the photon character of the field. This will lead us finally into a discussion of the coherence between photons, and of the important experiments of Hanbury-Brown and Twiss.

5.1 WHAT IS COHERENCE?

When we say that there is some coherence between two points P_1 and P_2 (with positions vectors $\mathbf{r}_1$ and $\mathbf{r}_2$ respectively) in an electromagnetic field, we mean (if we are thinking of the field as a classical field) that there is some *correlation in time* between the fluctuations in the electric fields $\mathbf{E}(\mathbf{r}_1, t)$ and $\mathbf{E}(\mathbf{r}_2, t)$ at the two points. Suppose for example that the x-components $E_x(\mathbf{r}_1, t)$ and $E_x(\mathbf{r}_2, t)$ are approximately sinusoidal, and that $E_x(\mathbf{r}_1, t)$ tends to be positive when $E_x(\mathbf{r}_2, t)$ is positive, and *vice versa*, as illustrated in Fig. 5.1(a). There is obviously some coherence between these two fields. They may for example have been produced by splitting a single beam of light at a semi-reflecting mirror, or they may be the fields at two different points in a laser beam. The two fields shown in Fig. 5.1(b) also have some coherence between them. In this case the fields tend to have opposite signs, but this in no way lessens the strong mutual correlation.

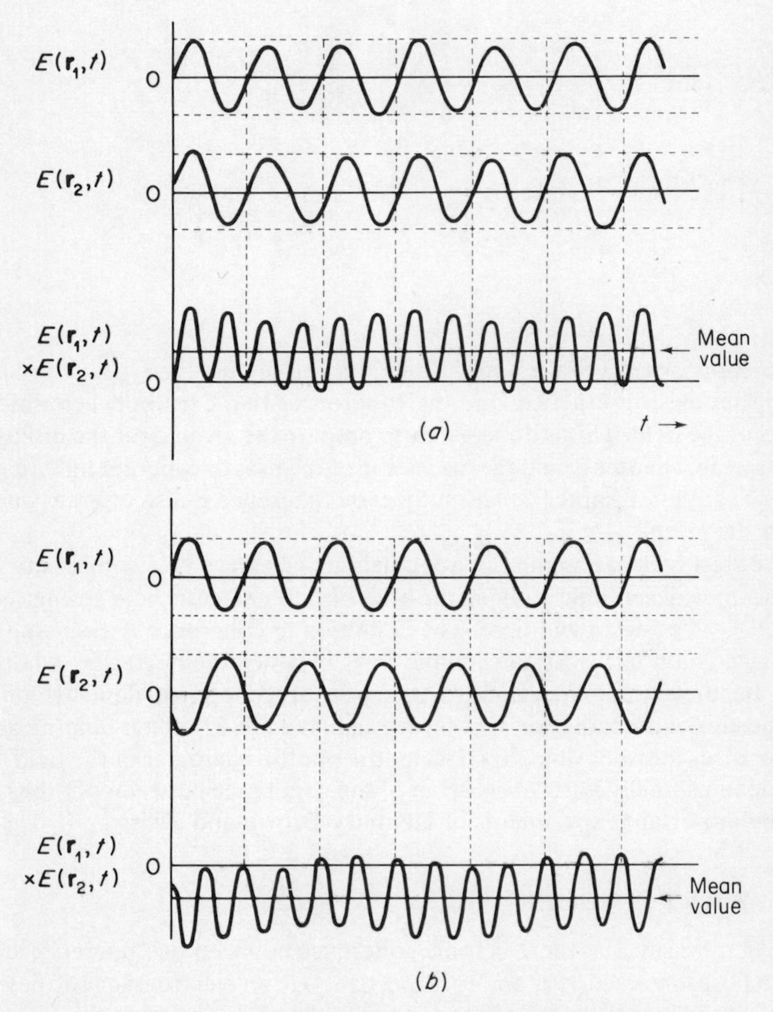

Fig. 5.1. (a) The approximately sinusoidal fields $E(\mathbf{r}_1, t)$ and $E(\mathbf{r}_2, t)$ are partially correlated; the mean value of their product is positive. (b) The two fields are still partially correlated, but now the mean value of their product is negative.

The coherence between two fields is significant only if it exists over a time interval long enough for it to be observed and measured, and so one way in which we might attempt to specify the degree of coherence is to take the *time average* of the product $E(\mathbf{r}_1, t) \times E(\mathbf{r}_2, t)$,

$$\text{degree of coherence} \propto \langle E(\mathbf{r}_1, t)E(\mathbf{r}_2, t)\rangle = \frac{1}{T}\int_0^T E(\mathbf{r}_1, t)E(\mathbf{r}_2, t)\,\mathrm{d}t. \qquad (5.1)$$

Here the time average is denoted by the brackets $\langle\ \rangle$; it is defined as shown, where T is the time interval over which the coherence is being observed and measured (and T is of course very much longer than the period of oscillation of E). The field strength E represents one of the components (that is, one direction of polarization) of $\mathbf{E}$, and so this definition applies to each component separately. For the two examples shown in Fig. 5.1 the time average is respectively positive and negative.

The definition given by Eq. (5.1) is too restrictive, because we can easily think of two fields $E(\mathbf{r}_1, t)$ and $E(\mathbf{r}_2, t)$ which are certainly correlated, and yet which have a zero value of $\langle E(\mathbf{r}_1, t)E(\mathbf{r}_2, t)\rangle$. An example is shown in Fig. 5.2. It is also too restrictive to compare the field at a point P_1 and time t with the field at another point P_2 only at the *same* time t. A satisfactory definition is obtained by expressing the field strengths in the complex notation (Section 2.2), and by including the possibility of correlation between the field $E(\mathbf{r}_1, t+\tau)$ and the field $E(\mathbf{r}_2, t)$ where τ is an arbitrary, adjustable time difference. In this way we obtain the *mutual coherence function*, defined to be

$$\Gamma(\mathbf{r}_1, \mathbf{r}_2, \tau) = \langle E(\mathbf{r}_1, t+\tau)E^*(\mathbf{r}_2, t)\rangle. \qquad (5.2)$$

As a first example of the calculation of the mutual coherence function, we consider the case in which the points P_1 and P_2 are both situated in the field

$$E = E_0\,\mathrm{e}^{\mathrm{j}(\omega t - kz)}$$

of a classical plane wave travelling in the z-direction. The mutual coherence function is

$$\Gamma(\mathbf{r}_1, \mathbf{r}_2, \tau) = E_0^2\,\mathrm{e}^{\mathrm{j}(\omega\tau + \phi_{12})},$$

where

$$\phi_{12} = k(z_2 - z_1)$$

(and E_0 is assumed to be real). We see that Γ is a complex number; the modulus E_0^2 gives information about the intensity of the field and about the correlation in the amplitudes of the field fluctuations, while the phase $(\omega\tau + \phi_{12})$ gives information about the relative positions in time of the maxima in the two fields. For the two fields shown in Fig. 5.2, for which $\phi_{12} = -\pi/2$,

$$\Gamma(\mathbf{r}_1, \mathbf{r}_2, \tau) = E_0^2\,\mathrm{e}^{\mathrm{j}(\omega\tau - \pi/2)}.$$

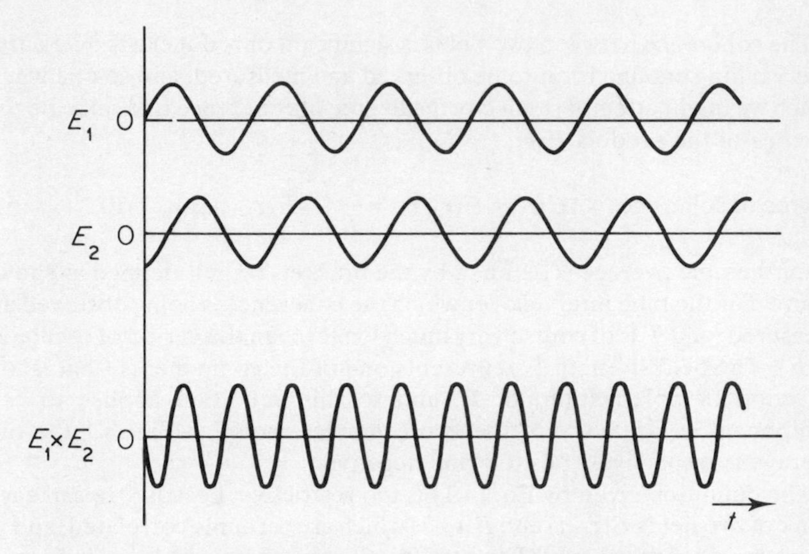

Fig. 5.2. The sinusoidal waves E_1 and E_2 are separated by one-quarter of a period. The time average of their product is zero.

If the fields are measured at the same time ($\tau = 0$), Γ is purely imaginary ($-\mathrm{j}\, E_0^2$), but if the field at P_1 is measured one-quarter of a period later than that at P_2 ($\omega\tau = \pi/2$) the measured fields are exactly in phase, and Γ is real ($+E_0^2$).

A function which shows the degree of coherence independently of the absolute intensities is the *normalized mutual coherence function*

$$\gamma(\mathbf{r}_1, \mathbf{r}_2, \tau) = \frac{\Gamma(\mathbf{r}_1, \mathbf{r}_2, \tau)}{(I_1 I_2)^{1/2}}, \tag{5.3}$$

where $I_i (i = 1 \text{ or } 2)$ is proportional to the intensity at point P_i and is given by

$$I_i = \langle E(\mathbf{r}_i, t) E^*(\mathbf{r}_i, t) \rangle = \Gamma(\mathbf{r}_i, \mathbf{r}_i, 0). \tag{5.4}$$

For the example of two points in the same plane wave of angular frequency ω,

$$\gamma(\mathbf{r}_1, \mathbf{r}_2, \tau) = \mathrm{e}^{\mathrm{j}(\omega\tau + \phi_{12})}.$$

We see that γ has a modulus of unity. In this example the maximum amount of correlation exists between the fields at the two points: when the field is known at one of the points it is also completely known at the other point. The coherence is said to be *complete*. When $|\gamma|$ is between 0 and 1 (as in the two examples shown in Fig. 5.1), the fields are said to be *partially coherent*. When $\gamma = 0$ the fields have completely independent time behaviours over long intervals of time, and are said to be *incoherent*.

One further generalization enables us to quantify the correlation between the field at the point P_1 at time t with the field at the *same* point but at a different time $t+\tau$. We can still use the function Γ, which now becomes

$$\Gamma(\mathbf{r}_1, \mathbf{r}_1, \tau) = \langle E(\mathbf{r}_1, t+\tau)E^*(\mathbf{r}_1, t)\rangle, \tag{5.5}$$

and is re-named the *autocorrelation function* of the field.

An example of a field for which the autocorrelation function is finite only over a limited range of values of τ is the wave-packet illustrated in Fig. 5.3. The time interval Δt shown in the figure is an approximate measure of the time-width of the wave-packet. The autocorrelation function has a maximum value when $\tau = 0$, is still significant when $\tau \lessdot \Delta t$, but is very small or zero when $\tau \gg \Delta t$. The time interval Δt (which we need not define exactly at this stage) is known as the *coherence time* of the wave-packet.

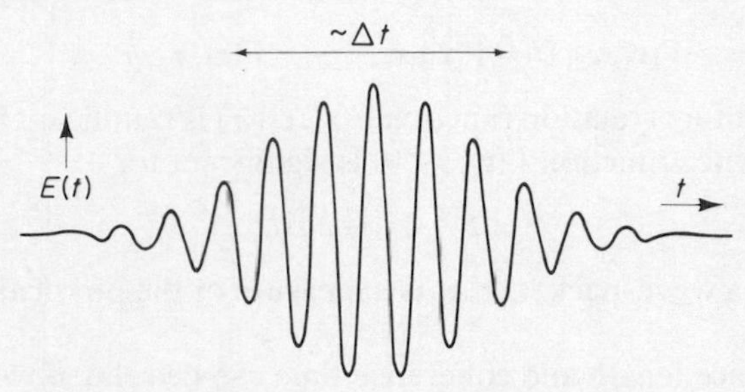

Fig. 5.3. A wave-packet of coherence time Δt.

The coherence time of a *continuous* beam of fluctuating intensity is found by plotting $\Gamma(\mathbf{r}, \mathbf{r}, \tau)$ against τ, and measuring the width of the distribution. If we can regard the beam as a random succession of independent wave-packets all having the same coherence time, the only non-zero contributions to the autocorrelation function are those arising from the correlation between the field $E(\mathbf{r}, t+\tau)$ of a wave-packet with the field $E(\mathbf{r}, t)$ of the *same* wave-packet, because the wave-packets are independent of each other. The autocorrelation function therefore has the same dependence on τ (apart from a different absolute magnitude) as that of a single wave-packet, and so the coherence time of the beam is that of a single wave-packet.

5.2 TEMPORAL AND SPATIAL COHERENCE

Two points P_1 and P_2 between which coherence exists may be separated longitudinally (with respect to the direction of the incident beam), or transversely, or both. In this section we start by considering the separate effects of longitudinal and transverse separations, and then consider the combined effect.

Temporal coherence

When P_1 and P_2 have a longitudinal separation in the direction of the beam (that is, when $z_1 \neq z_2$), but no transverse separation ($x_1 = x_2$, $y_1 = y_2$), the field at P_1 at time t is the field at P_2 at time $t + \tau_{12}$, where

$$\tau_{12} = (z_2 - z_1)/c;$$

hence,

$$E(\mathbf{r}_1, t) = E(\mathbf{r}_2, t + \tau_{12}). \tag{5.6}$$

Because of this connection between space and time, the associated longitudinal coherence is usually called *temporal coherence*. The fact that the beam is partially coherent over the coherence time Δt implies that it is also partially coherent over the length $c\,\Delta t$, called the *coherence length* of the beam. We can see this connection in another way by using Eq. (5.6) to give

$$\Gamma(\mathbf{r}_1, \mathbf{r}_2, 0) = \Gamma(\mathbf{r}_1, \mathbf{r}_1, \tau_{12}) = \Gamma(\mathbf{r}_2, \mathbf{r}_2, \tau_{12}), \tag{5.7}$$

Because the autocorrelation function $\Gamma(\mathbf{r}_1, \mathbf{r}_1, \tau)$ is significant for $\tau \lesssim \Delta t$, the mutual coherence function $\Gamma(\mathbf{r}_1, \mathbf{r}_2, 0)$ is significant for

$$|z_2 - z_1| \lesssim c\,\Delta t. \tag{5.8}$$

In the case of a wave-packet, $c\,\Delta t$ is a measure of the physical length of the packet.

The coherence length and coherence time can be related to the spread of frequencies in the beam. A classical field in which there is a range of angular frequencies can be expressed as

$$E(t) = \int a(\omega) e^{j[\omega t + \phi(\omega)]}\, d\omega, \tag{5.9}$$

where the amplitude a and phase ϕ both depend on ω (and we have suppressed the dependence of E on position). If we know the form of $E(t)$ we can calculate the amplitudes $a(\omega)$ by Fourier analysing $E(t)$. In this way we obtain the frequency spectrum of the beam. For example the wave-packet shown in Fig. 5.3 has an envelope of Gaussian shape, and by Fourier analysing the wave-packet we would find that the dependence of $|a(\omega)|^2$ on ω is also Gaussian (see Problem 5.2). An important relationship exists between the width $\Delta\omega$ of this frequency spectrum and the coherence time Δt. It is

$$\Delta t\, \Delta\omega \sim 1. \tag{5.10}$$

This does not depend on which particular definitions are used for Δt and $\Delta\omega$: any physically reasonable definition of the time width of the wave-packet and the frequency width of its frequency spectrum would give a product $\Delta t\, \Delta\omega$ in the range from about $\frac{1}{2}$ to 2π. The relationship applies also to beams consisting of a random succession of wave-packets, or more generally to any type of irregular beam that might exist in practice. It applies for example to the frequency spread $\Delta\nu$ of the light emitted when an atomic state of lifetime τ decays to the ground state, since for this light (see Eq. (3.56))

$$\tau\Delta\nu = \tfrac{1}{2}\pi.$$

The coherence time is therefore given in general by

$$\Delta t \sim 1/\Delta\omega, \tag{5.11}$$

and the coherence length by

$$l \sim c/\Delta\omega. \tag{5.12}$$

Another connection (which we shall need later) also exists between the time correlation and frequency spectrum of a beam. A classical monochromatic wave has the autocorrelation function

$$\Gamma(\mathbf{r}, \mathbf{r}, \tau) = E_0^2\, \mathrm{e}^{\mathrm{j}\omega\tau}. \tag{5.13}$$

If the intensity of a polychromatic wave is proportional to $I(\omega)\,\mathrm{d}\omega$ in the interval from ω to $\omega + \mathrm{d}\omega$, where

$$I(\omega) = |a(\omega)|^2,$$

then Eq. (5.13) can be generalized to give

$$\Gamma(\mathbf{r}, \mathbf{r}, \tau) = \int_0^\infty I(\omega)\, \mathrm{e}^{\mathrm{j}\omega\tau}\, \mathrm{d}\omega.$$

We see that the power spectrum $I(\omega)$ and the autocorrelation function $\Gamma(\mathbf{r}, \mathbf{r}, \tau)$ are related in the same way as a function and its Fourier transform. Therefore the inverse relationship

$$I(\omega) = (2\pi)^{-1} \int_{-\infty}^{+\infty} \Gamma(\mathbf{r}, \mathbf{r}, \tau)\, \mathrm{e}^{-\mathrm{j}\omega\tau}\, \mathrm{d}\tau \tag{5.14}$$

also exists. This provides a means of determining the power spectrum of a beam by measuring its autocorrelation function. The technique is particularly useful when $\Delta\omega$ is very small, since Δt is then correspondingly large and hence more easily measurable (the methods of measurement will be discussed in the next section).

Spatial coherence

The other type of spatial separation mentioned above is the transverse separation of the points P_1 and P_2 with respect to the beam direction. This gives rise to *spatial coherence*. In this case we have to consider the lateral (or transverse) distance over which a continuous beam is correlated. To do this it is helpful to consider the *wavefront* of the beam. This is the locus of all points having the same phase, although not necessarily the same amplitude. For example the idealized plane wave has a planar wavefront while a Hertzian dipole produces a wavefront of spherical shape. But real beams have uneven wavefronts, and at a given point in space the form of the unevenness changes with time. This is illustrated in Fig. 5.4. The phases at two fixed points P_1 and P_2 (where $z_1 = z_2$) are correlated if the points are separated by less than the *lateral coherence distance d*, but not if their lateral separation is much greater than this. Put in another way,

$$\gamma(\mathbf{r}_1, \mathbf{r}_2, 0) \ll 1, \quad \text{if } (x_1 - x_2)^2 + (y_1 - y_2)^2 \gg d^2.$$

Combined temporal and spatial coherence

In general P_1 and P_2 are separated longitudinally *and* laterally, and so the coherence between them involves both the temporal and spatial coherences.

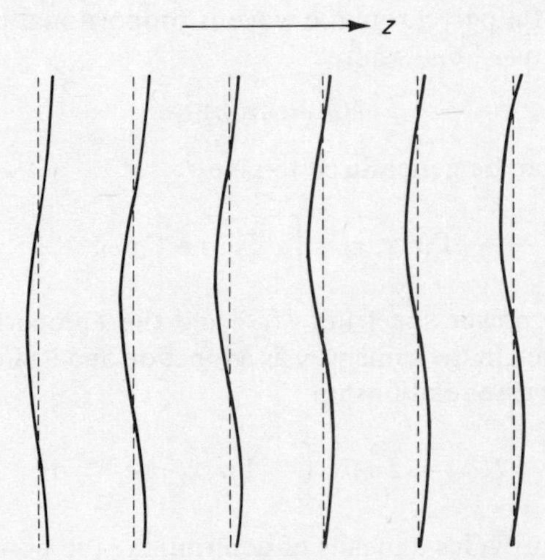

Fig. 5.4. The solid curves show the positions at which the phase of the electric field is zero. They represent the wave-fronts of the beam. The broken lines represent the wave-fronts of an idealized plane wave. At a fixed point the difference between the phase of the beam and that of the plane wave fluctuates with time. The relative phase at two fixed points is random (i.e. the fields at the two points are incoherent) if the lateral separation of the points is much larger than the lateral coherence distance d.

In considering the combined effect it is helpful to express the electric field strength in the form

$$E(\mathbf{r}, t) = E_0(\mathbf{r}, t)\, \mathrm{e}^{\mathrm{j}[\bar{\omega}t + \phi(\mathbf{r}, t)]}, \tag{5.15}$$

where the amplitude E_0 and phase ϕ are both slowly varying functions of position and time, and $\bar{\omega}$ is the mean frequency of the beam. The mutual coherence function then becomes

$$\Gamma_{12}(\tau) = \langle E_0(\mathbf{r}_1, t+\tau) E_0^*(\mathbf{r}_2, t)\, \mathrm{e}^{\mathrm{j}[\phi(\mathbf{r}_1, t+\tau) - \phi(\mathbf{r}_2, t)]} \rangle\, \mathrm{e}^{\mathrm{j}\bar{\omega}\tau},$$

where we now use a simplified notation for Γ. Since any complex number can be written as $A \exp(\mathrm{j}\beta)$, where A and β are real, we can express $\Gamma_{12}(\tau)$ in the form

$$\Gamma_{12}(\tau) = |\Gamma_{12}(\tau)|\, \mathrm{e}^{\mathrm{j}\alpha_{12}(\tau)}\, \mathrm{e}^{\mathrm{j}\bar{\omega}\tau}. \tag{5.16}$$

The real amplitude $|\Gamma_{12}(\tau)|$ and the real phase $\alpha_{12}(\tau)$ both depend on $\mathbf{r}_1$, $\mathbf{r}_2$, and τ, and both vary significantly only in times of the order of the coherence time Δt. The strongest dependence of $\Gamma_{12}(\tau)$ on τ comes therefore from the factor $\exp(\mathrm{j}\bar{\omega}\tau)$. This is the only important factor when $\tau \ll \Delta t$, in which case the mutual coherence function becomes simply

$$\Gamma_{12}(\tau) \simeq |\Gamma_{12}(0)|\, \mathrm{e}^{\mathrm{j}[\alpha_{12}(0) + \bar{\omega}\tau]} = \Gamma_{12}(0)\, \mathrm{e}^{\mathrm{j}\bar{\omega}\tau}. \tag{5.17}$$

The normalized mutual coherence function is similarly simplified to

$$\gamma_{12}(\tau) \simeq \gamma_{12}(0)\, \mathrm{e}^{\mathrm{j}\bar{\omega}\tau}. \tag{5.18}$$

This has the same form as $\gamma_{12}(\tau)$ for a plane wave, except that $\gamma_{12}(0)$ is not equal to unity in general.

The distinction that we have drawn between temporal and spatial coherence is useful in introducing the concept of coherence, and also in discussing particular experiments involving coherence, but is artificial in the sense that the two forms of coherence are not independent. For example a large lateral coherence distance can be achieved only if the light has a high spectral purity, in which case the longitudinal coherence length is also large. We must remember also that the field $E(\mathbf{r}, t)$ propagates in space and time according to the wave equation (2.18), which leads us to expect that the spatial and temporal coherences are intimately connected with each other. By developing this further it can be shown that the rates of change of Γ_{12} with respect to space and time are connected by the equation

$$\nabla_i^2 \Gamma_{12}(\tau) = \frac{1}{c^2} \frac{\partial^2 \Gamma_{12}(\tau)}{\partial \tau^2}, \tag{5.19}$$

where $i = 1$ or 2. We see that this also has the form of a wave equation, a fact which leads to many similarities between the calculation of Γ and the calculation of intensity and diffraction patterns.

5.3 THE MEASUREMENT OF COHERENCE

To make an experimental measurement of the coherence between $E(\mathbf{r}_1, t+\tau)$ and $E(\mathbf{r}_2, t)$ some way must be found to sample these field strengths and to bring them together at the *same* point and the *same* time. The two fields can then *interfere*. If they have an average phase difference of zero the interference is constructive, and the combined intensity is larger than the sum of the separate intensities, while if their average relative phase is π they interfere destructively. By measuring the interference we measure the coherence between the two fields.

Measurement of spatial coherence

An interference method of measuring spatial coherence is illustrated in Fig. 5.5. The incident beam is intercepted by an opaque screen having

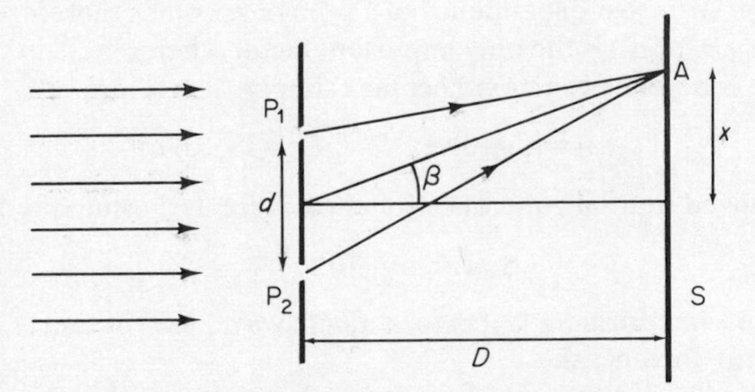

Fig. 5.5. An interference method of measuring the spatial coherence of the fields at the points P_1 and P_2. The two apertures act as point sources, producing an interference pattern on the screen *S*. In an alternative arrangement the light from P_1 and P_2 passes through a lens, and S is placed at the focal plane of the lens.

pin-hole apertures centred at the points P_1 and P_2 (having the vector positions $\mathbf{r}_1$ and $\mathbf{r}_2$ respectively). In practice the apertures would often be replaced by slits, giving the *Young's double slit* experiment, but the principle is the same. If the apertures have diameters much smaller than the wavelength of the light then each may be regarded as a point source which radiates approximately isotropically. At a distance s from the point $P_i (i = 1$ or 2) the field due to the radiation from P_i is proportional to $s^{-1}E(\mathbf{r}_i, t - s/c)$. The distance dependence s^{-1} reflects the fact that the intensity is proportional to s^{-2}, and the retarded time $(t - s/c)$ is used because the light takes the time s/c to travel the distance s. At a point A on the screen S (see figure) the combined field due to the two pin-holes is therefore

$$E(\mathbf{r}_A, t) \propto (P_1A)^{-1}E(\mathbf{r}_1, t - P_1A/c) + (P_2A)^{-1}E(\mathbf{r}_2, t - P_2A/c). \tag{5.20}$$

To simplify expression (5.20) we now assume that d and x are both much smaller than D (all these dimensions are defined in the figure, together with the angle β). Then

$$P_2A - P_1A \simeq d \sin\beta \simeq \frac{xd}{D}.$$

This allows us to put

$$(P_1A)^{-1} \simeq (P_2A)^{-1},$$

giving

$$E(\mathbf{r}_A, t + P_2A/c) \propto E(\mathbf{r}_1, t + \tau) + E(\mathbf{r}_2, t),$$

where

$$\tau = \frac{P_2A - P_1A}{c} \simeq \frac{xd}{cD}.$$

The intensity at A is therefore proportional to

$$\begin{aligned} I(x) &= \langle E(\mathbf{r}_A, t)E^*(\mathbf{r}_A, t)\rangle \\ &= \langle E(\mathbf{r}_1, t)E^*(\mathbf{r}_1, t)\rangle + \langle E(\mathbf{r}_2, t)E^*(\mathbf{r}_2, t)\rangle \\ &\quad + 2\,\mathrm{Re}\,[\langle E(\mathbf{r}_1, t+\tau)E^*(\mathbf{r}_2, t)\rangle] \\ &= I_1 + I_2 + 2\,\mathrm{Re}\,[\Gamma_{12}(\tau)] \\ &= I_1 + I_2 + 2(I_1I_2)^{1/2}\,\mathrm{Re}\,[\gamma_{12}(\tau)]. \end{aligned}$$

If I_1 and I_2 are the same, as if often the case, then the intensity pattern on the screen is given by

$$I(x) \propto 1 + \mathrm{Re}\,[\gamma_{12}(\tau)].$$

Assuming that the incident light is approximately monochromatic and that its coherence length is much larger than $c\tau$, we can now use Eq. (5.18), obtaining

$$I(x) \propto 1 + |\gamma_{12}(0)| \cos\left[\alpha_{12}(0) + \left(\frac{2\pi d}{D\bar{\lambda}}\right)x\right]. \tag{5.21}$$

The intensity maxima correspond to constructive interference, when the average phase difference between the light from P_1 and P_2 is zero (as in Fig. 5.1(a)). Conversely the intensity minima correspond to destructive interference, when the average phase difference is π (Fig. 5.1(b)).

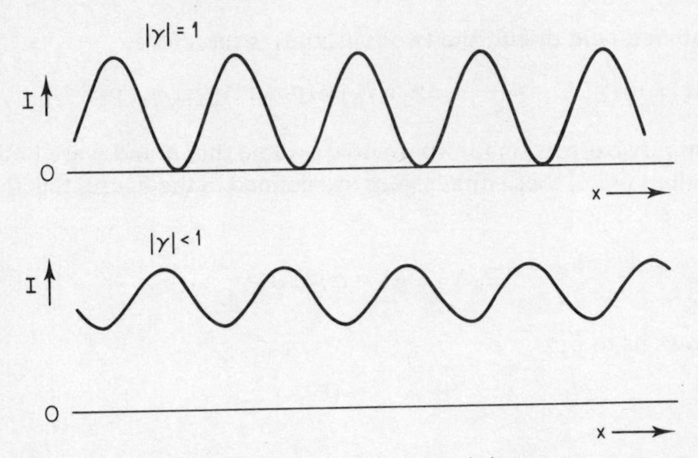

Fig. 5.6. Interference patterns for $|\gamma|=1$ (complete coherence), and $|\gamma|<1$ (partial coherence).

Two examples of this type of interference pattern are shown in Fig. 5.6. The first occurs only when there is complete coherence between P_1 and P_2 ($|\gamma|=1$) and when the two radiated intensities are the same ($I_1=I_2$). The second occurs either (i) when $I_1=I_2$, and $|\gamma|<1$, or (ii) when $I_1 \neq I_2$, and $|\gamma| \leq 1$. When the intensities are unequal the equation

$$I(x) \propto 1+\beta|\gamma_{12}(0)| \cos\left[\alpha_{12}(0)+\left(\frac{2\pi d}{D\bar{\lambda}}\right)x\right]$$

must be used, where

$$\beta=\frac{2(I_1 I_2)^{1/2}}{I_1+I_2}<1.$$

Confining our attention to the case of equal intensities, we see that the ratio of the maximum to minimum value of $I(x)$ is determined only by $|\gamma|$. We find from Eq. (5.21) that

$$|\gamma_{12}(0)|=\frac{I_{\max}-I_{\min}}{I_{\max}+I_{\min}}. \tag{5.22}$$

The quantity on the right-hand side of this equation is called the *visibility* of the interference pattern.

We see now how simple it is to measure the spatial coherence of a beam of light. The visibility of the Young's double slit interference pattern, with the slits separated by a distance d, gives the modulus $|\gamma|$ of the spatial coherence corresponding to this distance. If we want to know γ completely we must go further and find the phase α by measuring the position of the fringes with respect to the position $x=0$.

Measurement of temporal coherence

The temporal coherence $\Gamma(\mathbf{r}, \mathbf{r}, \tau)$ of a beam can be measured by sampling the field strength at the same point $\mathbf{r}$ at different times, and bringing the signals together at the *same* time and place, to interfere with each other. This is achieved with the *Michelson spectral interferometer*, shown in Fig. 5.7. If

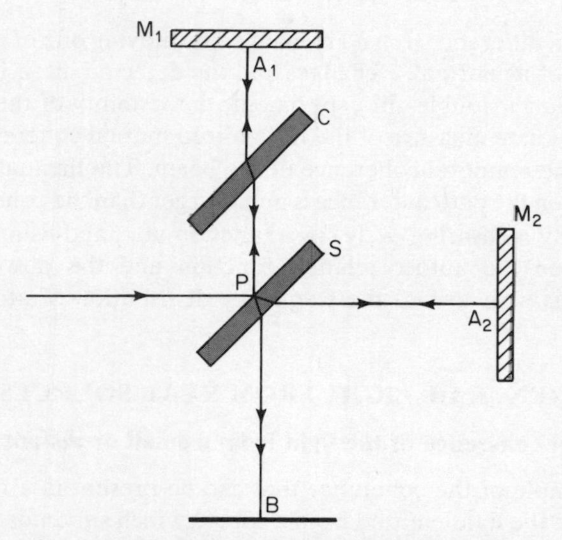

Fig. 5.7. Schematic diagram of a Michelson spectral interferometer, used to determine the temporal coherence of a beam. The correcting plate C ensures that the beams transverse the same total thickness of glass in the two arms of the interferometer.

the incoming beam is regarded as a series of parallel rays, each of these is split into two parts at a point P by the semi-reflecting front face of the block S. The two parts travel to mirrors M_1 and M_2 respectively, and then return to P, from where they travel along the same path towards a point B on the detector screen. The presence of the correcting plate C ensures that the same total thickness of glass is traversed in both arms of the interferometer, thus minimizing the effects of temperature fluctuations. The superposed beams have the same intensity when they arrive at the detector screen, because each has undergone one transmission and one reflection at the semi-reflecting surface of S.

If the path lengths $PA_1PB(=l_1)$ and $PA_2PB(=l_2)$ differ by

$$l_1 - l_2 = c\tau,$$

the detected intensity is proportional to

$$I = \langle |E(t+\tau)+E(t)|^2\rangle = 2\langle |E|^2\rangle + 2\,\mathrm{Re}\,[\Gamma_{11}(\tau)]$$
$$= 2\langle |E|^2\rangle\{1+\mathrm{Re}\,[\gamma_{11}(\tau)]\}.$$

Using Eq. (5.16) this becomes

$$I \propto 1 + |\gamma_{11}(\tau)|\cos[\alpha(\tau)+\bar{\omega}\tau].$$

When the path difference $(l_1 - l_2)$ is changed by moving one of the mirrors in the direction of its normal, τ changes and the detected intensity fluctuates. As in the Young's double-slit experiment, the visibility of the fluctuations gives an immediate measure of the normalized mutual coherence function, this time of the temporal coherence of the beam. The fluctuations become very weak when the path difference is much larger than the coherence length of the light. By measuring $\gamma_{11}(\tau)$ as a function of τ, and using the relation (5.14) between the autocorrelation function and the power spectrum, Michelson was able to find the frequency distribution of atomic spectral lines.

5.4 COHERENCE OF LIGHT FROM REAL SOURCES

5.4.1 Spatial coherence of the light from a small or distant source

As an example of the coherence that can be present in a real radiation field, consider the light emitted from a source which subtends a small angle at the observer. The source may for example be an illuminated pinhole placed a few cms from the observer, or it may be a distant star. The problem is simplified if we assume that the source is equivalent to a uniformly illuminated circular disc, and that it emits light of mean angular frequency $\bar{\omega}$ and width $\Delta\omega$, where $\Delta\omega \ll \bar{\omega}$. We assume also that each point of the source emits light quite independently of all other points. This means that there is no coherence in the source itself, but it does not mean, as we shall soon see, that there is no coherence in the radiation field far from the source.

The degree of spatial coherence in the light from the source can be found by a Young's slit experiment. Fig. 5.8(a) shows the disc source, and it shows also the interference pattern that would be produced by the light from a single point source S_a situated at one extremity of the source. There would be an intensity maximum at the point A on the screen, because the path lengths S_aP_1A and S_aP_2A are the same (provided that $\theta \ll 1$, and $d \ll D$). The x-coordinate of the point A is therefore

$$x_A = -D\theta/2.$$

The distance Δx between the intensity maxima can be deduced from Eq. (5.21); it is

$$\Delta x = D\lambda/d.$$

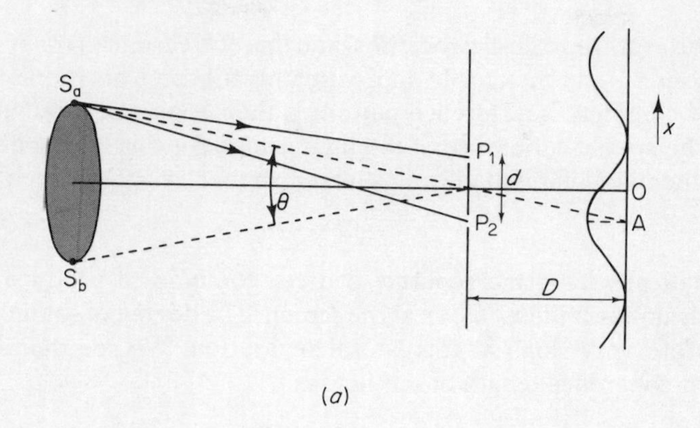

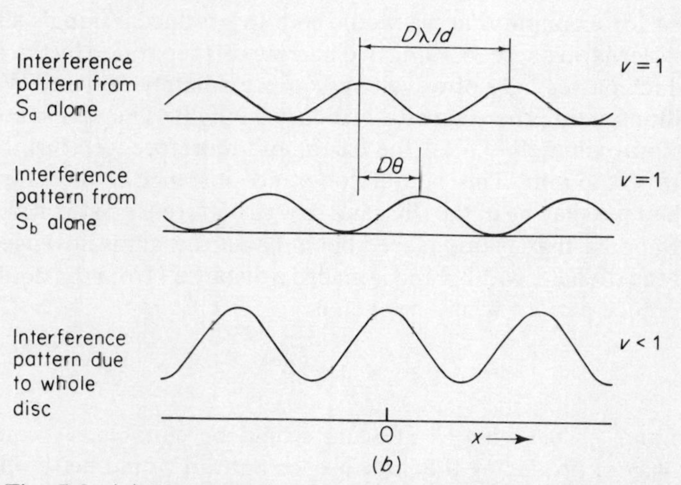

Fig. 5.8. (*a*) Young's slit experiment, showing the interference pattern produced on the screen when only the light from the point source S_a passes through the apertures at P_1 and P_2. (*b*) The interference pattern due to S_a or S_b alone have a visibility of unity; the combined pattern due to the whole disc has a visibility less than unity.

The visibility of the interference pattern due to the point source is unity. This interference pattern is shown again in Fig. 5.8(*b*), together with the pattern produced by the light from a point S_b at the opposite extremity. The central maxima of these two interference patterns are separated by $D\theta$.

If the slit separation d is made small, so that Δx is much larger than $D\theta$, which requires that

$$d \ll \lambda/\theta,$$

the patterns from these two points, and therefore from all other points on the disc source, have maxima in approximately the same positions. The visibility of the complete interference pattern is then approximately unity, showing that the spatial coherence of the light is high for this lateral distance d. On the other hand if d is made much larger, so that $\Delta x \ll D\theta$, which requires that

$$d \gg \lambda/\theta,$$

the patterns from the separate sources combine to produce an approximately uniform illumination at the screen. The degree of spatial coherence is therefore very small at this lateral separation. We see therefore that the lateral coherence length of the light is

$$d \sim \lambda/\theta. \tag{5.23}$$

The coherence extends in both transverse (x and y) directions.

Suppose for example that we would like to produce Young's slit interference patterns on a screen, using two narrow slits separated by 0.5 mm and a filter which passes light of wavelength approximately 500 nm. We might start by illuminating the slits with ordinary sunlight. The sun subtends an angle of approximately 0.5° at the earth, and therefore λ/θ has the value 6×10^{-5} m $= 0.06$ mm. This lateral coherence distance is therefore much smaller than the spacing of the slits, and so an interference pattern would not be seen. Suppose that a lamp placed behind a narrow slit is now used as the source. If the slit has a width s and is placed a distance l from the double slits, an interference pattern would be seen if

$$\theta = \frac{s}{l} \leqslant \frac{\lambda}{d} = 10^{-3}.$$

For example, $s = 0.1$ mm, $l = 100$ mm would be suitable. A completely different way of producing the interference pattern would be to illuminate the double slits with laser light. A clear pattern is then easily obtained, even if the angle subtended by the laser exit aperture is much larger than λ/d. This happens because the source is almost completely coherent across the whole of its area, as is the laser beam itself, and so Eq. (5.23) no longer applies. The coherence properties of laser light will be treated in more detail in Chapter 7.

To see how the mutual coherence function Γ can be evaluated quantitatively, we consider the flat source S shown in Fig. 5.9(a). If the element of area $d\sigma$ emits light of intensity $I(\sigma)\, d\sigma$ then the electric field strength at a point P_1 a distance R_1 away is

$$E_1 \propto [I(\sigma)\, d\sigma]^{1/2}/R_1,$$

provided that the line from $d\sigma$ to P_1 is approximately normal to the plane of $d\sigma$, as in the figure. The degree of coherence between two points P_1 and P_2,

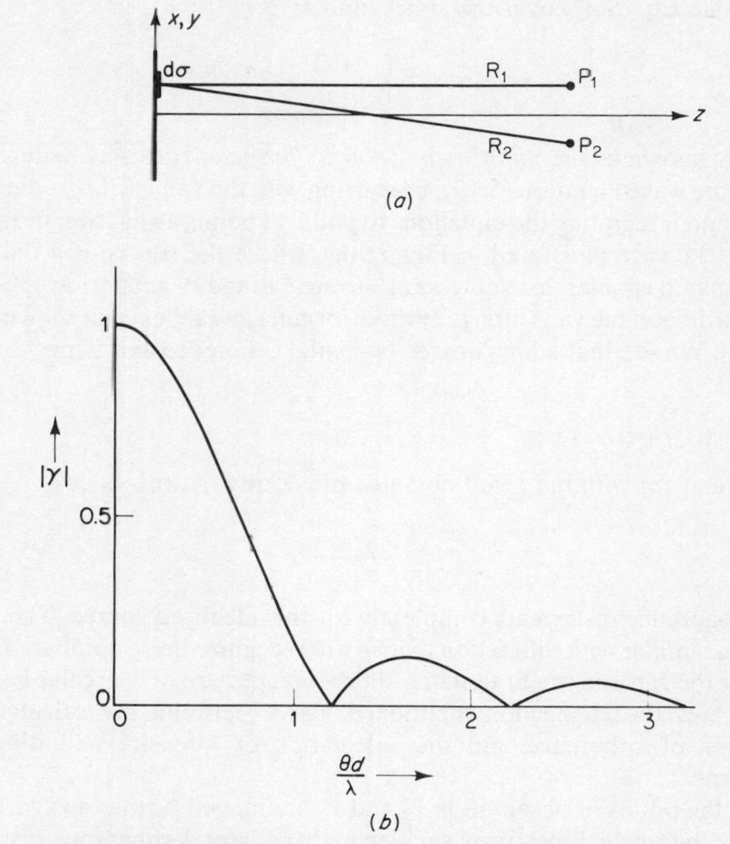

Fig. 5.9. (a) Co-ordinates used in the van Cittert–Zernike formula for the coherence at the points P_1 and P_2 caused by a flat source S. (b) The magnitude of the normalized mutual coherence function when the source subtends the angle θ and is circular and uniformly illuminated, and P_1 and P_2 are separated by the distance d and have the same z co-ordinate.

due to the light emitted from $d\sigma$ only, is therefore

$$\Gamma_{d\sigma}(\mathbf{r}_1, \mathbf{r}_2, \tau) \propto (R_1R_2)^{-1}\, e^{j\bar{\omega}[\tau+(R_2-R_1)/c]} I(\sigma)\, d\sigma, \tag{5.24}$$

assuming that the light is nearly monochromatic ($\Delta\omega \ll \bar{\omega}$). The proportionality constant omitted from this equation depends on the definition of I, and on the connection between I and E, and is of no interest if we eventually intend to calculate γ for the source, rather than Γ.

If the separate areas $d\sigma$ emit independently (i.e. if they are mutually incoherent), the total degree of coherence for the light at P_1 and P_2 is obtained by summing the separate contributions. We must therefore integrate Eq. (5.24) over the areas, obtaining

$$\Gamma(\mathbf{r}_1, \mathbf{r}_2, \tau) \propto e^{j\bar{\omega}\tau} \int_S \frac{I(\sigma)}{R_1 R_2} e^{j\bar{\omega}(R_2 - R_1)/c} \, d\sigma. \tag{5.25}$$

This is known as the *van Cittert–Zernicke formula*. It can be deduced also from the wave equation (5.19), by starting with the values of Γ at the source itself and integrating the equations to find Γ at points away from the source.

For the case illustrated in Fig. 5.9(a), where the source is a uniformly illuminated circular disc subtending an angle θ, and P_1 and P_2 have the same z-coordinate, the van Cittert–Zernike formula gives the result shown in Fig. 5.9(b). We see that a high degree of spatial coherence exists for

$$d \leqslant \frac{\lambda}{\theta}, \tag{5.26}$$

in agreement with the result obtained previously. At the values

$$\frac{d\theta}{\lambda} = 1.22, 2.23, \ldots$$

the coherence disappears completely for this idealized source. The reader who is familiar with diffraction theory will recognize these numbers as those giving the zeros in the Fraunhofer diffraction pattern of a circular hole; this illustrates the connection mentioned above between the calculation of degrees of coherence and the calculation of intensity and diffraction patterns.

As the points of observation P_1 and P_2 are moved further away from the disc S the angle θ becomes smaller and the lateral coherence distance d becomes larger. We see therefore that although there is no spatial coherence at the surface of the source, there is a high spatial coherence far from the source. We shall consider this behaviour again later, in terms of the coherence between the photons observed at P_1 and P_2. Use can be made of this behaviour to *deduce* the angle θ subtended by the source, by *measuring* the spatial coherence. The technique was first employed by Michelson, to find the angular sizes of distant stars. He placed mirrors at the points P_1 and P_2, instead of apertures, as shown in Fig. 5.10. This arrangement is known as the *Michelson stellar interferometer*. The distance d is varied by moving the outer two mirrors, while the inner mirrors and the apertures A_1 and A_2 remain fixed. The light from the star is spread by diffraction at the apertures, and appears as an extended image at the focal point of the lens. The light from the two apertures interferes, as in the arrangement of Fig. 5.5, and so the image has the form of interference fringes of small spacing Δx (because d is large). The visibility of the fringes gives the spatial coherence $|\gamma|$ and by measuring this coherence as a function of d, the van Cittert–Zernike formula

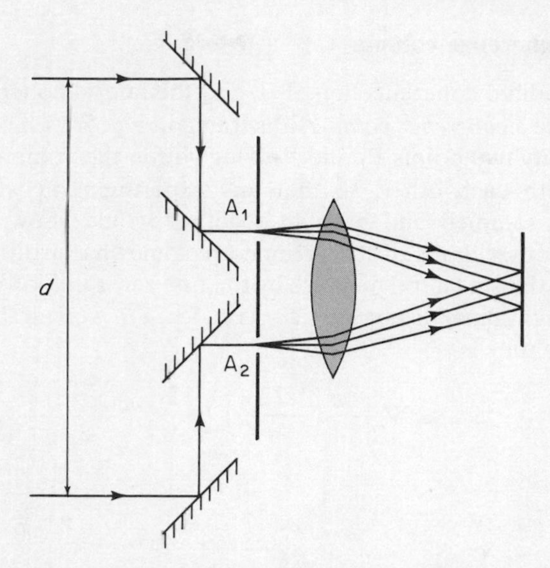

Fig. 5.10. Schematic diagram of a Michelson stellar interferometer, used to determine the angular size of stars.

(5.25) can be used to deduce the angular diameter of the star. In Michelson's early experiments he was able to measure angular diameters as small as 0.02 seconds of arc (10^{-7} rad), an angle equal to that subtended by a small coin at a distance of 100 km!

5.4.2 Temporal coherence

The temporal coherence of the light from a source depends primarily on the spread of frequency, and not on the size of the source (see Eq. (5.12)). A simple example is the light emitted by atoms in an excited state of lifetime τ. Neglecting collision and Doppler broadening (see Section 6.3), the radiation has a Lorentzian frequency distribution (see Fig. 3.17), with a half width given by

$$\Delta\omega = 1/\tau.$$

The coherence length is therefore

$$l_c \sim c/\Delta\omega = c\tau. \tag{5.27}$$

For example a typical atomic transition in the visible region has $\tau \sim 10^{-8}$ s and therefore $l_c \sim 3$ m. In practice collision and Doppler broadening are usually important and the coherence length is typically two orders of magnitude smaller, in the region of a few cms.

5.4.3 The coherence volume

The longitudinal coherence length l_c and the lateral coherence length d together define a *coherence volume*, illustrated in Fig. 5.11. The electric field strengths at any two points P_1 and P_2 lying within the volume are partially correlated with each other, so that any experiment in which the field strengths are sampled and brought together would show partial interference. The outer limits of the volume are somewhat arbitrary, but if we define the maximum lateral distance by the first zeros in Fig. 5.9(*b*), and the maximum longitudinal distance by $2c\tau$ (Eq. (5.27)), we find that the magnitude of the volume is

$$V_c \sim \frac{\pi}{4}\left(\frac{1.22\,\lambda}{\theta}\right)^2 \times \frac{2c}{\Delta\omega}. \tag{5.28}$$

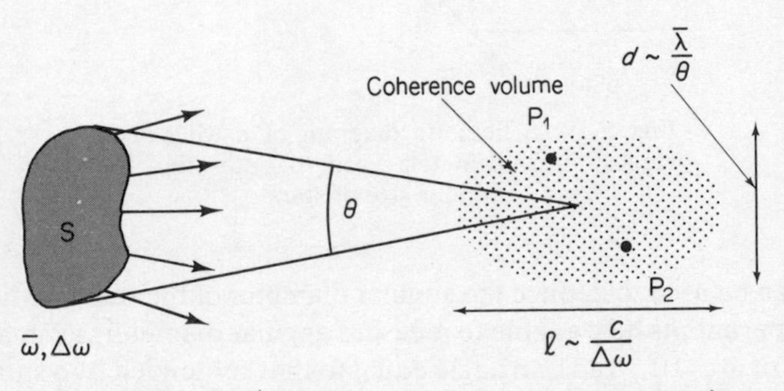

Fig. 5.11. Partial coherence exists between any two points lying within the coherence volume.

Another important volume is the volume of *momentum space* occupied by the observed light. The longitudinal momentum of the emitted photons has the mean value $\hbar\omega/c$, and therefore the total spread in their transverse momentum is

$$\Delta p_{x,y} = \frac{\hbar\omega\theta}{c}. \tag{5.29}$$

If we take the total range of angular frequencies to be $\sim 2\,\Delta\omega$, the spread in longitudinal momentum is

$$\Delta p_z \sim \frac{2\hbar\,\Delta\omega}{c}. \tag{5.30}$$

The total volume of momentum space occupied by the observed photons is therefore

$$V_{\text{mom}} = \Delta p_x\,\Delta p_y\,\Delta p_z \sim \frac{2\hbar^2\theta^2\omega^2\,\Delta\omega}{c^3}. \tag{5.31}$$

This volume of momentum space is proportional to $\theta^2 \Delta\omega$, whereas the volume of co-ordinate space is inversely proportional to $\theta^2 \Delta\omega$. A narrow range of photon momentum implies a large coherence volume, and *vice versa*. By taking the product of V_c and V_{mom} we obtain the volume of *phase-space* occupied by the observed partially coherent photons. Clearly this is independent of θ or $\Delta\omega$. It has the value

$$V_{\text{phase-space}} = \Delta x\ \Delta y\ \Delta z\ \Delta p_x\ \Delta p_y\ \Delta p_z \sim 4(1.22)^2\pi^3\hbar^3 \simeq h^3. \quad (5.32)$$

Equation (5.32) is an important result because it allows us to interpret the classical concept of coherence in terms of the behaviour of photons. We saw in Chapter 4 that the number of normal modes in the interval from ν to $\nu + d\nu$ in a cubic cavity of volume a^3 is (Eq. (4.5))

$$n = a^3\rho(\nu)\,d\nu = \frac{4\pi a^3\nu^2\,d\nu}{c^3}$$

(this differs from Eq. (4.5) by a factor 2, since for the present purposes we are considering only one direction of polarization). The volume of momentum space occupied by these modes is the volume between spheres of radius $(h\nu/c)$ and $(h\nu/c)+d(h\nu/c)$, namely

$$V_{\text{mom}} = 4\pi(h/c)^3\nu^2\,d\nu.$$

The average momentum space occupied by each mode is therefore

$$\frac{V_{\text{mom}}}{n} = \left(\frac{h}{a}\right)^3,$$

and the average phase space per mode is

$$V_{\text{phase-space}} = h^3. \quad (5.33)$$

Each mode may of course contain more than one photon, but all the photons in the same mode have the same frequency and direction, and the phase space occupied by the 'multiple photon' remains the same as that of a single photon. With this in mind we can look again at Fig. 5.11 and Eq. (5.32), and interpret the coherence volume as the volume of co-ordinate space occupied by a single or multiple photon. When the momentum space of the photon is small the coherence volume is large, and *vice versa*. The existence of partial coherence within the coherence volume is now easily understood. If P_1 and P_2 are within the volume of one photon their electric field strengths are correlated because of the correlation between the different parts of a photon wave-packet. On the other hand if they are further apart they are governed by uncorrelated photons and so the corresponding electric field strengths are uncorrelated.

Before leaving the subject of the phase space occupied by photons, let us go a little further and consider the concept of the *polychromatic photon*. In the discussion of Section 4.1, we regarded the photon as the quantum of a normal mode of the radiation field in a cavity. As such it has a well-defined frequency ν and wavenumber $\mathbf{k}$, but it may be situated at any point $\mathbf{r}$ inside the cavity. In other words it occupies a very small volume of momentum space but a very large volume (that of the whole cavity) of co-ordinate space. On the other hand the wave-packet illustrated in Fig. 5.3 is composed of waves from many normal modes of the cavity, and hence occupies a much larger volume of momentum space, but is much more localized in co-ordinate space. To make this photon localized in the longitudinal direction it must be composed of modes having a range of frequencies, and hence must be polychromatic. To make it localized in the transverse direction it must be composed of modes having a range of directions $\mathbf{k}$, and hence it is no longer unidirectional.

At this stage we may ask what type of photon actually exists in a radiation field, for example a field in thermal equilibrium with the cavity walls. In fact the different types are just different descriptions of a uniquely defined system. The field may be regarded either as a set of normal modes, each of which may contain zero, one or more unlocalized but monochromatic photons, or it may be regarded as a collection of localized and polychromatic wave-packets, each composed of a suitable superposition of the normal modes. When photons are created (i.e. emitted) or annihilated (i.e. detected), the type of photon depends on the creation or annihilation process. For example if a detection event occurs in a known, short time interval, and at a known, limited location, then the detected photons must be highly polychromatic and multidirectional, while if on the contrary the time and position of the detection event are not well known but the frequency and direction of detection are highly specific, then the detected photons must be unlocalized and may possibly be single mode photons.

5.5 INTENSITY CORRELATIONS

So far we have dealt with the correlation or coherence between the electric field strengths $E(\mathbf{r}_1, t+\tau)$ and $E(\mathbf{r}_2, t)$. It is possible also for correlations to exist between the intensities

$$I(\mathbf{r}_1, t+\tau) = |E(\mathbf{r}_1, t+\tau)|^2$$

and

$$I(\mathbf{r}_2, t) = |E(\mathbf{r}_2, t)|^2.$$

Of course a trivial correlation exists in the sense that the *mean* intensities

$$\bar{I}(\mathbf{r}_i) = \langle I(\mathbf{r}_i, t)\rangle$$

(where $i = 1$ or 2) are usually proportional to each other, but our present concern is with the correlations in the intensity *fluctuations*, given by

$$\Delta I(\mathbf{r}_i, t) = I(\mathbf{r}_i, t) - \bar{I}(\mathbf{r}_i). \tag{5.34}$$

The intensity correlations are of special interest in the case of light from thermal sources. We start by treating the problem classically, and then go on to see how the correlations are modified by the existence of photons.

5.5.1 Chaotic light

To simulate the intensity fluctuations in the light from a thermal source we can suppose that each atom of the source radiates independently and that the electric field of the emitted light is composed of a large number of independent waves,

$$E = \sum_n a_n \, \mathrm{e}^{\mathrm{j}(\omega_n t + \phi_n)}, \tag{5.35}$$

where the ϕ_n are random phases (i.e. they are in no way related to each other). This type of light is usually called *chaotic light*. If the bandwidth $\Delta\omega$ is small compared with the mean frequency $\bar{\omega}$ it is convenient to separate out the main time dependence, exp $(\mathrm{j}\bar{\omega}t)$, and write

$$E = E_0 \, \mathrm{e}^{\mathrm{j}\bar{\omega}t}, \quad \text{where } E_0 = \sum_n a_n \, \mathrm{e}^{\mathrm{j}[(\omega - \bar{\omega})t + \phi_n]}.$$

Figure 5.12 shows how the summation can be represented as a vector sum on an Argand diagram. The electric field E_0 is given by the length OA, and because the terms have been given the same amplitude a but random phase

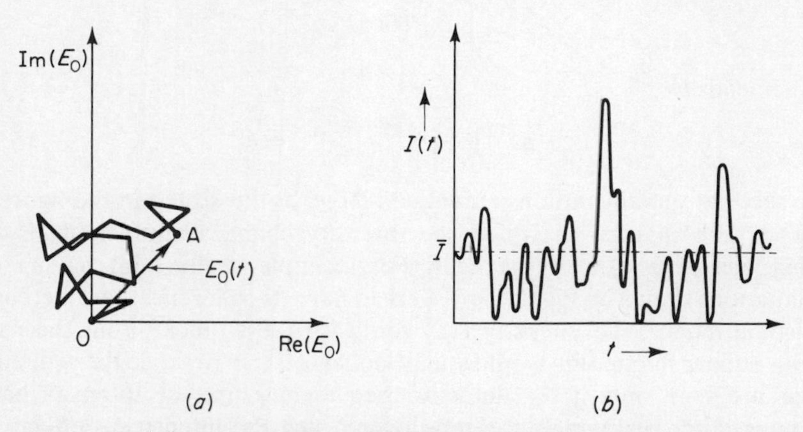

Fig. 5.12. (*a*) Representation of the electric field of a chaotic wave as a sum of a large number of independent terms of different frequencies and random phase. (*b*) Fluctuations in the intensity $I(t) = |E_0(t)|^2$.

ϕ_n, the point A is reached from O by a *random walk* of equal steps. At a later time $t+\Delta t$ the steps will have rotated through their respective angles $(\omega_n - \bar{\omega})\,\Delta t$, and their phases may also have changed, which means that a different random walk leads to a different point A, and hence to different values of E_0, E, and I. Figure 5.12(b) shows how the intensity changes with time. It is far from uniform, with occasional excursions to values much greater than the mean.

To find the probability distribution of I we can start by using random walk theory to find the probability distribution of E_0. The result (which is the same for terms of equal or unequal magnitude) is that the phase of E_0 is uniformly distributed from 0 to 2π, and that the probability that $|E_0|$ lies in the range from $\mathscr{E}$ to $\mathscr{E}+\mathrm{d}\mathscr{E}$ is

$$P(\mathscr{E})\mathrm{d}\mathscr{E} = \frac{2\mathscr{E}}{\overline{E_0^2}} \exp\,(-\mathscr{E}^2/\overline{E_0^2})\,\mathrm{d}\mathscr{E}. \tag{5.36}$$

This probability distribution is illustrated in Fig. 5.13(a). The most probable value of E_0 is zero. The figure also shows two other distributions, both of them quite different from that of the chaotic wave. The classical sinusoidal wave has a constant phase angle and amplitude, while the laser beam has a phase angle which varies slowly and randomly with time, but with E_0 approximately constant (see Chapter 7). Returning to the chaotic wave, since $I = \mathscr{E}^2$ we find that the probability distribution for I is

$$P(I)\,\mathrm{d}I = (\bar{I})^{-1} \exp\,(-I/\bar{I})\,\mathrm{d}I, \tag{5.37}$$

where $\bar{I}$ is the mean value. To find the root-mean-square fluctuation we need to evaluate

$$\overline{I^2} = \int I^2 P(I)\,\mathrm{d}I = 2\bar{I}^2,$$

which leads to

$$[\overline{(\Delta I)^2}]^{1/2} = [\overline{I^2} - \bar{I}^2]^{1/2} = \bar{I}. \tag{5.38}$$

We see that the fluctuations are indeed large, as illustrated in the figure.

The time interval in which the intensity changes appreciably is the coherence time (Δt) of the light. For example if the light having the fluctuations shown on Fig. 5.12(b) were to have its coherence time increased by some means, the intensity $I(t)$ would vary less quickly (but the root-mean-square fluctuation would still be equal to $\bar{I}$). In practice these fluctuations are seen only if the detector used to measure the intensity has a response time T (namely the time over which the intensity is effectively averaged) which is shorter than Δt; otherwise the observed root-mean-square fluctuation is less than $\bar{I}$. When T is much larger than Δt the fluctuation is given by

$$\overline{(\Delta I)^2} = \bar{I}^2 \frac{\Delta t}{T}. \tag{5.39}$$

A further source of fluctuations arises from the existence of photons in the beam. Suppose that a detector is used which counts the number of photons arriving in the time interval T. If the mean counting rate is $\bar{n}$, and if the photons arrive randomly (i.e. independently of each other), the actual number n arriving in a given time interval T is governed by Poisson statistics. The fluctuation in n is given by

$$\overline{(\Delta n)^2} = \bar{n}. \tag{5.40}$$

This source of fluctuation would exist even in a beam for which the classical intensity fluctuation is small or zero (such as those depicted in Fig. 5. 13(b) and (c)).

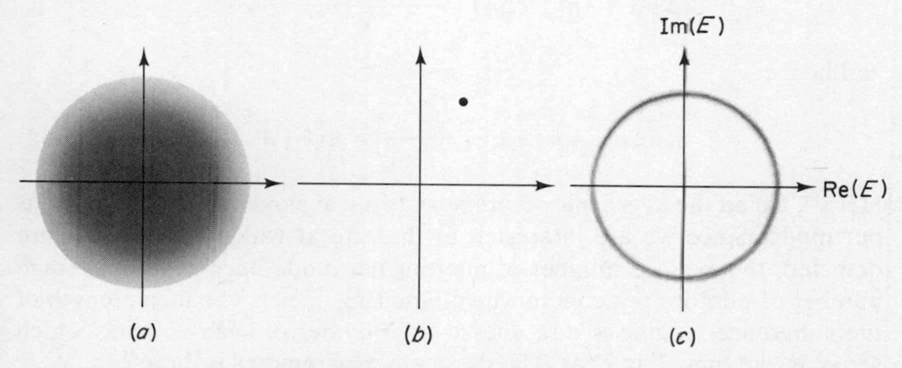

Fig. 5.13. Probability distributions of the amplitude and phase of the electric field in three different types of beam: (a) chaotic beam, (b) classical sinusoidal beam, and (c) laser beam. In part (a) the probability is represented by the depth of shading.

Let us now combine the fluctuation given by Eq. (5.39) with the fluctuation given by Eq. (5.40) for a chaotic beam. We firstly convert from I to n by putting

$$I = nh\nu/T,$$

giving

$$\overline{(\Delta n)^2}_{\text{class}} = \bar{n}^2 \frac{\Delta t}{\tau}.$$

In the same way that independent errors are combined by adding their squares, we now obtain

$$\overline{(\Delta n)^2} = \bar{n} + \bar{n}^2 \frac{\Delta t}{T}. \tag{5.41}$$

This equation gives the mean-square *photon fluctuation* of a beam of chaotic light.

Equation (5.41) can be derived also by considering the radiation energy U contained in a volume V in thermal equilibrium at temperature θ, in the frequency interval from ν to $\nu+\mathrm{d}\nu$. The mean value of U is (Eq. (4.12))

$$\bar{U} = h\nu\bar{n} = \frac{8\pi h\nu^3 V\,\mathrm{d}\nu}{c^3\,(\mathrm{e}^{h\nu/k\theta}-1)}.$$

The energy fluctuation for *any* system in thermal equilibrium is given by

$$\overline{(\Delta U)^2} = k\theta^2 \frac{\partial \bar{U}}{\partial \theta}.$$

In the present case this leads to

$$\overline{(\Delta U)^2} = (h\nu)^2\,\overline{(\Delta n)^2} = \frac{8\pi h^2\nu^4 V\,\mathrm{e}^{h\nu/k\theta}\,\mathrm{d}\nu}{c^3(\mathrm{e}^{h\nu/k\theta}-1)^2},$$

and hence

$$\overline{(\Delta n)^2} = \bar{n}\left(1+\frac{1}{\mathrm{e}^{h\nu/k\theta}-1}\right) = \bar{n}(1+\delta). \tag{5.42}$$

Here δ, called the *degeneracy parameter*, is the average number of photons per mode. Since we are interested in the rate at which the photons are detected, the average number of photons per mode becomes the average number of photons per coherence volume (Fig. 5.11). The mean length of the coherence volume is $c\Delta t$, and so the number of such volumes which arrive in the time T is $T/\Delta t$. The degeneracy parameter is therefore

$$\delta = \bar{n}\Big/\left(\frac{T}{\Delta t}\right). \tag{5.43}$$

Equations (5.42) and (5.43) now give Eq. (5.41).

5.5.2 Photon correlations

An important experiment was carried out by Hanbury-Brown and Twiss in 1956. Their apparatus is shown schematically in Fig. 5.14. They used two independent photomultipliers (see Chapter 9), each at the focus of a concave collecting mirror, to observe the light from the star Sirius. The signals were amplified, then the mean values were effectively subtracted and the remaining fluctuations were multiplied together in a correlator. Finally the output of the correlator was averaged over times of a few minutes. The resulting signal is proportional to the *cross-correlation function* $\overline{\Delta n_1\,\Delta n_2}$, where Δn_1 and Δn_2 are the fluctuations in the mean numbers of photons arriving at the

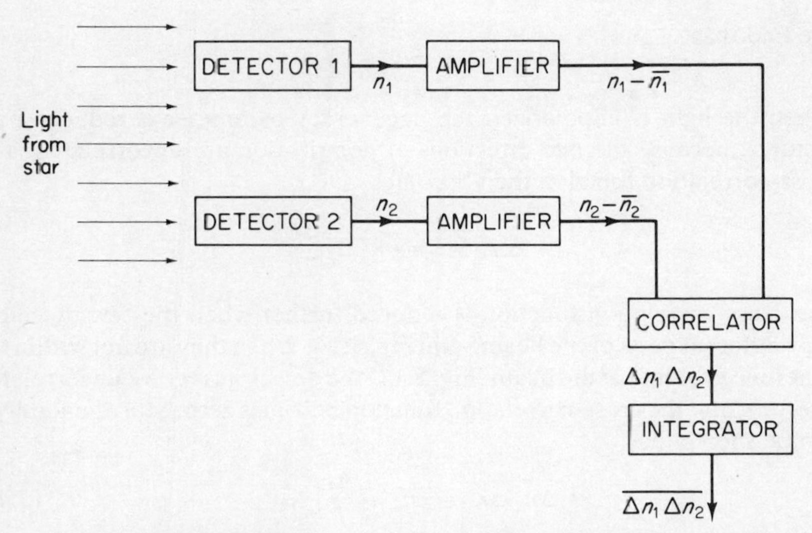

Fig. 5.14. Schematic diagram of the intensity interferometer used by Hanbury-Brown and Twiss to show the existence of photon correlations, and to measure the angular diameter of Sirius.

photomultipliers during the response time T of the detector system. They found that $\overline{\Delta n_1\,\Delta n_2}$ is non-zero, thus showing that a *photon correlation* exists between the beams of light striking the two detectors. In other words the photons do not arrive at the two detectors independently: instead the arrival of a photon at one detector is partially correlated with the arrival of a different photon at the other detector.

To show that this result is a consequence of Eq. (5.41) it is helpful to start by considering a simpler experiment in which the two detectors view the *same* beam of light (this experiment was also carried out by Hanbury-Brown and Twiss, splitting the light from a mercury source by a semi-reflecting mirror, and directing the resulting two beams to two photomultiplier detectors). In this case we can put

$$n = n_1 + n_2$$

where n is the total number of photons detected in the response time T. The relationships that we need are

$$\overline{(\Delta n)^2} = \bar{n}(1+\bar{n}\ \Delta t/T),$$

$$\overline{(\Delta n_i)^2} = \bar{n}_i(1+\bar{n}_i\ \Delta t/T),$$

and

$$\overline{(\Delta n)^2} = \overline{(\Delta n_1 + \Delta n_2)^2} = \overline{(\Delta n_1)^2} + \overline{(\Delta n_2)^2} + 2\ \overline{\Delta n_1\ \Delta n_2}.$$

We find that

$$\overline{\Delta n_1 \, \Delta n_2} = \overline{n_1} \, \overline{n_2} \, \Delta t / T.$$

When the light is unpolarized the degeneracy parameter is reduced by a factor 2, because the two directions of polarization are uncorrelated. The cross-correlation function then becomes

$$\overline{\Delta n_1 \, \Delta n_2} = \tfrac{1}{2} \, \overline{n_1} \, \overline{n_2} \, \frac{\Delta t}{T}.$$

The cross-correlation function is reduced further when the two detectors view different parts of the beam, as in Fig. 5.14. When they are not within the coherence volume of the beam (Fig. 5.11) the detectors receive uncorrelated photons, and the cross-correlation function becomes zero. More generally it can be shown that

$$\overline{\Delta n_1 \, \Delta n_2} = \tfrac{1}{2} \, \overline{n_1} \, \overline{n_2} \, \frac{\Delta t}{T} \, |\gamma_{12}|^2, \tag{5.44}$$

where γ_{12} is the normalized mutual coherence function for the two detectors.

Hanbury-Brown and Twiss measured the cross-correlation function at four different transverse separations d of the two detectors. The result is shown in Fig. 5.15. A non-zero correlation was found even at a separation as

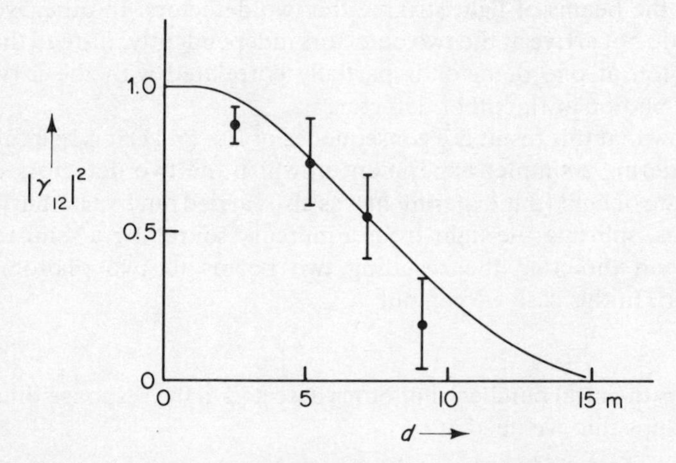

Fig. 5.15. The experimental points show the dependence on detector separation of the cross-correlation function for the light from Sirius. The function $|\gamma_{12}|^2$ is obtained by normalizing the results to unity, at $d = 0$. d is the transverse separation of the detectors. (Adapted from Hanbury-Brown and Twiss, *Nature*, **178** (1956), 1048.)

large as 9 m. The angular diameter of the star can be estimated if we use Fig. 5.9(b) and make the simplifying assumption that the detected light is approximately monochromatic with a wavelength of approximately 500 nm. We see that $|\gamma| = 0.5$ when $d \simeq 10$ m (Fig. 5.15) or when $\theta d/\lambda \simeq 0.7$ (Fig. 5.9(b)). Therefore,

$$\theta \simeq \frac{0.7\,\lambda}{10\,\mathrm{m}} \simeq 3.5 \times 10^{-8}\ \mathrm{rad} \simeq 0.007''.$$

An accurate calculation, shown by the curve in the figure, gives the result 0.0069 sec of arc.

This experiment of Hanbury-Brown and Twiss was important not only in demonstrating the existence of intensity and photon correlations, but also in showing that measurements of these correlations can give useful results. Intensity interferometers of this type are now routinely used to measure angular diameters, often by observing at radio frequencies. High angular resolution is obtained by using radio telescopes separated by very large (even intercontinental) distances. Another very useful application of photon correlations is in determining the spectral distribution of laboratory sources. The light from a source is detected by one photomultiplier, and the fluctuations $\Delta n(t)$ occurring at time t are correlated with the fluctuations $\Delta n(t+\tau)$ at a later time $t+\tau$. Equation (5.44) then enables the autocorrelation function to be deduced, from which the spectral distribution $I(\omega)$ of the light can be calculated (Eq. (5.14)). The technique is called *photon correlation spectroscopy*.

The experiment and interpretation of Hanbury-Brown and Twiss aroused some controversy at the time. The detection process involves the detection of single photons when they cause photoelectric ejection at the cathode of the photomultiplier, and it was asked, how can the detection of a photon at one photomultiplier be correlated with the detection of a different photon at the other photomultiplier, if the photons move independently in the beam? The analysis given above provides one answer. A more qualitative explanation is that the photons have a tendency to *bunch* together in chaotic light, causing the fluctuations to be greater than if they were completely independent (Eq. (5.40)). This happens because when two or more photons lie within the same coherence volume they must have essentially identical position and time co-ordinates. Because the photons are bosons there is no restriction on the total number of photons that can occupy a single coherence volume, provided that the average density of photons is high enough. When this type of bunch of 'multiple photon' is observed by two detectors there is some chance of a correlated response, giving a non-zero value of $\Delta n_1 \Delta n_2$.

PROBLEMS: CHAPTER 5

(Answers to selected problems are given in Appendix E.)

5.1 In an observation of Betelgeux using a Michelson stellar interferometer it was found that the interference bands first disappeared when the outer mirrors were 3 m apart. Estimate the angular diameter of this star, assuming a mean wavelength of 500 nm.

5.2 Show that the Fourier transform of the wave packet

$$E(t) = E_0\, e^{j\bar{\omega}t}\, e^{-(t-t_0)^2/4T^2}$$

is proportional to

$$a(\omega) = e^{-(\omega-\bar{\omega})^2T^2}$$

Find the standard deviations Δt and $\Delta\omega$ (i.e. the root-mean-square deviations) of the functions $|E(t)|^2$ and $|a(\omega)|^2$ respectively, and show that

$$\Delta t \Delta\omega = \tfrac{1}{2}.$$

5.3 Three monochromatic sources of equal intensity are equally spaced along a straight line. A distant screen is placed parallel to the line of the sources. Calculate the one-dimensional intensity distribution that will be observed along a line parallel to the sources if

(a) all three sources are coherent and in phase,
(b) one of the outer sources is not coherent with the other two,
(c) the centre source is not coherent with the outer two.

5.4 Estimate the coherence area of the light from a pinhole of diameter 0.5 mm, at a distance of 1 m from the pinhole, assuming that it is illuminated by incoherent light.

5.5 Estimate the degeneracy parameter for

(a) light of wavelength 600 nm in thermal radiation of temperature 2000 K
(b) radio waves of wavelength 1 m in thermal radiation of temperature 10^4 K.

5.6 Show that for the plane wave having the time-dependence

$$E = \sum_{n=1}^{N} e^{j\omega_n t}$$

the cross-correlation function $\langle \Delta I(t+\tau)\Delta I(t)\rangle$ measured in a Hanbury-Brown–Twiss experiment is proportional to $|\gamma(\tau)|^2$.

CHAPTER 6

Lasers

Until the 1950's laboratory built sources of electromagnetic radiation had utilized either (i) the spontaneously emitted radiation from excited atoms, molecules, and nuclei, or (ii) the thermal radiation from hot bodies, or (iii) the classical radiation produced by accelerated charges (e.g. microwave sources). The need for more intense beams of radiation had often been felt before then, particularly for spectroscopic studies, but as text-books of the period usually pointed out, the brightness of a beam cannot exceed the brightness of the original source. For example, if the radiation from the sun is focussed to give the brightest possible spot, the temperature at the spot cannot exceed that of the surface of the sun (effectively 5800 K), and the intensity in the beam producing the spot cannot exceed the intensity at the surface of the sun. Non-equilibrium sources offer the possibility of higher intensities at the wavelengths of atomic emission lines, but as with equilibrium sources the radiation is emitted in all directions from the source and the maximum intensity that can be focussed and utilized is still limited by the maximum power that can be used to excite the source.

The key to obtaining higher intensities lies in the process of stimulated emission of radiation (see Section 4.2). This occurs when an atom in an excited state of energy $\mathscr{E}_2$ is stimulated to decay to a lower state of energy $\mathscr{E}_1$ either by action of a single photon,

$$A(\mathscr{E}_2)+h\nu \rightarrow A(\mathscr{E}_1)+2h\nu,$$

or by action of n coherent photons,

$$A(\mathscr{E}_2)+nh\nu \rightarrow A(\mathscr{E}_1)+(n+1)h\nu,$$

where $h\nu = \mathscr{E}_2 - \mathscr{E}_1$. The extra photon has of course the *same frequency* as the stimulating photons. Another very important property can be derived from the quantum-mechanical description of the process, where it is usual to consider a closed cavity, so that the photons have the frequencies of the normal modes of the cavity. The stimulated emission event is then described as the *creation* of a photon in a normal mode. The rate of creation is proportional to the number of photons already in *that mode* (see Appendix C), and so the created photon has the *same direction of propagation*, as well as the same frequency, as the stimulating photons. It also has the same phase and the same direction of polarization. These statements are not limited to closed cavities, and are also true when an unenclosed travelling beam of light induces stimulated emission.

The fact that the direction is the same is extremely important. It gives the possibility of concentrating the power available at the frequency ν in one direction, instead of spreading it in all directions. Another important fact is that the rate for the process is proportional to $u(\nu)$, the density of radiation at the frequency ν. This implies that if $u(\nu)$ is already high the atoms in the state of energy $\mathscr{E}_2$ are more likely to decay to the state of energy $\mathscr{E}_1$, contributing a further photon at the frequency ν, than to decay spontaneously to some other state with the emission of some other frequency. This channelling of energy into the frequency ν makes further stimulated emission events even more likely, giving a run-away effect. If these properties are utilized in a non-equilibrium source, and it is arranged that the emitted radiation is not re-absorbed before leaving the source, we see that it becomes possible to channel a large part of the available power into only one frequency and only one direction. In this way it is possible to create a beam having a narrow frequency spread and a value of $u(\nu)$ many orders of magnitude higher than that attainable with a laboratory thermal source.

These properties of stimulated emission had of course been known since the work of Einstein in 1916, and the properties of non-equilibrium electrically-excited gas discharges were also reasonably well understood by the end of the 1920s, but it is one of the quirks of history that the necessary theoretical and technical knowledge was not effectively used until 1954 (although there had been some speculation since about 1940). But as soon as the first experimental use of stimulated emission had been published, the rate of progress in the subject was explosive.

In the first device Townes, Gordon, and Zeiger (1954) used stimulated emission between the lowest two levels of the ammonia molecule to amplify microwave radiation of wavelength 12.6 mm (this will be described in Section 6.4.1). They called the device a *maser*, an acronym for '**m**icrowave **a**mplification by the **s**timulated **e**mission of **r**adiation'. Once this had been invented it was an easier step to speculate that it might be possible to amplify shorter wavelengths, perhaps in the infra-red and visible regions. The basic ideas were published by Schawlow and Townes in 1958. The race to produce the first device was won in mid-1960 by Maiman, who used two levels in the ruby crystal (see Section 6.2.1) to produce red light of wavelength 694.3 nm. Later in the same year Javan, Bennett, and Herriott published details of a successful device using a gas discharge in a mixture of He and Ne (see Section 6.4.2). These devices were at first often described as optical masers, but have since almost universally come to be known by the acronym *lasers*, from 'light amplification by the stimulated emission of radiation'. At the present time a wide variety of lasers exists in the infrared, visible and ultraviolet regions, and some of these will be described in Section 6.4.

As well as having directionality and high intensity, the beams produced by lasers can have the two further valuable properties of being coherent (because the stimulated and stimulating photons are in phase) and of having a very narrow spread of wavelengths. Alternatively, in some types of laser, the beams can consist of trains of extremely short pulses of radiation. These and other properties will be discussed in Chapter 7.

In this chapter we start by considering a hypothetical two-level system and then go on to discuss multi-level systems, using the ruby laser as an example. Then we look at the line widths and laser modes in more detail, and finish the chapter with a discussion of the physical principles and modes of operation of four particular types of laser systems.

6.1 A SIMPLE TWO-LEVEL SYSTEM

Let us start by considering a hypothetical laser in which a gas of atoms having only the two levels shown in Fig. 6.1(*a*) is enclosed in the box shown in Fig. 6.1(*b*). The two levels have energies $\mathscr{E}_1$ and $\mathscr{E}_2$, and degeneracies g_1 and g_2 respectively, and the box contains N_1 atoms in level 1 and N_2 in level 2. The end walls of the box consist of two plane mirrors parallel to each other and separated by the distance L.

We have seen in Section 4.2 that the N_1 atoms in the lower level absorb photons of energy

$$h\nu_0 = \mathscr{E}_2 - \mathscr{E}_1$$

at the rate

$$R_{\text{abs}} = B_{12} N_1 u(\nu_0),$$

where B_{12} is the Einstein coefficient for absorption and $u(\nu_0)\,\mathrm{d}\nu$ is the energy density in the frequency interval of width $\mathrm{d}\nu$, centred at ν_0. The N_2 atoms in the upper level are stimulated to emit photons at the rate

$$R_{\text{stim}} = B_{21} N_2 u(\nu_0),$$

where B_{21} is the Einstein coefficient for stimulated emission. They also

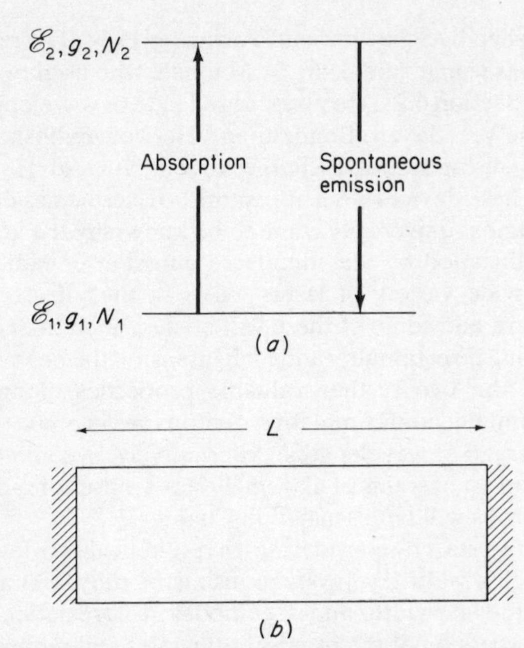

Fig. 6.1. (*a*) A two-level system. The energies, degeneracies and populations are denoted by $\mathscr{E}$, g, and N respectively. (*b*) Box containing the two-level atoms. The end walls are flat parallel mirrors.

decay spontaneously at the rate

$$R_{\text{spont}} = A_{21}N_2 = \tau_2^{-1}N_2,$$

where A_{21} is the Einstein coefficient for spontaneous decay and τ_2 is the lifetime of the upper level. Using Eqs. (4.30) and (4.31) we find that

$$B_{21} = \frac{c^3}{8\pi h\nu_0^3\tau_2}, \tag{6.1}$$

and as before (Eq. (4.29)),

$$B_{12} = \frac{g_2}{g_1}B_{21}. \tag{6.2}$$

The thermal motion of the atoms causes the emitted and absorbed photons to have a range of frequencies about the central frequency ν_0, as shown in Fig. 6.2. This broadening of the line is referred to as Doppler broadening, and will be discussed in more detail in Section 6.3. Other causes, such as pressure broadening and the natural width of the line, may

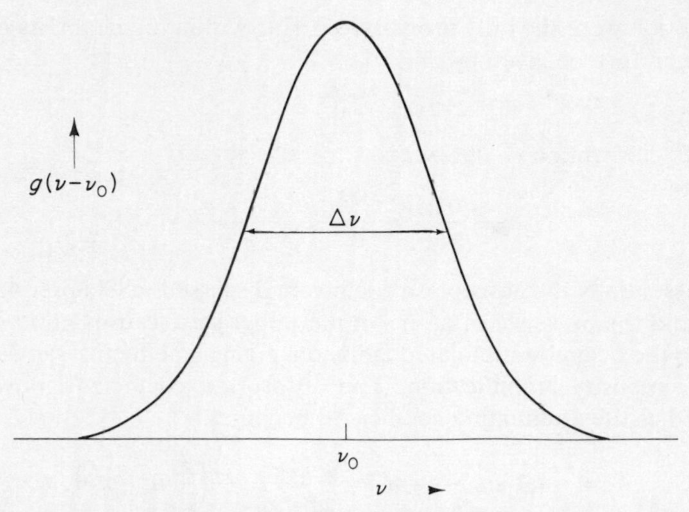

Fig. 6.2. Shape of the absorption and emission lines. $\Delta\nu$ is the full width at half maximum.

also contribute to this line shape. For the present we can suppose that the line shape is dominated by Doppler broadening, and that it is represented by the function $g(\nu-\nu_0)$, where the area under the line is normalized to unity,

$$\int g(\nu-\nu_0)\,\mathrm{d}\nu = 1. \tag{6.3}$$

Expression (6.1) refers to the whole line. When we are interested in a specific frequency ν it becomes

$$B_{21}(\nu) = \frac{c^3 g(\nu-\nu_0)}{8\pi h \nu_0^3 \tau_2}. \tag{6.4}$$

Intensity amplification

Let us consider now a monochromatic beam of frequency ν travelling in the $+z$ direction, perpendicular to the two mirrors of Fig. 6.1(b), with an intensity

$$I(z) = cu(\nu, z).$$

If the density of atoms in the lower level is n_1, the number of photons absorbed in one second within a slab of depth $\mathrm{d}z$ and unit area is $B_{12}(\nu)n_1\,\mathrm{d}z\,u(\nu, z)$. Each absorption leads to the removal of an energy $h\nu$ from the beam, and so

$$\mathrm{d}I = -h\nu B_{12}(\nu)n_1\,\mathrm{d}z\,u(\nu, z) = -I(z)\frac{h\nu B_{12}(\nu)n_1}{c}\,\mathrm{d}z.$$

If absorption were the only process occurring within the beam, its intensity would therefore be given by

$$I(z) = I(0)\,\mathrm{e}^{-\mu z},$$

where the absorption coefficient is

$$\mu = \frac{h\nu B_{12}(\nu) n_1}{c} = \frac{g_2 c^2 g(\nu - \nu_0) n_1}{g_1 8\pi \nu_0^2 \tau_2}.$$

This is essentially the absorption coefficient discussed in Chapter 4. On the other hand the presence of atoms in the upper state causes photons to be *added* to the beam by stimulated emission, giving an effective 'anti-absorption' or intensity amplification. The absorption coefficient, now better described as the attenuation coefficient, becomes

$$\mu = \frac{h\nu}{c}(B_{12} n_1 - B_{21} n_2) = \frac{c^2 g(\nu - \nu_0)}{8\pi \nu_0^2 \tau_2}\left(\frac{g_2}{g_1} n_1 - n_2\right). \tag{6.5}$$

This can be positive or negative, depending on the relative magnitudes of $g_2 n_1 / g_1$ and n_2. (The process of spontaneous emission also contributes to the intensity I, but since the photons are emitted isotropically the increase in I in the specific direction z is negligible and so we can ignore the process for the present purposes.)

For a system in thermal equilibrium the attenuation coefficient is always positive, and the beam intensity decreases with distance. This follows because for each atomic level the population of each of its sub-levels (the number of which is the degeneracy of the atomic level, see Section 4.2) is proportional to the Boltzmann factor. Therefore

$$n_{1,2} \propto g_{1,2}\,\mathrm{e}^{-\mathscr{E}_{1,2}/kT}, \tag{6.6}$$

and

$$\frac{g_2}{g_1} n_1 - n_2 = n_2(\mathrm{e}^{(\mathscr{E}_2 - \mathscr{E}_1)/kT} - 1) = n_2(\mathrm{e}^{h\nu/kT} - 1) > 0,$$

thus making $\mu > 0$.

Population inversion

The attenuation coefficient can be made negative by setting up a non-equilibrium system in which the condition

$$\Delta n = n_2 - \frac{g_2}{g_1} n_1 > 0 \tag{6.7}$$

is satisfied. There is then said to be a *population inversion* of the levels 1 and 2. The population inversion is created by a suitable *pumping mechanism*, and some examples of these will be discussed in later sections, in describing particular laser systems. The condition (6.7) has occasionally been described as the levels 1 and 2 having a *negative temperature*, since Eq. (6.6) gives an effective temperature

$$T_{\text{eff}} = -\frac{h\nu}{k \ln (n_2 g_1 / n_1 g_2)}$$

which is less than zero.

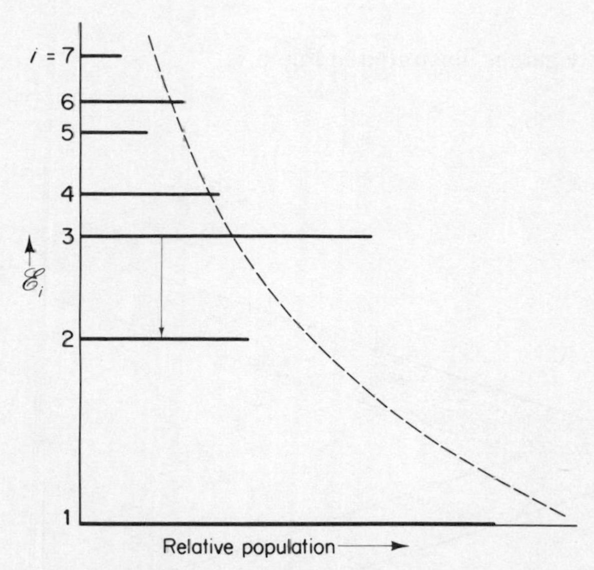

Fig. 6.3. Relative populations of the levels of a hypothetical atom acting as a laser medium. A population inversion exists between levels 2 and 3, which are the active levels of the laser. Population inversions exist also between other levels (5 and 6) not being directly used to produce laser light. The broken line represents the Boltzmann factor $\exp(-\mathscr{E}_i/kT)$.

Figure 6.3 shows a system in which a population inversion exists between the levels 2 and 3 and also between the levels 5 and 6. The broken line shows the Boltzmann distribution $\exp(-\mathscr{E}_i/kT)$, where kT is the average excitation energy of the atoms.

When a population inversion exists the beam intensity increases with distance,

$$I_\nu(z) = I_\nu(0)\, e^{\kappa_\nu z},$$

where the *gain coefficient* κ_ν is

$$\kappa_\nu = -\mu_\nu = \frac{c^2 g(\nu - \nu_0)}{8\pi\nu^2\tau_2}\Delta n. \tag{6.8}$$

In traversing the length L of the laser the intensity therefore increases by the factor

$$\left(\frac{I_\nu(L)}{I_\nu(0)}\right)_{\text{gain}} = e^{\kappa_\nu L}. \tag{6.9}$$

This intensity gain is illustrated in Fig. 6.4.

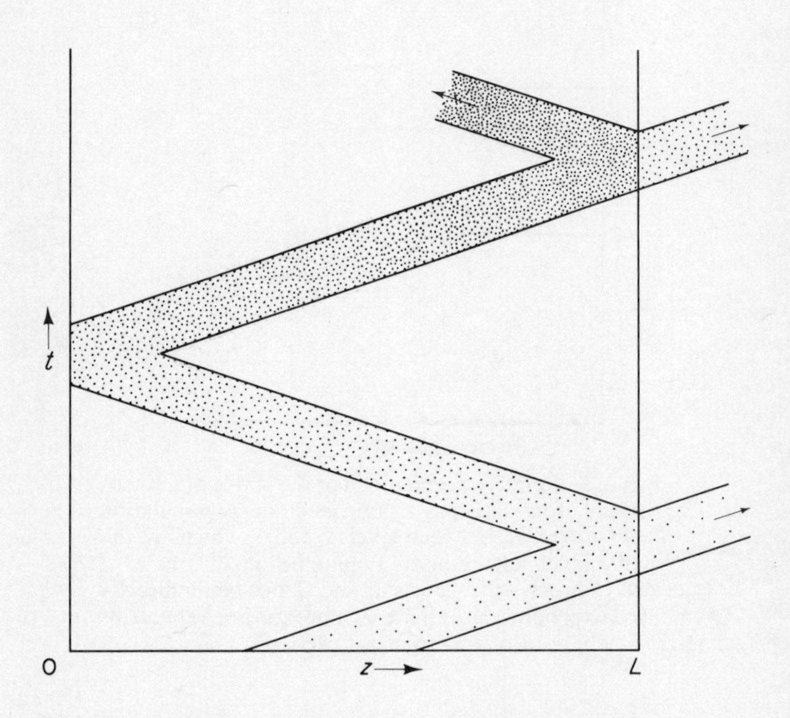

Fig. 6.4. Schematic illustration of the intensity gain in a laser. The initial event (spontaneous emission) occurs (in this example) approximately midway between the mirrors (situated at $z = 0$ and L) of the laser. The increasing depth of shading represents the increase with time of the intensity of the beam as it traverses the space between the mirrors. A small part of the beam passes through the partially transmitting right-hand mirror, forming the externally observed laser beam. The increase with time of the coherence and length of the wavepacket is not indicated in this figure.

Loss processes

Apart from absorption and stimulated emission there are various other processes which cause changes in the intensity of the beam, all of them resulting in a loss of intensity. The most important of these occurs in those lasers in which the beam must be reflected many times between the two end mirrors in order to obtain a significant intensity gain. If the successively reflected waves are in phase with each other there is no loss in intensity when they interfere, but if they are out of phase they interfere destructively, eventually cancelling each other out after many reflections. This cancellation occurs also on the Fabry-Perot interferometer, and we see that the laser cavity that we are considering (with flat and parallel end faces) is indeed one form of Fabry-Perot interferometer, with a rather large separation of the mirrors. As in the Fabry-Perot interferometer, the intensity loss due to destructive interference in the laser cavity is zero only if the condition

$$2L = n\lambda \tag{6.10}$$

is satisfied, where n is an integer. With this condition the electromagnetic field oscillates as a standing wave (see also Section 4.1), of frequency

$$\nu_n = n\frac{c}{2L}. \tag{6.11}$$

This mode of oscillation is a *normal mode* of the cavity (c.f. the discussion in Section 4.1). The mode number n is usually very large; for example if the cavity length is 0.6 m and the wavelength is 600 nm, n is 2×10^6. The frequency spacing of these modes is

$$\Delta\nu_n = \frac{c}{2L}. \tag{6.12}$$

Another source of loss comes from the non-unity reflection coefficients R_1 and R_2 of the end mirrors. The attenuation due to this loss after a return trip inside the laser is R_1R_2, and if R_1 and R_2 are both nearly unity this can be expressed as

$$\left[\frac{I_\nu(2L)}{I_\nu(0)}\right]_{\text{refl loss}} = R_1R_2 \simeq e^{-(1-R_1R_2)}.$$

Other sources of loss are associated with diffraction effects (see Section 6.3) and with scattering in the laser medium, and these can be jointly characterized by a coefficient β given by

$$\left[\frac{I_\nu(L)}{I_\nu(0)}\right]_{\text{diff loss}} = e^{-\beta}.$$

Combining this with the reflection loss gives

$$\left[\frac{I_\nu(L)}{I_0(L)}\right]_{\text{loss}} = \mathrm{e}^{-\alpha L},$$

where

$$\alpha = \frac{1}{L}\left[\tfrac{1}{2}(1 - R_1 R_2) + \beta\right].$$

The more usual way of expressing this is in terms of the loss in intensity with time,

$$\left[\frac{I_\nu(t)}{I_\nu(0)}\right]_{\text{loss}} = \mathrm{e}^{-\alpha c t} = \mathrm{e}^{-t/t_c}, \tag{6.13}$$

where

$$t_c = \frac{L}{c\left[\tfrac{1}{2}(1 - R_1 R_2) + \beta\right]} \tag{6.14}$$

is the *decay time of the cavity*, or the mean time for which a photon would bounce back and forth between the mirrors before being lost. For example, for a laser of length 0.6 m, whose mirrors both have reflection coefficients of 99%, and with a diffraction loss of 1% per transit,

$$t_c = \frac{0.6}{3 \times 10^8 (10^{-2} + 10^{-2})} \simeq 10^{-7}\ \text{s}.$$

The losses can also be characterized by the *quality factor* or *Q-value* of the cavity, defined by

$$Q = 2\pi \frac{\text{energy stored}}{\text{energy lost per period of oscillation}}$$

$$= \frac{2\pi I_\nu}{\left(-\frac{\mathrm{d}I_\nu}{\mathrm{d}t}\Big/\nu\right)} = 2\pi \nu t_c. \tag{6.15}$$

This has a very high value, $\sim 10^8$ or 10^9 for visible lasers. This is several orders of magnitude larger than the Q-values of more familiar oscillating systems, such as pendulums or electrical oscillators.

Threshold condition

Figure 6.5 shows the magnitudes of the gains and losses for our hypothetical laser, as a function of frequency, for a narrow range of frequency covering an absorption line. The absorption and stimulated emission pro-

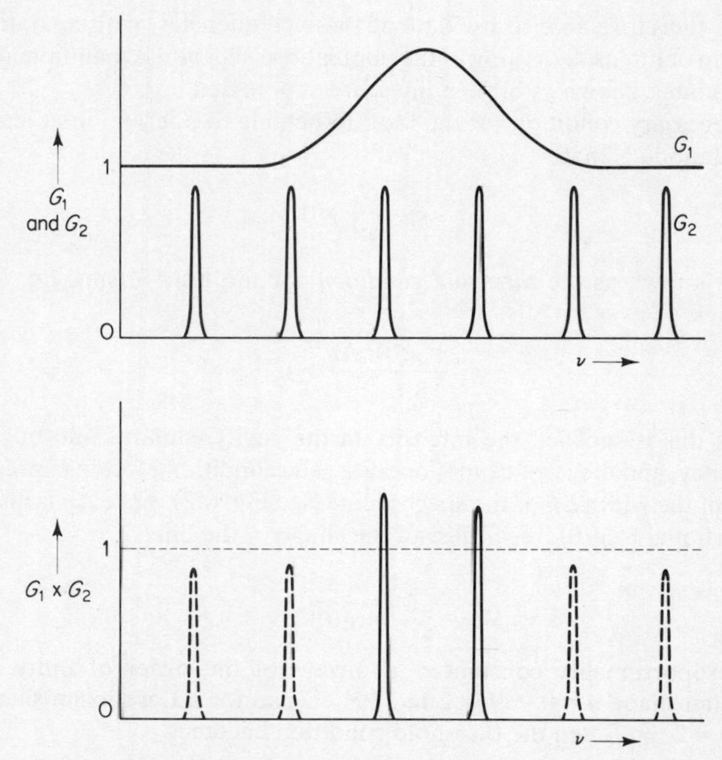

Fig. 6.5. G_1 is the average gain per pass due to the combined effects of absorption and stimulated emission. $G_1 \geqslant 1$ when a population inversion exists. G_2 is the gain per pass due to the various loss processes. The laser can oscillate only when $G_1G_2 > 1$.

cesses give an average intensity gain per pass of

$$G_1 = \mathrm{e}^{\kappa_\nu L}.$$

Assuming that a population inversion exists in the cavity, $\kappa_\nu > 0$ and $G_1 \geqslant 1$, as drawn in the figure (otherwise the peak in G_1 becomes a dip). On the other hand the loss processes give an effective gain G_2 which is always less than 1. Its maximum value is

$$G_2^{\max} = \mathrm{e}^{-L/(ct_c)},$$

which occurs when condition (6.10) is satisfied. $G_2^{\max}$ is independent of ν over the narrow frequency range shown in the figure. Two such maxima in G_2 occur within the region of the spectral line, for the example shown.

The lower part of the figure shows the product G_1G_2. We see that $G_1G_2 > 1$ at two different normal mode frequencies in this example. The laser is therefore able to oscillate at these frequencies, with a continuous build-up of intensity (as long as the population inversion is maintained). The other modes, shown as broken lines, are suppressed.

A necessary condition for the laser to be able to operate on at least one normal mode is that

$$\kappa_{\nu_0} - \frac{1}{ct_c} > 0. \tag{6.16}$$

This is known as the *threshold condition* for the laser. Using Eq. (6.8) it becomes

$$\frac{c^2 g(0) t_c \Delta n}{8\pi \nu_0^2 \tau_2} > 1. \tag{6.17}$$

Unless this is satisfied the intensity in the cavity cannot build up at any frequency, and the laser cannot operate. The condition is often expressed in terms of the width $\Delta\nu$ of the spectral line (see Fig. 6.2), since $\Delta\nu$ is inversely proportional to $g(0)$, regardless of the shape of the line,

$$\Delta\nu = \frac{a}{g(0)}. \tag{6.18}$$

The proportionality constant a is always of the order of unity. For a Gaussian shape $a = 0.939 (= 2(\ln 2/\pi)^{1/2})$ and for a Lorentzian shape $a = 0.637 (= 2/\pi)$. Then the threshold condition becomes

$$1 < \frac{ac^2 t_c \, \Delta n}{8\pi \nu_0^2 \tau_2 \, \Delta\nu} \simeq \frac{c^2 t_c \, \Delta n}{8\pi \nu_0^2 \tau_2 \, \Delta\nu}. \tag{6.19}$$

The threshold condition is a necessary condition to be satisfied before a laser beam can be produced, but what other conditions must also be satisfied? One obvious requirement is that the beam to be amplified must originate at some time. Usually this happens by the spontaneous decay of an excited atom, producing a single photon. A second requirement is that this initial photon should travel in a direction parallel, or nearly parallel, to the axis of the laser. Then the beam can be reflected several times, enabling the intensity to increase. Because spontaneous emission is an isotropic process there are always some photons emitted in the required direction. The number of reflections of the beam being amplified, and the necessary degree of parallelism to the axis, depend on the value of G_1. If this is high then only a small number of transits is necessary and the angle of the beam is not critical. In extreme conditions G_1 can be so high that an intense beam is produced by a single transit and no feedback mirrors are needed; this is described as *amplified spontaneous emission*.

More usually the mean number of reflections is arranged to be of the order of the number that occur within the cavity decay time t_c, namely ct_c/L, which is usually of the order of 100. This helps to ensure that the output beam is as parallel and coherent as possible (see Chapter 7). The requirement of a large number of reflections also introduces the further requirement that the end mirrors must be accurately parallel to each other if they are flat (as in the present example) or else the successive reflections cause the beam to move steadily sideways, eventually missing the mirrors. This is called *walk-off*. This requirement is relaxed when curved mirrors are used, which is more usual (see Section 6.3).

As the population inversion in a laser cavity is slowly increased through the threshold value the distribution of energy in the cavity modes undergoes a drastic change. Below threshold all the modes within the Doppler profile $g(\nu-\nu_0)$ are populated with photons. As the population inversion is increased a point is reached at which one mode with a large value of t_c satisfies the threshold condition. All the excited atoms which can emit into this mode are then quickly forced to do so, by stimulated emission. The spontaneously emitted photon which triggers this process is polychromatic and multidirectional (see Section 5.4.3), but from it is selected the monochromatic and unidirectional part that corresponds to the single mode being amplified. The energy contained in the mode suddenly increases to a level many orders of magnitude higher than that in the other modes, and a high intensity of light is emitted in the direction of the mode, along the laser axis.

Power requirements

Let us consider now how difficult it is to satisfy the threshold condition in practice. We have seen already that $t_c \sim 10^{-7}$ seconds, and in the case of optical frequencies we can take $\tau_2 \sim 10^{-8}$ s and $\lambda \sim 600$ nm. If $\Delta\nu$ is caused by the Doppler effect only, and the effective temperature of the laser medium is not too far from room temperature, then (see Section 6.3)

$$\Delta\nu \sim 10^{-6}\nu,$$

and the threshold condition becomes

$$\Delta n \geqslant \frac{8\pi\tau_2 \times 10^{-6}\nu^3}{c^3 t_c} \sim 10^{13}\ \mathrm{m}^{-3}.$$

This population inversion is very small compared with n_1 (at atmospheric pressure, $n_1 \simeq 3\times 10^{25}/\mathrm{m}^3$). But maintaining it for continuous operation of the laser is not easy. Since the density of atoms in the upper level is given by

$$n_2 = \Delta n + \frac{g_2}{g_1} n_1$$

(Eq. (6.7)), n_2 is at least equal to Δn and is generally very much larger (for a two-level system). Furthermore, the atoms can exist in the upper level only for the mean time τ_2 (or less, if the laser is operating and stimulated emission occurs). Re-excitation to the upper level requires an energy $h\nu$ if the re-excitation process is perfectly efficient (but much more energy in practice), and so the threshold power required to populate the upper level is

$$P \gtrsim \frac{n_2 h\nu}{\tau_2} \gg \frac{\Delta n\, h\nu}{\tau_2}.$$

With the values of t_c, ν_0, and $\Delta\nu$ used previously this gives $P \gg 300\ \mathrm{W/m^3}$. One of the main difficulties in laser operation is therefore that of feeding in, and then dissipating, the high power needed to sustain a large enough population inversion. This is one reason why many lasers are operated in the pulsed mode, as we shall see.

The most important single factor in making the population inversion and power requirements as easy as possible is the working frequency of the laser, since with the assumptions we have made the critical population inversion Δn is proportional to ν^3 and the critical power is proportional to ν^4. It is essentially for this reason that an X-ray laser is yet to be built (see Problem 6.5).

6.2 MULTI-LEVEL SYSTEMS

Although the ammonia maser (to be described in Section 6.4.1) is a two-level system, nearly all other maser and laser systems use more than two levels. Some examples of 3 and 4-level systems are shown in Fig. 6.6.

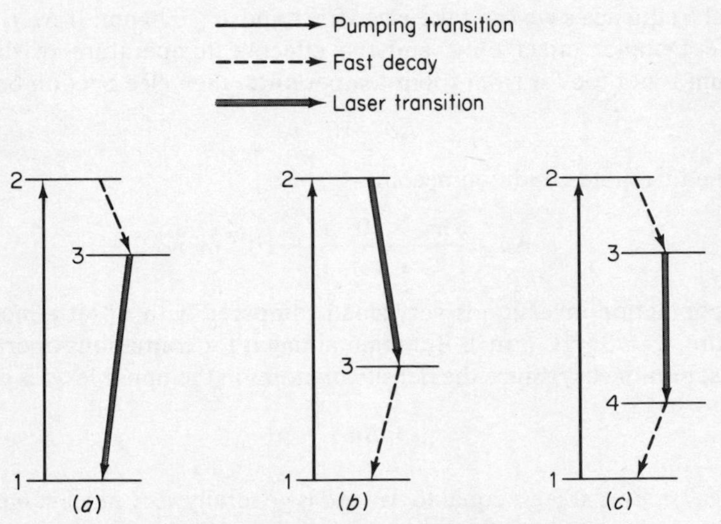

Fig. 6.6. Laser systems involving more than two levels.

Part (*a*) of the figure shows a 3-level system. Level 1 may be the ground-state of an atom or molecule, and transitions are made from this level to the excited level 2 by some suitable *pumping mechanism*. This may for example take the form of photon absorption, or electron–atom or atom–atom collisions. Examples of these and other pumping mechanisms will be given later. Level 2 decays quickly and preferentially to level 3 (with only a small probability of decaying back to the ground state or other lower states). If level 3 is long-lived (metastable) and the pumping transitions $1 \rightarrow 2$ are sufficiently frequent, it is possible for the population n_3 to exceed n_1, and then laser action can ensue. Less than half the atoms or molecules must be left in the ground state (since $n_1 < n_2 < n_2 + n_3$), which requires a large pumping power and a high pumping efficiency.

A more convenient system is shown in part (*b*) of the figure. For continuous operation the transition $3 \rightarrow 1$ has to be faster than the laser transition $2 \rightarrow 3$, in order to maintain the inversion condition $n_2 > n_3$. Both n_2 and n_3 are usually very much smaller than n_1, allowing a lower pumping power to be used. The 4-level system shown in Fig. 6.6(*c*) differs only in having an indirect route to the upper laser level.

Other systems also exist, in great variety. It has been said, perhaps flippantly, that *any* type of medium can be used for laser action, provided that sufficient pumping power is available. Many of those which have been made to work cannot be conveniently described in terms of discrete levels of a single atom, molecule or ion. In some cases for example the laser transition occurs between a discrete level and a continuum of levels, or even between two continua. In other cases the pumping mechanism involves a different atom from that in which the laser transition occurs. We shall meet a few examples of these other types of laser systems in later sections.

6.2.1 The ruby laser

Before going further into the theoretical aspects of lasers, let us see how the first laser worked. This is the ruby laser, built by Maiman in 1960. Improved designs are still used at the present time.

Ruby consists of crystalline aluminium oxide (Al_2O_3, also known as corundum), in which a small fraction of the aluminium has been replaced by chromium. The chromium exists in the crystal in the form of Cr^{3+} ions, and these are active in absorbing any green or violet light travelling through the crystal. The absorbed energy re-appears as red fluorescent light. The mechanism through which this happens can be seen from the simplified energy level diagram shown in Fig. 6.7. The details of the spectroscopic notation of the levels need not concern us, but we can see that two broad levels exist which are capable of absorbing light in the region of 560 and 410 nm, the oscillator strengths of these transitions being of the order of

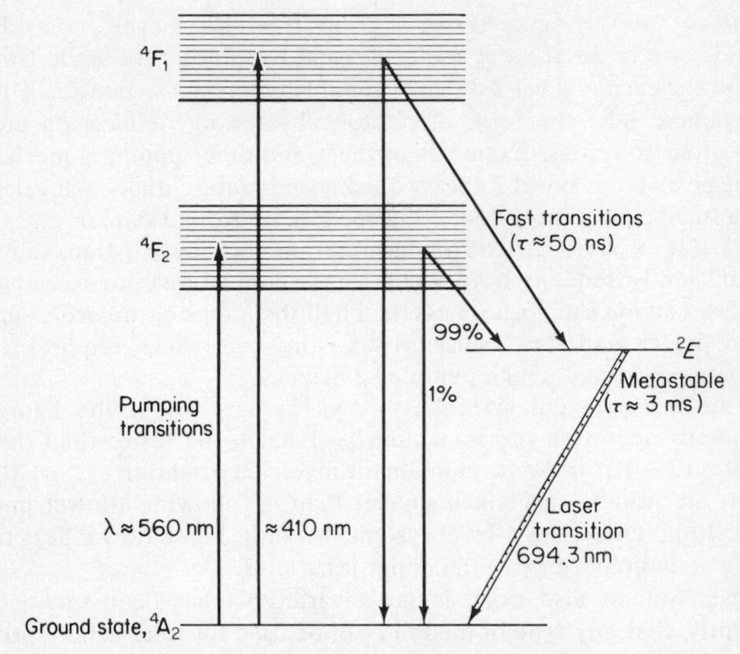

Fig. 6.7. Simplified energy level diagram for the chromium ion in ruby.

10^{-3}. The oscillator strengths for the transitions from these bands to the metastable (^{2}E) level are much larger however, and so when these bands are excited they decay preferentially to this level rather than back to the ground state, with a lifetime of approximately 50 ns. The oscillator strength connecting the metastable level to the ground state is very small, $\sim 10^{-6}$, which is of course the reason why it is metastable (its lifetime is approximately 3 ms).

The crystal used by Maiman for the first laser contained approximately 0.05% by weight of Cr_2O_3 in Al_2O_3, and was in the form of a cylindrical rod 20 mm long and 7 mm in diameter. It therefore contained approximately 10^{19} Cr^{3+} ions. The experimental arrangement is shown in Fig. 6.8. The ends of the rod are ground flat and parallel. One end is coated to make it reflecting, while the other end is made partially reflecting. The rod is irradiated by a xenon flash lamp. This consists of a glass tube filled with xenon at a pressure of about 150 torr, and activated by discharging a capacitor bank through the gas. The electrical energy is converted to light energy with an efficiency of the order of 10%, and the resulting flash of white light lasts approximately 1 ms. The surrounding reflector helps to direct this light onto the ruby rod, which absorbs some of it in being excited to the bands shown in Fig. 6.7. The bands decay quickly to the metastable (^{2}E)

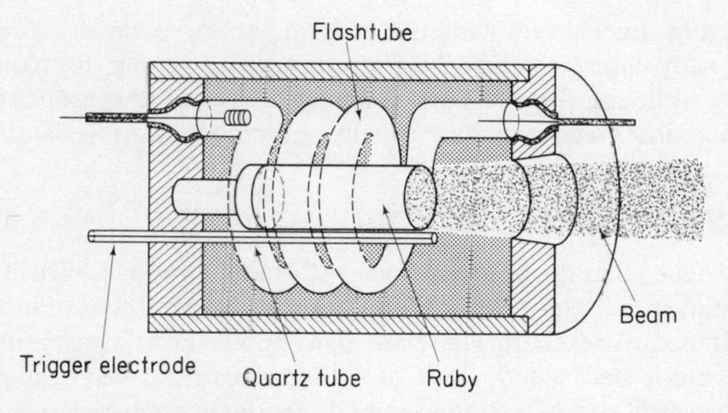

Fig. 6.8. Schematic diagram of the first ruby laser, approximately full size. (Taken from T. H. Maiman *et al.*, *Physical Review*, **123** (1961), 1151.)

level, which has a high probability of remaining excited for the duration of the flash. This results in a decrease in the number N_1 of chromium ions in the ground state and a corresponding increase in the number N_2 in the ^2E level (we use N to denote the total number in the crystal, and n to denote the number per unit volume).

A population inversion is created in this system only when at least half the ground state ions have been excited, which requires that enough light be absorbed from the pumping flash to excite approximately 5×10^{18} ions in the crystal. Since each excitation requires an energy of approximately 4×10^{-19} J ($= hc/\lambda$, where $\lambda \simeq 560$ or 410 nm), a total of a least 2 J must be absorbed by the crystal. Some part of this energy (depending on the efficiency of the laser) re-appears as the energy of the laser beam. The remaining energy from the flash lamp is dissipated as heat.

The ^2E to ground state transition has a width $\Delta\nu$ which is typically about 50 times larger than the frequency separation $\Delta\nu_n$ given by Eq. (6.12). Because the population inversion ΔN required to satisfy the threshold condition at the centre of the broadened line is of the order of 10^{16}, an excess which is small compared with the total number of 5×10^{18} ions that must be excited in the crystal, the threshold condition is also usually satisfied over much of the width of the line, and many different modes are excited at the same time. The number can be reduced by the mode-selection techniques discussed in the next section.

To run the ruby laser continuously (usually described as CW operation of the laser) would require a pumping power high enough to keep more than half the chromium ions in excited states. From the figures given above we see that more than about 20 J of electrical energy is required every millisecond for this crystal, a power greater than 20 kW. The problems of supplying this power and of quickly removing the excess heat are obviously formidable, and the ruby laser is clearly not a convenient system for continuous operation, although this has been achieved. For similar reasons there are many other laser systems that are usually operated in the pulsed mode only.

6.3 LEVEL WIDTHS AND MODE SELECTION

We have seen that the line shape $g(\nu-\nu_0)$ and the width $\Delta\nu$ are important in determining the laser threshold condition and the number of modes that are simultaneously excited. There are many applications which require that only *one* mode be excited, since only then does the laser light have the ultimate spectral width, coherence and directionality of which it is capable. Some of these applications will be described in the next chapter.

We start the present section by considering the line shape $g(\nu-\nu_0)$ in more detail. After this the different types of modes and their properties are considered, and finally we discuss the techniques for selecting a single mode.

Level widths

The three main contributions to the width of a line come from (i) the natural width of the line, due to the finite lifetime of the atomic excited states, (ii) pressure broadening, caused by interactions between atoms, and (iii) Doppler broadening, caused by the thermal motion of the atoms. The first of these has been discussed in Chapters 3 and 4: an excited state with a lifetime τ for spontaneous decay to the ground state gives an emission (or absorption) line having the frequency spectrum

$$g_{\text{nat}}(\nu-\nu_0) \propto \frac{1}{(\nu-\nu_0)^2+(1/4\pi\tau)^2}. \tag{6.20}$$

This has a Lorentzian shape, and its width (by which we always mean the full width at half the maximum height) is

$$\Delta\nu_{\text{nat}} = \frac{1}{2\pi\tau}.$$

The energy $h\,\Delta\nu$ can be regarded as an uncertainty in the energy of the level, essentially caused by the uncertainty τ in the time of the decay process. When the upper and lower levels of a transition both have a finite lifetime, the two energy uncertainties both contribute to the width of the line, giving

$$\Delta\nu_{\text{nat}} = \frac{1}{2\pi}\left(\frac{1}{\tau_1}+\frac{1}{\tau_2}\right). \tag{6.21}$$

Pressure (or collision) broadening arises from the interactions between the atoms of a gas when they collide or pass close to each other. These encounters disturb the time dependence exp (jωt) of the excited state, introducing extra Fourier components into the time dependence and hence causing an increase in the width of the line. If the mean time between encounters is T_c it can be shown that the line profile is still Lorentzian,

$$g_{\text{Press}}(\nu-\nu_0) \propto \frac{1}{(\nu-\nu_0)^2+(\frac{1}{2}\Delta\nu_{\text{Press}})^2}, \tag{6.22}$$

with a width

$$\Delta\nu_{\text{Press}} = \frac{1}{2\pi}\left(\frac{1}{\tau_1}+\frac{1}{\tau_2}+\frac{2}{T_c}\right). \tag{6.23}$$

A similar type of broadening exists in solid or liquid media, due to the interaction of an atom or ion with the neighbouring atoms and also with the phonons (quanta of vibrational energy) of the lattice.

The third main effect, Doppler broadening, is caused by the fact that the light emitted by an atom moving with a velocity component v_z in the direction of an observer appears to be shifted in frequency by an amount proportional to v_z,

$$\nu' = \nu\left(1+\frac{v_z}{c}\right). \tag{6.24}$$

For a gas in thermal equilibrium the probability distribution $P(v_z)$ is proportional to exp $(-av_z^2)$, where a depends on the temperature T of the gas and the mass M of the atoms. The emission (or absorption) line therefore has a Gaussian line shape,

$$g_{\text{Doppler}}(\nu-\nu_0) \propto \mathrm{e}^{-b(\nu-\nu_0)^2}, \tag{6.25}$$

where b also depends on T and M. The width of the line is given by

$$\Delta\nu_{\text{Doppler}} = \frac{2\nu_0}{c}\left(\frac{2kT\ln 2}{A}\right)^{1/2} = 7.16\times10^{-7}\nu_0\left(\frac{T}{A}\right)^{1/2}, \tag{6.26}$$

where A is the atomic or molecular weight of the emitting or absorbing particles.

Homogeneous and inhomogeneous broadening

There is one important respect in which Doppler broadening is fundamentally different from natural or pressure broadening. It is possible in principle (and also in practice, see Chapter 7), to remove some or all of the excited atoms having a particular value of v_z, thus creating a 'hole' in the velocity distribution $P(v_z)$, and hence a hole in the frequency distribution $g_{\text{Doppler}}(\nu-\nu_0)$. This happens for example at the frequency of each mode for which the threshold condition is satisfied (see Fig. 6.5). The photons at these frequencies stimulate the emission of those excited atoms that happen to have the appropriate values of v_z (assuming for the moment that other sources of broadening are negligible), thus depopulating these values of v_z. This is illustrated in Fig. 6.9. The process is often described as *burning a hole* in the velocity distribution. The depopulation stops of course as soon as the number of atoms having a particular value of v_z is reduced below the level at which the threshold condition is satisfied. The important point is that the atoms can be distinguished by the frequency at which they absorb light from the beam, and that the absorption profile of the gas is an inhomogeneous sum of separate and independent absorption lines. For this reason Doppler broadening is described as *inhomogeneous broadening.*

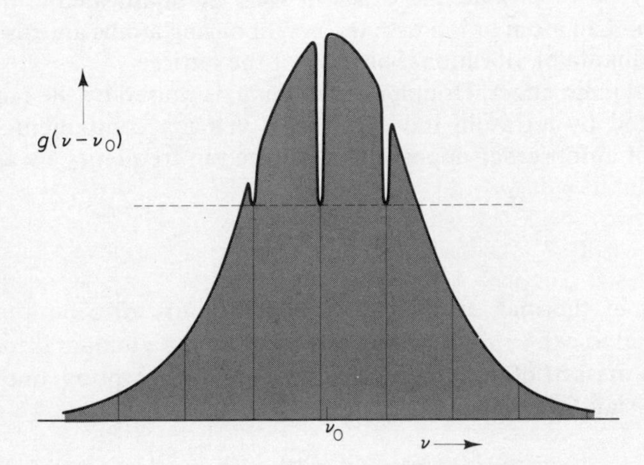

Fig. 6.9. Holes burnt in the atomic velocity distribution, and hence in the line profile $g(\nu - \nu_0)$, by those modes for which the threshold condition is satisfied. At the low points of the holes (broken line) the gain by stimulated emission is exactly balanced by the cavity losses.

By contrast, when Doppler broadening is negligible and only pressure broadening and natural broadening are significant, the shape of the frequency distribution is not altered by the selective removal of atoms. The absorption and emission profile of the macroscopic assembly of atoms is the same as that of each of the individual atoms. Stimulated emission at a particular frequency changes the number of excited atoms but does not alter g_{Press} at that frequency. Pressure and natural line shape broadening are therefore described as *homogeneous broadening.*

In practice line shapes have contributions from inhomogeneous and homogeneous broadening, the former usually being much the larger. The line shape is therefore usually approximately Gaussian.

Laser modes

In considering a simple two-level system in Section 6.1 we assumed that the end mirrors of the laser cavity are flat, and parallel to each other. The losses in this cavity are small only when the length $2L$ (where L is the cavity length) contain a whole number of wavelengths. The electromagnetic field then oscillates in one of the normal modes of the cavity, with the frequency (Eq. (6.11))

$$\nu_n = q\frac{c}{2L}.$$

We now consider other shapes and configurations of the end mirrors, and other types of normal mode.

The modes in which we are interested are those in which the electric and magnetic fields are transverse to the axis of the cavity. These are called the TEM modes. In the case of a rectangular box-like cavity the modes are labelled by the numbers m, n, and q, giving the designation TEM_{mnq}. Here m and n are the numbers of nodes in the transverse x and y directions, and $q+1$ is the number of nodes (including those at the mirrors) in the longitudinal z direction. As we saw earlier, q is usually very large, $\sim 10^6$, but the values of m and n which are relevant for laser operation are small or zero. When the mirrors are flat the frequency of a TEM_{mnq} mode is independent of m and n (to the first order in m/q and n/q) and is given by Eq. (6.11), namely

$$\nu_{mnq} = q\frac{c}{2L}. \tag{6.27}$$

In practical laser systems one or both of the mirrors is usually curved, as shown in Fig. 6.10. This makes alignment of the mirrors much easier, and gives a more stable operation of the laser, as we shall see later. In all the configurations shown a ray launched at a small angle to the axis, or parallel to but away from the axis, is successively reflected by the mirrors without ever travelling far from the axis, as the reader can verify qualitatively by sketching ray paths. The configurations which have this stability are those for which the space between each mirror and its centre of curvature contains either the centre of curvature of the other mirror, or the other mirror itself, but not both. Expressed mathematically this *stability criterion* is

$$(R_1+R_2-L)(R_1-L)(R_2-L) \geqslant 0, \tag{6.28}$$

where R_1 and R_2 are the radii of curvature of the mirrors, and L is their separation. This can also be written (see Problem 6.6) in the form

$$0 \leqslant \left(1-\frac{L}{R_1}\right)\left(1-\frac{L}{R_2}\right) \leqslant 1. \tag{6.29}$$

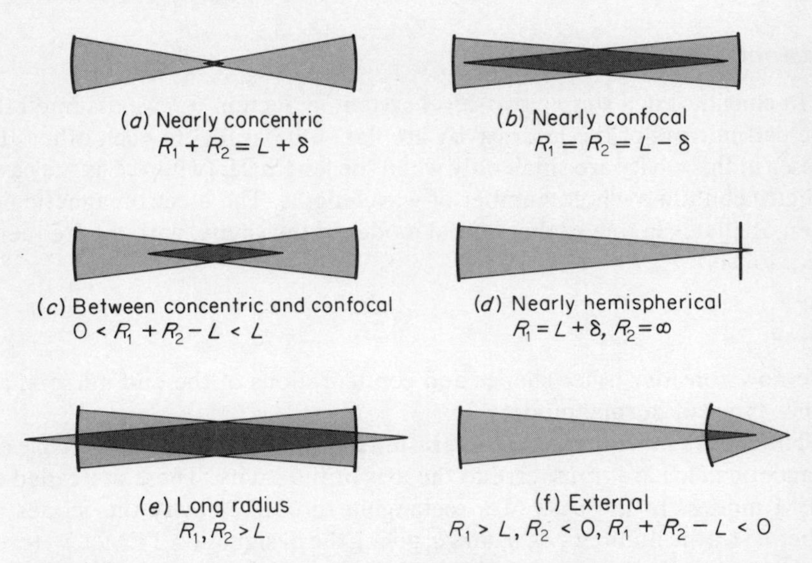

Fig. 6.10. Mirror configurations satisfying the stability criterion (Eq. (6.29)). The distance between the mirrors is L, and δ is arbitrary but small.

Of course it is not sufficient to consider only the geometrical optics of narrow rays, since the mirrors also cause diffraction of the reflected light. Because of this diffraction the minimum half-angle of divergence of a beam reflected from a circular mirror of radius a is of the order of λ/a. Hence the minimum radius of the reflected beam after travelling a distance L is $\sim L\lambda/a$. If this is less than the radius a of the opposite mirror most or all of the beam hits the mirror. The condition for this to happen is therefore

$$\frac{L\lambda}{a} \leqslant a.$$

If this is not satisfied an appreciable fraction of the beam misses the mirror. The beam is then said to suffer *diffraction losses*, and is reduced in intensity at each transit. This criterion is usually expressed in the form

$$F = \frac{a^2}{L\lambda} \gtrsim 1, \tag{6.30}$$

where F is the *Fresnel number* of the optical system.

When the criteria (6.29) and (6.30) are both satisfied, it is possible to set up a standing wave confined to the space between the mirrors. In other words the open two-mirror system then possesses normal modes (and these are independent of the position and shape of the side walls, provided that they are far enough away). These are again designated as TEM_{mnq} modes, with the same meanings for the numbers m, n, and q. It can be shown that the frequencies of these modes are

$$\nu_{mnq} = q\frac{c}{2L} + (m+n+1)\frac{c}{2\pi L}\cos^{-1}\left[\left(1-\frac{L}{R_1}\right)\left(1-\frac{L}{R_2}\right)\right]. \quad (6.31)$$

When the end mirrors are flat, R_1 and R_2 are both infinite and the frequency is independent of m and n, as given by Eq. (6.27). We can see also that the modes exist only when the stability criterion (6.29) is satisfied, since only then is the mode frequency a real quantity.

When the stability criteria (6.29) and (6.30) are not satisfied the losses become large and the cavity lifetime t_c, see Eq. (6.13) and (6.14), becomes short. This makes the threshold condition harder to satisfy, and so these modes are usually suppressed. However, there are some resonators, called *unstable resonators*, which violate the stability criteria and yet still result in laser action. These are exceptionally high gain systems in which only a few passes are required.

The uniphase mode

Of special importance are the TEM_{00q} modes. They have no nodes in the transverse direction and the electric field has a simple Gaussian dependence on the transverse displacement r,

$$E_{00q}(r, \theta) = E_{00q}(x, y) \propto e^{-r^2/w^2}, \quad (6.32)$$

where w is a characteristic length called the *beam radius*. The variation of intensity with r also has a Gaussian shape,

$$I(r) = I_0\, e^{-2r^2/w^2}. \quad (6.33)$$

This is shown in Fig. 6.11, together with the intensity distribution of the

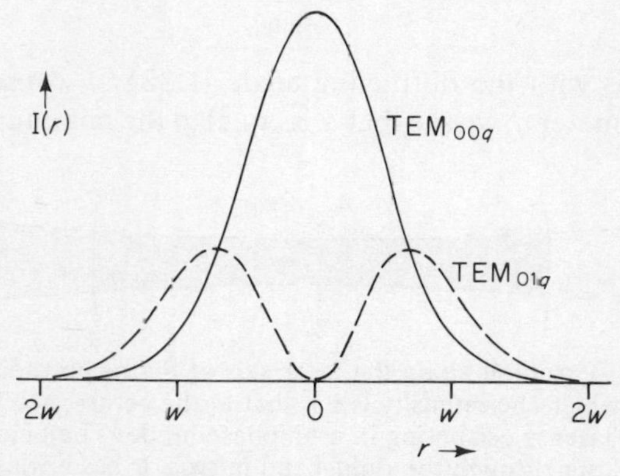

Fig. 6.11. Intensity variation in the transverse direction for the modes TEM_{00q} and TEM_{01q}.

TEM_{01q} mode. 85% of the total intensity in the TEM_{00q} mode is within the radius w. The phase of the electric field of a TEM_{00q} mode does not change across the wavefront of the mode, unlike the phase in modes having non-zero values of m and n. When one or more TEM_{00q} modes having different values of q are excited in a laser the wavefront still has a constant phase, and the laser is then often described as oscillating in a *uniphase* or *fundamental mode.* A single TEM_{00q} mode is then called a *single-frequency uniphase mode.*

The uniform phase and Gaussian intensity profile of the uniphase mode are both retained after the light has emerged from the laser cavity. The beam radius w changes, because the propagating beam spreads by diffraction, but the transverse intensity distribution remains Gaussian (readers who are familiar with diffraction theory will realize that this is because the Fourier transform of a Gaussian function is also a Gaussian). The dependence of w on z is shown in Fig. 6.12. There is a waist of radius w_0, at which the wavefront is planar, and at other positions the beam radius is given by

$$w(z) = w_0\left[1+\left(\frac{z}{z_0}\right)^2\right]^{1/2}, \tag{6.34}$$

where z is measured from the waist and where

$$z_0 = \frac{\pi w_0^2}{\lambda}. \tag{6.35}$$

The beam profile shown in Fig. 6.12 therefore has a hyperbolic shape, and the envelope of the beam is a hyperboloid of revolution.

It can be seen from Eq. (6.34) that at large distances the semi-angle of divergence of the uniphase beam is

$$\alpha = \frac{\lambda}{\pi w_0}. \tag{6.36}$$

Comparing this with the diffracting angle, $1.22\lambda/d$, caused by a circular aperture of diameter d, we see that α is equal to the minimum half-angle of a

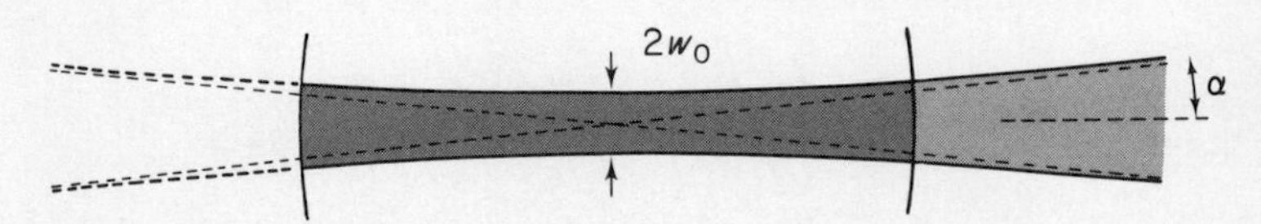

Fig. 6.12. Variation along the laser axis of the beam radius (i.e. the radius at which the intensity is e^{-2} that at the centre, see Eq. (6.33)) when the laser is oscillating in a uniphase mode. The beam emerges from the laser through the right-hand mirror. It has a waist of radius w_0, and at large distances it diverges with a semi-angle α. At the mirrors the beam profile is normal to the mirror surfaces.

beam emerging from an aperture of diameter $1.22\pi w_0$ ($\simeq 4w_0$). The value of w_0 is determined by the requirement that each mirror surface defines a wavefront of the beam (and therefore the envelope of the beam is normal to the mirror surfaces), and is given by

$$w_0 = \left(\frac{\lambda z_0}{\pi}\right)^{1/2} = \left(\frac{\lambda L}{\pi}\right)^{1/2}\left[\frac{(R_1-L)(R_2-L)(R_1+R_2-L)}{L(R_1+R_2-2L)^2}\right]^{1/4}. \quad (6.37)$$

It can be seen from this equation that w_0 is real only if the stability criterion (6.28) is satisfied, thus again justifying this criterion.

Diffraction losses of different modes

The diffraction losses depend on the type of mode and on the radii of the mirrors, as well as on the Fresnel number (Eq. (6.30)). For example the losses of the TEM_{mnq} modes are larger when m or n are non-zero than when m and n are both zero, because when m and n are non-zero the mean beam radius is always larger than that given by Eqs. (6.34) and (6.37). For example it can be seen from Fig. 6.11 that the TEM_{01q} mode extends to larger radii than the TEM_{00q} mode. The losses are also large when the mirror radii are such that w_0 (and also w at the mirrors) is large. This happens for example when R_1 and R_2 are both infinitely large (plane mirrors), since there is then an infinity of axes along which the beam could be localized (and Eq. (6.37) gives an infinite value of w_0). Figure 6.13 shows in greater detail the dependence on the Fresnel number of the diffraction losses for two mirror configurations (confocal and planar) and two modes (TEM_{00q} and TEM_{01q}).

Mode rejection

Having discussed the types of laser mode and their properties, let us now consider the techniques for rejecting unwanted modes. We see from Fig. 6.13 that in the case of the confocal configuration the diffraction losses of the uniphase and non-uniphase modes are substantially different for the same Fresnel number. One obvious technique for rejecting the non-uniphase modes is therefore to adjust the Fresnel number (usually by putting an aperture in front of one of the mirrors) so that the diffraction losses in the uniphase mode are low enough for the laser threshold condition to be satisfied, but at the same time are high enough in all the other modes to prevent their being excited. In practice the mirror surfaces have to be accurate to less than about $\lambda/20$, or else the preferred mode may be one in which the phase changes across the mirror. The mirrors must also be clean and free from blemishes, not only to reduce losses but also because an absorbing patch would cause a node at that point, thus excluding the uniphase mode.

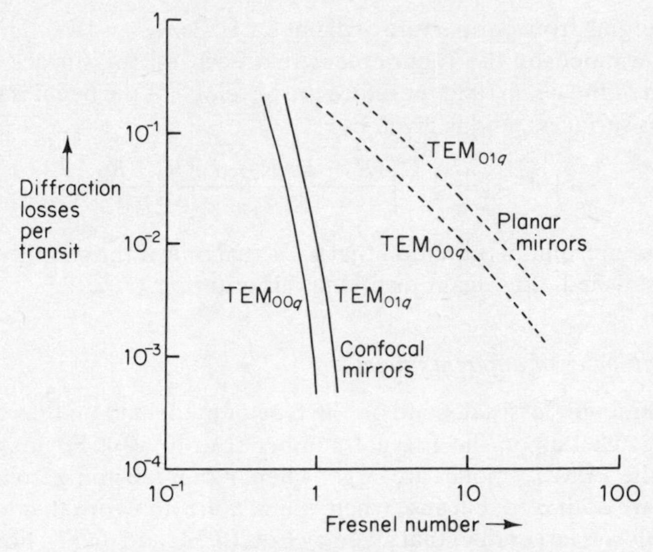

Fig. 6.13. Diffraction losses for 2 mirror configurations and 2 modes.

Assuming that the non-axial modes have all been excluded, the possible frequencies of oscillation are (Eq. (6.31))

$$\nu_{00q} = q\frac{c}{2L} + \frac{c}{2\pi L}\cos^{-1}\left[\left(1-\frac{L}{R_1}\right)\left(1-\frac{L}{R_2}\right)\right]^{1/2}.$$

These are spaced by $(c/2L)$, which is small enough in general, see Fig. 6.5, to allow several axial modes to be excited at the same time. The usual way of selecting one of these it to introduce another cavity length l into the system. One way of doing this is to put a Fabry–Perot etalon inside the laser cavity. If its optical thickness is l the frequencies which it can transmit are $q'c/2l$, where q' is an integer. The combination of the two frequency conditions can then be arranged to allow the excitation of only one mode, as illustrated in Fig. 6.14.

6.4 PARTICULAR MASERS AND LASERS

The technical development of lasers has been, and still is, extremely rapid, and a wide variety of laser media and pumping systems have been used. In this section a small selection of different laser systems is described, to illustrate some of the physical principles and some of the technical considerations. We start with the ammonia maser, for its historical interest, and then go on to gas discharge lasers, choosing the helium–neon laser as the main example because it is, and will no doubt remain for some time, the most

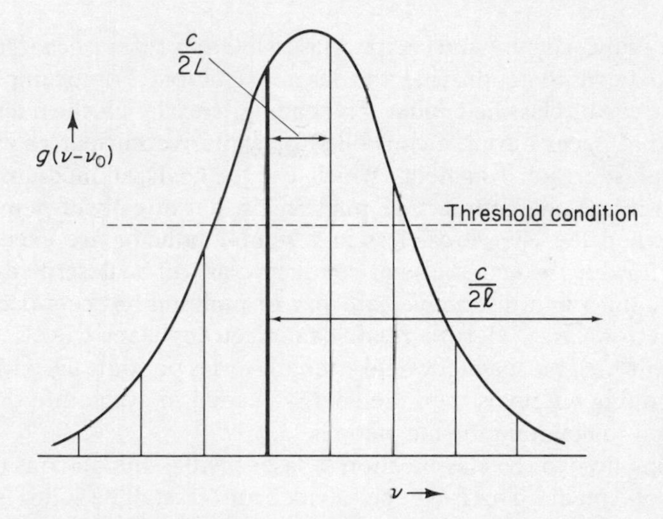

Fig. 6.14. The heavy vertical line shows a coincidence in the mode frequencies of the coupled cavities. i.e.

$$\nu' = \frac{qc}{2L} = \frac{q'c}{2l},$$

where q and q' are both integers. Only the single frequency ν' can be excited.

familiar of the small and cheap lasers. The CO_2, ion and excimer lasers will also be described. Next we deal with the dye lasers, which are important for their tunability, and then finish with the semi-conductor lasers, which are important for communication purposes.

Before describing these particular devices, let us see if it is possible to classify lasers in some way. This will help to show us the wide variety of different types of lasers, and it may also help us to understand more easily new types of lasers as they are invented. Three useful methods of classification are according to (*a*) the pumping mechanism, (*b*) the medium, and (*c*) the type of excited level.

Starting with the pumping mechanism, the type we have met already (in Section 6.2.1) is optical pumping. This can be subdivided into pumping with incoherent light (as in the ruby laser), or pumping with the coherent light from another laser. We shall see in Section 6.4.3 that the coherent form is useful for pumping dye lasers for example. A second important category of pumping is by charged particle impact. This usually takes the form of electron-impact excitation, which is the usual means of pumping for gas lasers (Section 6.4.2). The electrons either form a component of a gas discharge inside the laser medium (as in the He–Ne laser), or they are injected into the laser medium in the form of an externally accelerated beam (as is sometimes the case in the excimer and exciplex lasers). Other types of charged particle can also be used to provide the laser excitation. For example nuclear pumping can be classified under this heading, since in this the nuclear fission events that occur during a controlled or explosive nuclear chain reaction provide fast fission fragments which excite the laser medium. Another sub-division of charged-particle pumping is that of current pumping. This occurs when the charge carriers in a semi-conductor are excited by the current flowing through the semi-conductor, as will be described in Section 6.4.4. A third and distinctive category of pumping process occurs in the free-electron laser. Here a relativistic electron beam travels through a region in which the magnetic field strength varies periodically with position. The pumping energy is then the energy needed to accelerate the electron beam and to maintain the magnetic field.

Turning now to the classification of laser media, one obvious category is that of solid media. This can be subdivided into crystalline solids (such as the ruby crystal) or glasses (such as the neodymium-YAG laser) or semi-conductors (see Section 6.4.4). The next category is that of liquid media. These usually take the form of solutions of organic dye molecules (to be discussed in Section 6.4.3), but a few inorganic materials are sometimes also used as laser media. The third category in this progression is of course that of gaseous media. Some examples will be given in Section 6.4.2. Yet another distinctive type of medium is the electron beam and field of the free-electron laser, briefly described above.

The third way mentioned above for classifying lasers is in terms of the types of levels that give rise to the laser transition. At the longest wavelengths the upper and lower levels of the transition must be separated by an appropriately small energy. In the first device that we shall discuss below, the ammonia maser, the energy separation is very small (10^{-4} eV, corresponding to $\lambda = 12.6$ mm) and is caused by the tunnelling of the N atom through the plane formed by the three H atoms. Apart from special examples such as this, the smallest energies which can be employed for laser action are the rotational energies of polyatomic molecules. These energies are typically from about 10^{-3} to 10^{-2} eV (corresponding to $\lambda \sim 100$ μm to 1 mm, in the infrared). Next come molecular vibrational energies, typically in the region of 10^{-1} eV ($\lambda \sim 10$ μm). We shall meet an example of a laser using vibrational levels (the CO_2 laser) in Section 6.4.2. The next step in energy is provided by the excitation energies of the charge carriers of semi-conductors (~ 1 eV) and the excitation energies of electronic excited states of atoms and molecules (~few eV), giving the possibility of laser action in the near infrared, visible and near ultraviolet regions. Sections 6.4.2, 6.4.3, and 6.6.4 contain several examples of this type. For even higher energies (>few eV) it is necessary to use the electronic excited states of atomic or molecular ions (Section 6.4.2), or the energies of relativistic electrons (in the free-electron laser).

Although none of the three classification schemes is complete or exact, they do show us the very wide range and diversity of laser systems. Let us now go into greater detail on a few particular systems.

6.4.1 The ammonia maser

As mentioned at the beginning of the chapter, the details of the first maser, the ammonia maser, were published by Gordon, Zeiger, and Townes in 1954 (*Phys. Rev.*, **95**, 282). This maser is a two-level system, operating on the lowest two levels of the ammonia (NH_3) molecule.

The most stable geometry, in a classical sense, for this molecule is the pyramidal structure in which the three H atoms form an equilateral triangle, with the N atom displaced out of the plane of the triangle. The N atom can be on either side of the plane, as shown in Fig. 6.15(*a*), where the two possible positions are labelled 1 and 2. Both these configurations have the same energy, and both represent positions of minimum potential energy, about which the N and H atoms can vibrate. In fact the potential in these positions is only slightly lower than when the N atom is mid-way between these positions, coplanar with the 3 H atoms, and so the zero point energy of the vibration of the N atom is sufficient to allow it to pass through the H plane and to oscillate between the two local potential minima. The molecule is

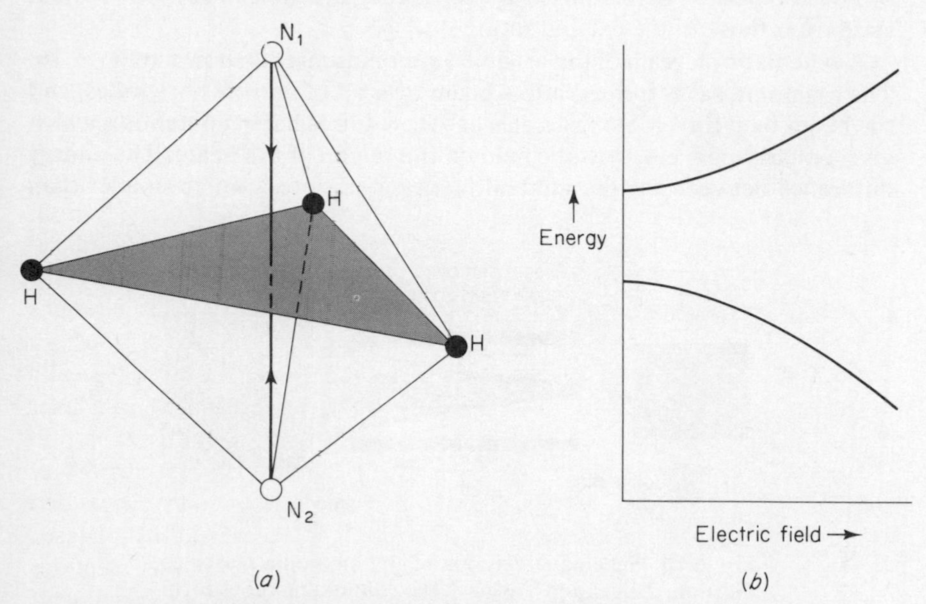

Fig. 6.15. (*a*) N_1 and N_2 are two possible positions of the nitrogen atom in the NH_3 molecule. In the two lowest energy states of the molecule the N atom vibrates between these two positions. (*b*) The dependence on electric field strength of the energy of the lowest two levels of the ammonia molecule.

therefore effectively in both configurations at the same time, with an equal probability of being in either. This implies that the wavefunction ψ of the molecule is a mixture of the wavefunctions ψ_1 and ψ_2 of the two separate configurations, the amplitude of each having the same magnitude, but with either the same or opposite signs,

$$\psi \propto \psi_1 \pm \psi_2.$$

The two different choices of sign give two energies that are slightly different, the positive sign giving the ground state of the molecule and the negative sign the first excited state, having an energy 9.86×10^{-5} eV higher, equivalent to a frequency of 23,870 MHz and a wavelength of 12.6 mm. It is this energy difference that is utilized in Townes' ammonia maser.

An important property of these lowest two levels, as far as the operation of the ammonia maser is concerned, is that when the ammonia molecule is in an electrostatic field the energy of the ground state decreases, while that of the excited state increases, as shown in Fig. 6.15(*b*). When placed in a non-uniform electrostatic field the molecules tend to move in the direction in which their total energy is minimized. Therefore the ground state molecules tend to move towards positions of higher electric field whereas the excited state molecules tend to move in the opposite direction, towards positions of zero field. This convenient difference in properties therefore provides a mechanism for physically separating the molecules in the ground state from those in the excited state.

A schematic diagram of the original ammonia maser is shown in Fig. 6.16. The ammonia gas is formed into a beam by a set of narrow bore tubes, and the beam then travels along an axis between 4 rods having potentials which give a quadrupole electrostatic field on the region of the beam. The energy difference between the ground and first excited state is much smaller than

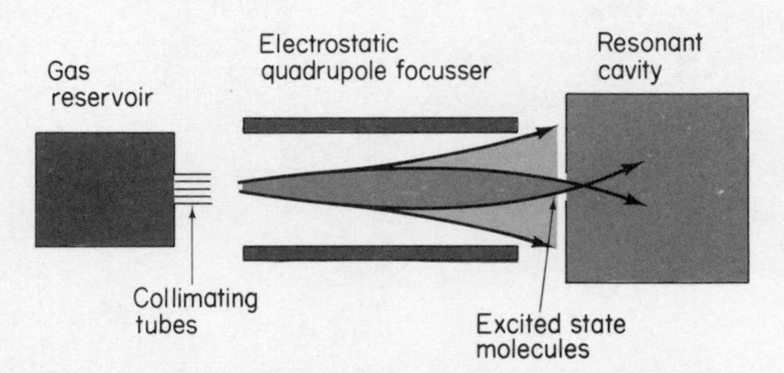

Fig. 6.16. Schematic diagram of the ammonia maser of Gordon, Zeiger and Townes. The collimating tubes form the ammonia gas into a beam which then passes through a quadrupole electric field. This field focusses the molecules in the higher energy state into a resonant cavity.

kT, and so approximately equal numbers of molecules are in these two states. The electric field strength is zero along the axis of the beam but large near the rods, which results in the excited state molecules staying near the axis while the ground state molecules move radially outwards. The excited state molecules then enter a cavity that has previously been tuned (by adjusting its dimensions) so that it resonates near the frequency 23,870 MHz.

If the density of the excited state molecules in the cavity is high enough for the threshold condition (6.19) to be satisfied then spontaneous emission can initiate a self-sustaining oscillation. The power is very low ($\sim 10^{-10}$ W) but the frequency is extremely precise, being constant to the order of 1 Hz. When used as a frequency standard or a clock the accuracy is therefore of the order of one part in 10^{10}, an accuracy seldom improved upon in the world of science.

6.4.2 Gas discharge lasers

Soon after the discovery of the ruby laser Javan, Bennett, and Herriott (*Phys. Rev. Letters*, **6**, 106 (1961)) suceeded in producing laser light (at 5 different wavelengths, ranging from 1.12 to 1.21 μm) from an electrical discharge in a mixture of helium and neon gases. The same laser can also be used to produce visible red light (at 632.8 nm), and in this form it is the familiar low-power, low-cost laser frequently used for alignment and test purposes.

Figure 6.17 shows the levels of the He and Ne atoms that are relevant to the production of the 632.8 nm line. The neon atoms are pumped into the upper level by a two-stage process. The first of these stages consists of the excitation of helium atoms into the $2\,^1S$ level by inelastic electron impact

$$e + \text{He (ground state)} \rightarrow \text{He* } (2\,^1S) + e.$$

This can occur at any incident electron energy above the threshold energy for the reaction. Indirect excitation into the $2\,^1S$ level is also possible, by cascading transitions from higher levels,

$$\begin{aligned} e + \text{He (ground state)} &\rightarrow \text{He* (higher excited states)} + e \\ &\qquad \hookrightarrow \text{He* } (2^1S) + h\nu. \end{aligned}$$

The $2\,^1S$ level is metastable, with a lifetime of 2×10^{-2} s, and so once excited in either of these ways the atoms stay excited until they collide with a wall or another atom.

The second stage of the pumping process occurs when the metastable helium atoms collide with the neon atoms that are also present in the laser discharge tube, since it is possible in these conditions for the excitation

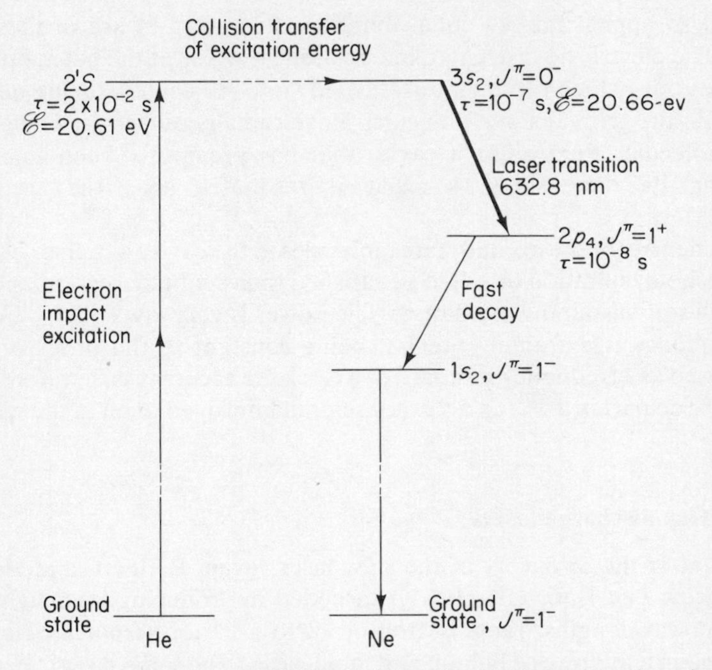

Fig. 6.17. The levels of He and Ne involved in the production of the laser line at 632.8 nm. (Several other energy levels also exist in this region of excitation energy.)

energy to be transferred from the helium to the neon atoms,

$$\text{He}^* \,(2\,^1S,\ 20.61\ \text{eV}) + \text{Ne (ground state)}$$
$$\rightarrow \text{He (ground state)} + \text{Ne}^* \,(3s_2,\ 20.66\ \text{eV}).$$

The cross-section for this process is large, because the energy inbalance (0.05 eV) can easily be provided by the thermal kinetic energy ($\sim kT$) of the atoms. The net effect of the two-stage pumping process is an efficient channelling into the Ne level of the energy originally present as the kinetic energy of the electrons of the discharge.

The Ne $3s_2$ level decays to the $2p_4$ level, with a lifetime of approximately 10^{-7} s. The $2p_4$ level is not populated by collisional transfer with the He atoms, and it has a shorter lifetime (approximately 10^{-8} s) than the $3s_2$ level, which implies that a population inversion can be maintained between the $3s_2$ and $2p_4$ levels. Therefore laser action can occur between them, giving the wavelength 632.8 nm. Similar mechanisms involving other levels lead to laser action at the infra-red wavelengths mentioned above.

The mechanical details of a simple low-power He–Ne laser are shown schematically in Fig. 6.18. The end mirrors are concave and are an integral

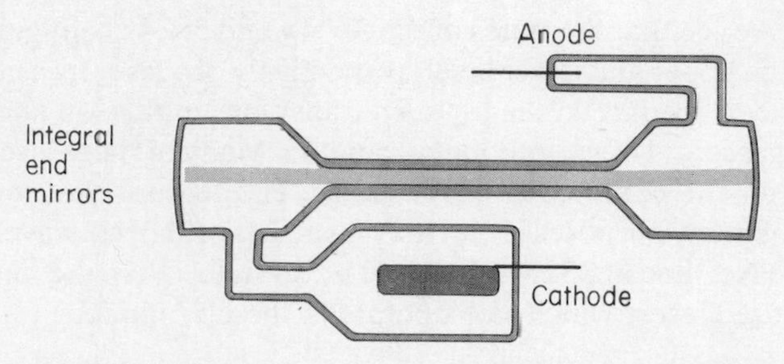

Fig. 6.18. Schematic diagram of a simple low-power He–Ne laser.

part of the laser cavity. The gas (at a pressure ~1 torr, of which ~10% is neon) is excited by a DC discharge (typically ~5 mA at ~1000 V) between an anode and a large area cold cathode, both in side-arms. The continuous power in the 632.8 nm line is typically ~1 mW, giving an overall efficiency $\sim 10^{-4}$.

The CO_2 laser

A much more powerful and efficient gas discharge laser is the CO_2 laser. It can provide a continuous power of up to about 1 MW (and a peak power of up to about 1 TW (10^{12} W) when operated in the pulsed mode). Under some conditions it can have an efficiency as high as about 20%. There are several possible CO_2 laser transitions, all of them occurring between states in which the molecule is both vibrationally and rotationally excited, and all of them in the infra-red. The transition for which the attainable power is highest has a wavelength of 10.57 μm. As in the case of the He–Ne laser the pumping mechanism consists of the excitation of another gas (this time N_2) by electron impact, followed by collisional energy transfer.

Ion lasers

Another type of gas discharge laser is that using atomic ions. For example in the argon-ion laser the two-step process

$$e + \text{Ar (ground state)} \rightarrow \text{Ar}^+ \text{ (ground or excited state)} + e + e,$$

$$e + \text{Ar}^+ \text{ (ground or excited state)} \rightarrow \text{Ar}^{+*} \text{ (excited state)} + e$$

results in the production of excited argon ions, some of which have the configuration $3p^4 4p$ for the outer shell of electrons. The decay route is

$$\text{Ar}^{+*}(3p^4 4p) \xrightarrow{\tau \sim 10^{-8}\,\text{s}} \text{Ar}^{+*}(3p^4 4s) \xrightarrow{\tau \sim 3\times 10^{-10}\,\text{s}} \text{Ar}^{+}(3p^5, \text{ground state}),$$

from which we see that the states of the $3p^4 4p$ and $3p^4 4s$ configurations are suitable as the upper and lower levels respectively of a laser transition. Laser action has been produced in many such transitions in this ion and in ions of other inert gases and of various metal vapours. Many of these lasers are able to operate on the continuous (CW) mode. The feature that makes them attractive for many purposes is that they give lines of shorter wavelength (for example the Ne^+ line at 332.4 nm, in the ultra-violet) than the lines of other commonly used lasers which also operate in the CW mode.

Excimer and exciplex lasers

It is not possible to form a stable Ar_2 molecule in its ground state because ground state Ar atoms repel each other at short distances, but in some circumstances it is possible to form an excited molecule (Ar_2^*) which can exist for a time of the order of 10^{-7} s. This excited molecule decays radiatively to the unstable ground state of the molecule, which then quickly dissociates into two atoms. Excited molecules of this type (e.g. Ar_2^*, Kr_2^*, Xe_2^*) are called *excimers*. The potential energy curves (i.e. the variation of the electronic energy with the internuclear separation) for the ground state of Xe_2 and the lower few states of Xe_2^* are shown in Fig. 6.19. When the constituent atoms are different from each other, as in the KrF^* molecule, the excited molecule is called an *exciplex*. These two types of molecule form the basis of an important class of gas lasers.

The laser transition occurs between the excimer or exciplex state, and the unstable ground state of the molecule. Because the lower state is unstable its population is essentially zero, and the emitted light is not re-absorbed. Another important consequence of the instability of the ground state is that the emitted light has a range of wavelengths, as we can see from Fig. 6.19. The laser can therefore be tuned over a broad range (a discussion of the techniques for tuning is deferred until the next section). The mean wavelengths are shorter than those of other gas lasers that we have discussed, for example 126 nm for Ar_2^*, 173 nm for Xe_2^* and 249 nm for KrF^*. The excimer and exciplex lasers are pumped typically by pulses of high energy (~ 1 MeV) electrons, or by a pulsed discharge, and so operate only in the pulsed mode.

6.4.3 Dye lasers

An important advance in laser design occurred in 1966 when Sorokin and Lankard (and soon afterwards several other groups of workers) showed that it is possible to use the properties of certain types of organic molecules to construct lasers for which the wavelength of the laser line can be controlled and varied over an interval that is typically from 5 to 50 nm wide. This

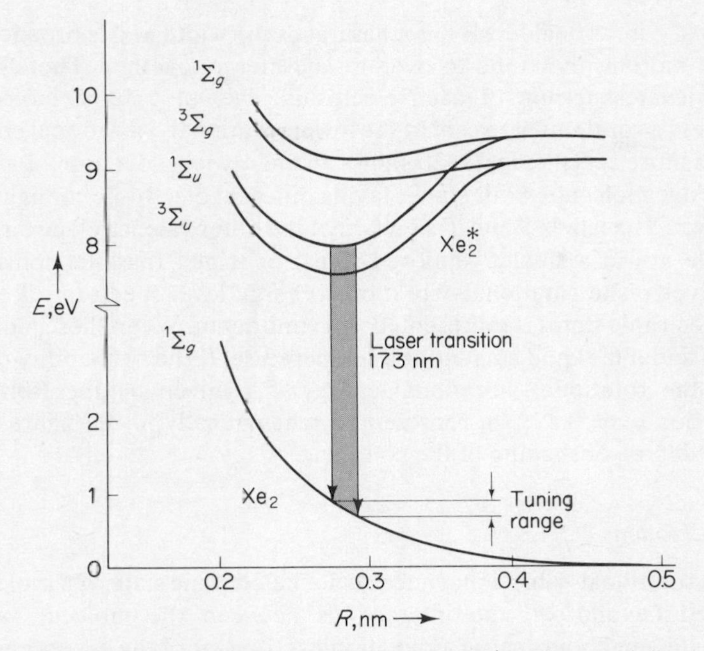

Fig. 6.19. Potential energy curves of the Xe_2^* excimer and the Xe_2 unstable ground state, showing the dependence of the electronic energy E on the internuclear separation R. The laser transition occupies a band of energies, and is the only optically allowed transition between these excimer states and the ground state.

tunability has since made possible many new experiments, leading to a greatly increased understanding of many aspects of atomic and molecular physics.

Energy levels of organic molecules

Large organic molecules have many different excited configur..ions of the electrons, as have atoms, but in addition they have many different modes in which the nuclear framework of the molecule can vibrate. A molecule of N atoms has up to $3N-5$ different vibrational modes of different frec ies, each of which can contain any number of vibrational quanta. There are no restrictions on the way in which the vibrational energies of the different modes can add together, and so the total number of different combinations of vibrational modes and quanta that can exist within a constant width ΔE increases quickly as the total vibrational energy E increases. Also each of the vibrational combinations possesses a set of rotational energy levels, causing the density of levels to be even larger. Because each of these rotational–vibrational levels has a natural decay width and is broadened by thermal motion, they tend to overlap and merge together. Therefore the energy level spectrum of each electronic level of a large molecule is effectively a continuum, except at the lower rotational–vibrational energies.

To be more specific, Fig. 6.20 shows an energy level diagram of a typical organic dye molecule. Four states having different electronic configurations are shown. The labels S and T signify that the outer valence electrons of the molecule are in a singlet (opposed spins) or triplet (parallel spins) state respectively. The rotational–vibrational energy level spectrum of each of these electronic states is represented as a continuum. When these molecules are in a suitable liquid solvent at a temperature T the probability of their having the rotational–vibrational energy $\mathscr{E}$ is given by the Boltzmann distribution $\exp(-\mathscr{E}/kT)$, represented schematically in the figure by the varying degree of shading of the continua.

Thermalization

If the rotational–vibrational energy of an electronic state of a molecule is increased beyond kT the interactions between the molecule and its immediate neighbours cause a radiationless transfer of the excess energy to the neighbouring molecules and eventually to the whole system, so that the molecule becomes *thermalized* (i.e., it regains the Boltzmann probability

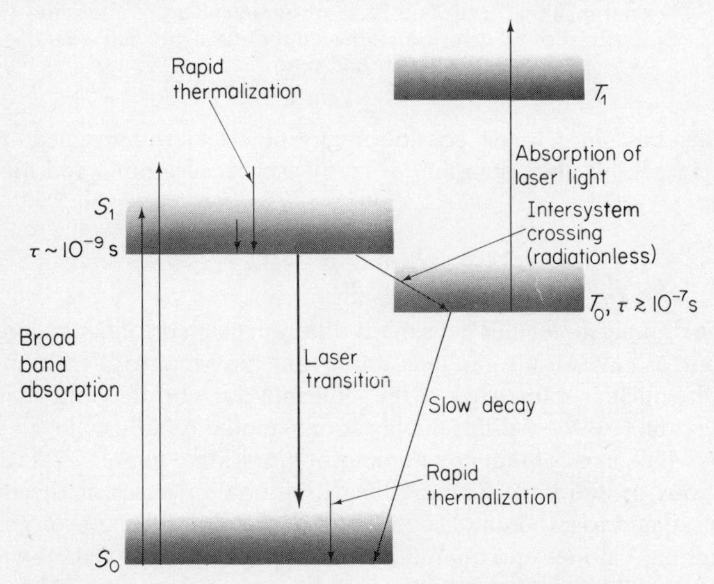

Fig. 6.20. Schematic diagram of some of the energy levels of a fluorescent organic dye molecule, showing the transitions that are relevant for laser action.

distribution of energy). This happens very quickly, in times of the order of 10^{-12} s. The thermalization occurs within electronically excited states (such as S_1, T_0, and T_1 in Fig. 6.18), as well as in the ground electronic state S_0.

Laser operation

With these facts to hand we are now able to understand how a dye laser works. The pumping mechanism is broad-band absorption of light to excite the S_1 state, as indicated in Fig. 6.20. The molecules for which this absorption has a high cross-section are those which are most effective in colouring other substances when added in small quantities, namely the dye molecules. The rapid thermalization to the lowest vibrational level of S_1, which acts as the upper laser level, ensures an efficient and rapid channelling of the absorbed energy into this level. The subsequent radiative decay to the S_0 state is efficient when this decay route does not compete seriously with alternative (and possibly non-radiative) decay routes, such as the route $S_1 \rightarrow T_0$ shown in the figure. This means that the molecule should fluoresce efficiently (compare the discussion of the ruby fluorescence, Section 6.2.1), and so the most suitable molecules tend to be the fluorescent dye molecules. The absorption and fluorescence curves of one of the most commonly used laser dyes, rhodamine 6G, are shown in Fig. 6.21.

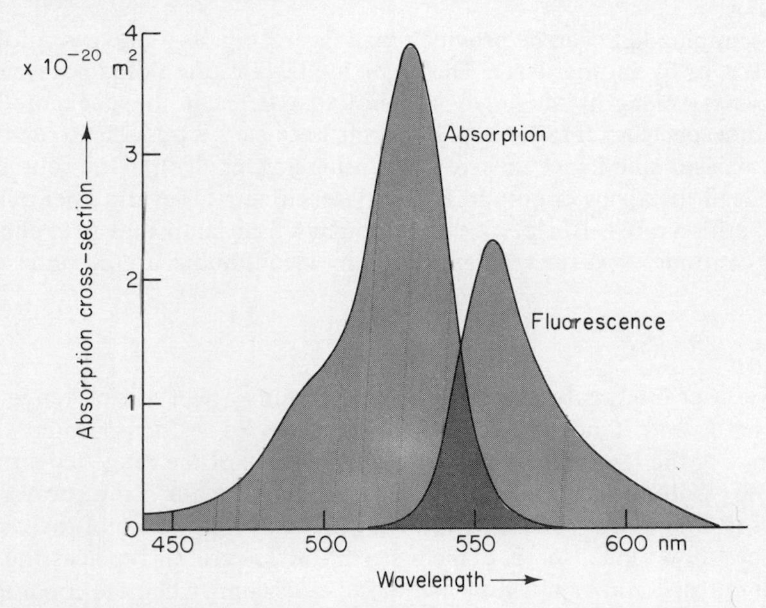

Fig. 6.21. Variation with wavelength of the absorption cross-section and fluorescence yield of rhodamine 6G in methanol (after C. V. Shank, *Rev. Mod. Phys.*, **47**, 649 (1975)).

The lower level of the laser transition can be anywhere in the S_0 continuum (except in the region in which the absorption and fluorescence curves overlap), thus allowing the laser wavelength to be varied over a wide range. In the case of rhodamine 6G the accessible range is from approximately 570 to 630 nm. Any of the excited molecules can be stimulated to emit anywhere in this range of wavelength, which implies that the fluorescence line broadening is homogeneous. This is an essential requirement if all the available energy is to be channelled into a single laser frequency. Finally, the lower level is rapidly depopulated by the thermalization process, thus completing all the necessary ingredients for an efficient 4-level laser system.

This efficiency is reduced by the existence of the electronically excited states T_0 and T_1 shown in Fig. 6.20. The spin selection rule $\Delta S = 0$ (Eq. (4.70)) prevents normal radiative transitions between the $S_{0,1}$ and $T_{0,1}$ systems, but there is a small transition probability for the radiationless intersystem crossing $S_1 \rightarrow T_0$. The T_0 state has a considerably longer lifetime that the S_1 state, which results in a continual increase in the T_0 population at the start of the laser action, thus reducing the number of molecules which can take part in further laser action. The T_0 population can be reduced by adding substances (such as COT, cyclooctatetrene) which reduce the T_0 lifetime, and also by flowing the dye solution through the laser cell to physically remove the T_0 molecules and replace them with ground-state molecules.

The pumping light can be provided by a flash lamp, as in the case of the ruby laser, or by another laser. The argon ion laser is often used because it produces wavelengths (down to 437 nm) that lie near the peak of the absorption spectrum of many laser dyes, and because it is possible to run the argon ion laser, and hence the dye laser (using flowing dye), in the continuous mode. Typical power outputs for dye lasers are up to several J per pulse (at a repetition rate $\sim$10 Hz) with excitation by a flash lamp, and up to about 1 W of continuous power with excitation by a continuous argon ion laser.

Wavelength tuning

A dye laser has a gain which is greater than unity over a wide range of wavelength, and if no wavelength selective device is incorporated the frequency of the laser light wanders over the centre of the range (covering $\sim$5 nm), resulting in an effectively polychromatic beam. Three ways of narrowing, and at the same time tuning, the wavelength range of the laser output are shown in Fig. 6.22. In the first a diffraction grating replaces one of the end mirrors, and light is returned to the cavity only when the condition

$$2d \sin \theta = n\lambda \tag{6.38}$$

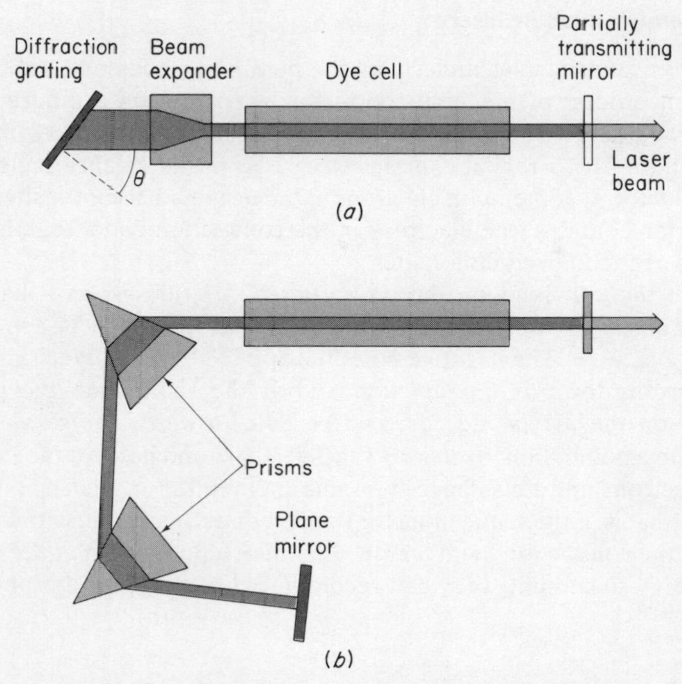

Fig. 6.22. Tuning of a dye laser with (*a*) a diffracting grating and (*b*) prisms.

is satisfied, where d is the grating spacing and n is the order of diffraction (usually unity). The wavelength of the laser beam is therefore controlled by the angle of the diffracting grating, and rotation of the grating allows the wavelength to be scanned over the width of the gain curve. The beam expander (a telescope) allows a larger area of the diffraction grating to be used, giving a higher resolution and also reducing the damage to the surface of the grating. The bandwidth of the laser beam is then of the order of 0.1 nm. Figure 6.22(*b*) shows an alternative arrangement in which wavelength selection is provided by the dispersion of one or more prisms. The angles of incidence at the prism faces can be made approximately equal to the Brewster angle (both on entering and on leaving the prisms, see Problem 6.11), thus reducing the reflection losses. In either arrangement the line width can be further narrowed by adding a Fabry–Perot etalon which has only one resonance frequency within the gain curve of the laser.

Lasing action has been achieved with many hundreds of dyes, covering the spectrum from the near infrared ($\simeq 1200$ nm) to the near ultra-violet ($\simeq 350$ nm).

6.4.4 Semi-conductor lasers

The physical principles underlying the pumping mechanism and photon generation process of the semi-conductor lasers are quite different from those of the lasers we have discussed so far. The laser action takes place at a p–n junction. The p region contains free holes in the valence band of the semiconductor, together with electrons at the occupied acceptor sites, while the n region contains free electrons in the conduction band, together with vacancies at the ionized donor sites.

To give the p–n junction the properties of a laser system it has to be *forward-biassed* by applying an external potential difference V_b across it, as shown in Fig. 6.23. The negative potential applied to the n-type side drives free electrons towards the junction and at the same time the positive potential on the p-type side drives free holes towards the same region, resulting in a population inversion of the electrons and holes at the junction. If the electrons and holes have the same momentum in the region of the junction (that is, if the semiconductor is of the direct-gap type, with the wave vector at the conduction band energy minimum equal to that at the valence band energy maximum) they can recombine directly to give a photon of energy

$$h\nu \simeq E_g.$$

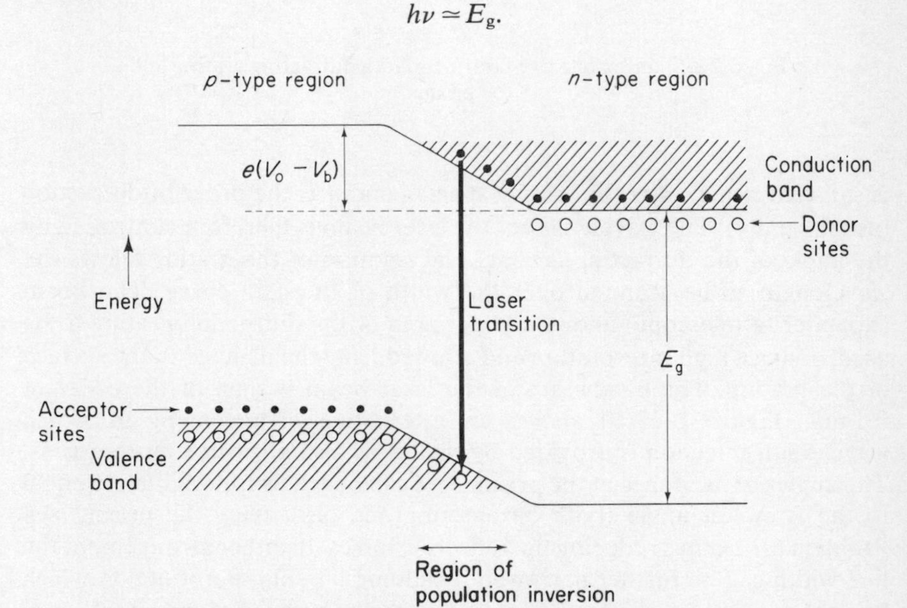

Fig. 6.23. Schematic illustration of the energy levels and their occupancy when a forward bias V_b is applied to a p–n junction. The symbol ○ represents a hole in the valence band or an ionized donor site; the symbol ● represents an electron in the conduction band or an occupied acceptor site. The applied potential difference V_b drives free electrons and free holes towards the junction.

Other recombination processes are also possible: unionized donor levels can combine with free holes, free electrons can combine with unoccupied acceptor levels, and unionized donor levels can combine with unoccupied acceptor levels. All these processes give photons of energy slightly lower than E_g.

A typical GaAs semiconductor laser is illustrated in Fig. 6.24. To give a large population inversion in the junction region the p and n-type regions have a high concentration ($\sim 0.1\%$) of the appropriate impurity atoms and a high current ($\geqslant 3 \times 10^8$ A/m^2 is necessary at room temperature) is driven through the junction by the forward bias. The laser wavelength is approximately 840 nm, corresponding to an energy of 1.47 eV. This is 0.04 eV less than the band gap E_g for GaAs, which implies that the unionized donors and unoccupied acceptors play a major role in the laser process. The width of the laser transition is typically 2 nm (caused by variations in E_g) and the axial mode spacing is typically 0.35 nm (for a mirror separation of 1 mm), and so several such modes are excited together. Many of the non-axial modes are also excited because their losses are small, which increases the angular divergence of the laser beam. This divergence is in any case large because of

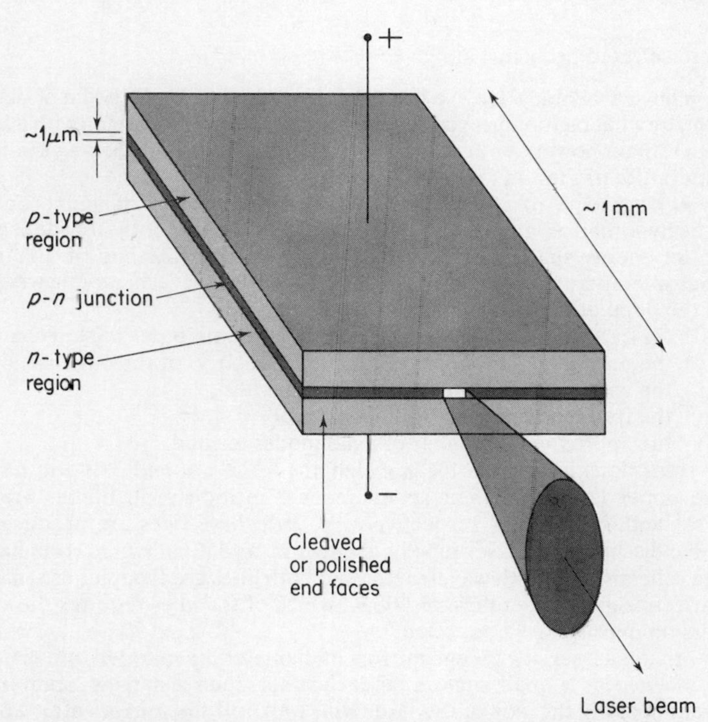

Fig. 6.24. Schematic diagram of a GaAs semi-conductor laser.

the thinness (~1 μm) of the junction region (c.f. Eq. (6.36)), and the narrow width (~5 to 20 μm) over which the laser action is confined.

The ratio of the number of photons generated to the number of electrons crossing the junction, called the external quantum efficiency, is typically 15% at room temperature. This is very high compared with the pumping efficiency of most other types of laser, and the overall efficiency (laser power/electrical input power) of semiconductor lasers is also correspondingly high, having values up to 10%. The semiconductor lasers can be operated in either the continuous or pulsed mode, although the latter is usually easier in practice.

A special feature of the semiconductor lasers is the very short lifetime of the upper laser state. The spontaneous recombination lifetime is typically $\sim 2\times 10^{-9}$ s, and when stimulated emission occurs this is reduced to $\sim 10^{-10}$ s. This short lifetime, together with the short drift times of the charge carriers, means that the power level of the laser can be changed very quickly, at frequencies up to about 10^{10} Hz. This makes the laser very suitable for communication purposes, as we shall see in Section 7.2.4.

PROBLEMS: CHAPTER 6

(Answers to selected problems are given in Appendix E.)

6.1 If you have available a He–Ne laser which produces a 1 mW beam of diameter 1 mm, by what factor must you reduce the diameter (by focussing with a lens) to give a radiant power per m^2 equal to that emitted by a body having the surface temperature of the sun (5800 K)?

6.2 A laser has end mirrors of reflectivity 99%, separated by 1 m, and it contains a gas of hypothetical atoms which have only two levels, both non-degenerate, with an energy spacing of 2 eV and an upper state lifetime of 10^{-8} s. The effective temperature of the gas is 400 K, the atoms have an atomic weight 20, and the population inversion is 10^{14} m^{-3}. Calculate:
   (i) the cavity lifetime (assuming that there are no other loss mechanisms),
   (ii) the natural and Doppler widths (in frequency) of the emission line,
   (iii) the gain per transit at the centre of the line,
   (iv) the frequency spacing of the axial modes, and
   (v) the approximate number of axial modes excited.

6.3 The transitions in neon at the wavelengths 632.8 nm and 3.39 μm have the same upper level, and their spontaneous transition probabilities are 1.4×10^6 s^{-1} and 9.7×10^5 s^{-1} respectively. If both these lines are produced in a He–Ne discharge in a laser tube of length 1 m, and if the two mirrors have the same reflectivity at both wavelengths and both lines are Doppler broadened at an effective gas temperature of 400 K, which of the lines requires the smaller inversion density for laser action?

6.4 Show that if a laser has flat end mirrors of diameter d, separated by a distance L and inclined at a small angle α to each other, then a narrow beam initially travelling along the axis of the laser will 'walk' off the mirrors after approximately $(2d/L\alpha)^{1/2}$ reflections. Show that α must be less than 0.4″ if more than 100 reflections are required and $d/L = 10^{-2}$.

6.5 Estimate the minimum power required to sustain continuous laser action at the X-ray wavelength of 5 nm, if the required gain coefficient is $10^{-1}\ \mathrm{m}^{-1}$ and the volume of the laser is $10^{-7}\ \mathrm{m}^3$. Assume that the upper level decays at the classical radiative rate, and that the Doppler width is negligible compared with the natural width of the line.

6.6 Use the identity

$$\left[\left(1-\frac{L}{R_1}\right)\left(1-\frac{L}{R_2}\right)\right]^{-1}=1+\frac{L(R_1+R_2-L)}{(R_1-L)(R_2-L)}$$

to show the equivalence of Eqs. (6.28) and (6.29).

6.7 A Fresnel number greater than 3 is required in a laser cavity of length 0.5 m. What is the minimum diameter of the end mirrors if the laser wavelength is (*a*) 632.8 nm, (*b*) 3.39 μm?

6.8 The end mirrors of a certain laser are nearly confocal (see Fig. 6.10(*b*)). They have a radius of curvature of 0.4 m and are separated by 0.5 m. The laser wavelength is 600 nm. What is the minimum beam radius at (*a*) the centre of the laser, (*b*) at the mirrors, and (*c*) 10 m away from the laser?

6.9 The wavelength of a dye laser is determined by the first order reflection in a diffraction grating used as an end mirror. At what rotational speed must the grating be turned if the wavelength is to be scanned at the rate 0.1 nm/s in the neighbourhood of 500 nm, and if the grating has 2000 lines/mm?

6.10 Comment briefly on the connection between the wave and particle aspects of electromagnetic radiation, as exemplified by the operation of a maser or laser.

6.11 A set of prisms is used to narrow the band width of a dye laser. Find the conditions for which the angles of incidence are equal to the Brewster angle at both the entrance and exit faces of the prisms.

6.12 Write a short essay on the proposition 'lasers could have been designed and built at any time from the early 1930s onwards'.

CHAPTER 7

The properties and uses of laser light

In Chapter 6 we considered the essential physical principles of laser action, and discussed the mode of operation of some specific lasers. In this chapter we concentrate on laser light and its uses, rather than on the way the light is produced. We start by considering the various properties of laser light in more detail. This is done without reference to which laser is being used to produce the light, and therefore it is not assumed that the reader has studied the section (6.4) on particular lasers. The first topic to be treated is the ultimate spectral line width of which lasers are capable, since the narrowness of the line is perhaps the most startling property possessed by lasers. We then go on to the other important and unique properties of high directionality and coherence. The first part of the chapter ends with a description of how lasers can be made to give ultra-short pulses of radiation.

In the second part of the chapter we discuss laser techniques and applications. This part is self-contained in the sense that it does not assume a knowledge of the first part of the chapter (although this knowledge would certainly be useful and would make some parts of the discussion easier to follow). We start with the technique of frequency doubling, and then go on to discuss the techniques of Doppler-free spectroscopy, of great importance for fundamental studies of atoms and molecules. This is followed by the more practical application of holography. The chapter finishes with several other practical applications of laser light, from the important topic of communications to the possibility of laser-induced fusion.

7.1 PROPERTIES OF LASER LIGHT

7.1.1 Spectral line width

The obvious starting point for a discussion of the spectral line width $\Delta\nu_L$ of a laser beam is to establish the theoretical lower limit to the width when the laser is operating in a single mode.

Let us suppose that the average number of photons in this mode is $\bar{n}$. When the laser is operating above threshold nearly all these photons will have been produced by the process of stimulated emission. As we saw at the beginning of the previous chapter, the stimulated photons are completely coherent with the stimulating photons and therefore most of the photons in the mode will be coherent with each other. The incoherent component is much smaller, and it exists only because the excited atoms of the laser also undergo spontaneous emission, sometimes emitting photons into the mode that we are considering. The relative numbers of stimulated and spontaneous photons in the mode can be found from the relationship (4.28) (see also the discussion following this equation): it is

$$\frac{\text{rate of stimulated emission into mode}}{\text{rate of spontaneous emission into mode}} = \bar{n}.$$

Therefore in equilibrium,

$$\text{mean number of coherent photons in mode} = \bar{n}\left(\frac{\bar{n}}{\bar{n}+1}\right) \simeq \bar{n} - 1.$$

$$\text{mean number of incoherent photons in mode} \simeq 1.$$

The number $\bar{n}$ is usually very large. To calculate it we can use the fact that the average survival time for a photon is the cavity decay time t_c. A laser operating in a single mode and producing a beam of power P therefore has an average mode energy given by

$$E = \bar{n}h\nu \simeq Pt_c. \tag{7.1}$$

For example a He–Ne laser producing 1 mW of continuous power, and having $t_c \sim 10^{-7}$ s, has

$$\bar{n} \sim \frac{10^{-3} \times 10^{-7}}{3 \times 10^{-19}} \simeq 3 \times 10^8.$$

If there were only coherent photons in the mode the number n at any one time would be governed by Poisson statistics (see Section 5.5.1), and the root-mean-square fluctuation about $\bar{n}$ would therefore be

$$\Delta n = \bar{n}^{1/2}. \tag{7.2}$$

The mean electric field strength $\bar{E}$ in the cavity is proportional to (cavity energy)$^{1/2}$ and hence to $\bar{n}^{1/2}$, and so the magnitude of E would fluctuate by ΔE, where

$$\frac{\Delta E}{\bar{E}} = \frac{1}{\bar{E}} \frac{dE}{dn} \Delta n = \frac{1}{\bar{n}^{1/2}} \tfrac{1}{2} \bar{n}^{-1/2} \Delta n = \frac{1}{2\bar{n}^{1/2}}. \tag{7.3}$$

The phase ϕ of the field would also fluctuate, by an amount given by the uncertainty relationship (see Appendix C)

$$\Delta n \, \Delta\phi \sim 1. \tag{7.4}$$

Therefore

$$\Delta\phi \sim \bar{n}^{-1/2}. \tag{7.5}$$

Another source of fluctuations is the continual addition of the incoherent (spontaneously emitted) photons to the mode, at the rate t_c^{-1} (since the mean number is 1, and they survive for the average time t_c). Each of these adds a field of magnitude $\bar{E}/\bar{n}^{1/2}$, with a random phase angle. They do not change $\bar{E}$ significantly, since this is kept within the bounds given by Eq. (7.1) by the energy balance of the laser, but they do change the mean phase angle $\bar{\phi}$, by an amount

$$\Delta\phi_r \sim \bar{n}^{-1/2}. \tag{7.6}$$

This is a more important source of fluctuation than those given by Eqs. (7.3) and (7.5), since it causes a continual and random change in $\bar{\phi}$ (whereas the other two sources do not change $\bar{\phi}$ or $\bar{E}$). This is illustrated in Fig. 7.1. The

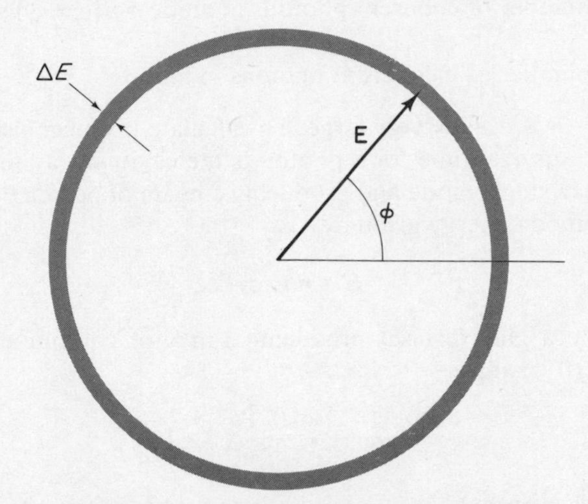

Fig. 7.1. The magnitude and phase of the electric field of a laser oscillating above threshold. The phase angle ϕ 'diffuses' around the circle with a characteristic diffusion time τ_p. The uncertainty ΔE in the amplitude is greatly exaggerated.

tip of the electric vector performs a 'random walk', and diffuses around the circle. Each step has a length $\sim\Delta\phi_r$, and the number of steps in time t is t/t_c. Using a result from diffusion theory we find that the root-mean-square phase change is

$$\Delta\phi(t) \sim (t/t_c)^{1/2} \times \Delta\phi_r,$$

and that the time taken for the phase to change by one radian is

$$\tau_p \sim t_c(\Delta\phi_r)^2 \sim \bar{n}t_c. \quad (7.7)$$

This is called the *phase diffusion time.*

The random changes in phase spoil the sinusoidal behaviour of the beam, and so add other frequency components. The phase diffusion time is effectively the coherence time discussed in Chapter 5, and so the frequency width $\Delta\nu_L$ of the beam is given by (Eq. (5.10))

$$\Delta\nu_L \sim \frac{1}{2\pi\tau_p}.$$

Using Eq. (7.1) to express $\bar{n}$ in terms of the power P of the laser, and using Eq. (3.56) or (5.10) to give the relationship between the time uncertainty t_c and the frequency uncertainty (linewidth) $\Delta\nu_c$ of the cavity mode,

$$\Delta\nu_c = \frac{1}{2\pi t_c}, \quad (7.8)$$

the value of $\Delta\nu_L$ can now be expressed as

$$\Delta\nu_L \sim \frac{2\pi h\nu(\Delta\nu_c)^2}{P}.$$

This is the minimum width that the laser line can possess.

A more exact treatment shows that $\Delta\nu_L$ depends also on the number N_2 of atoms in the upper laser level, and on the population inversion

$$\Delta N = N_2 - \frac{g_2}{g_1}N_1, \quad (7.9)$$

where g_1 and g_2 are the degeneracies of the lower and upper levels respectively. It is found that

$$\Delta\nu_L = \frac{2\pi h\nu(\Delta\nu_c)^2}{P}\frac{N_2}{\Delta N}. \quad (7.10)$$

The frequency width given by Eq. (7.10) is much smaller than the experimentally measured widths of laser lines. For example a He–Ne laser working at 632.8 nm, with a power of 1 mW, a cavity decay time of 10^{-7} s, and with $N_2 \rightarrow \Delta N$, has, using Eq. (7.10), the theoretical limit

$$\Delta\nu_L \sim 10 \text{ Hz},$$

but a measured line width which is ~ few kHz in the best case. The origin of this discrepancy between theory and experiment lies in the stability of the mode frequency ν. Any changes in the cavity length L, caused by thermal expansion or contraction of the spacer between the end mirrors, or by mechanical vibration of the laser, cause corresponding changes in ν. To maintain ν constant to 10 Hz in a cavity of length 1 m would require

$$\frac{\Delta L}{L} \leqslant \frac{\Delta\nu_L}{\nu} \sim 2 \times 10^{-14}, \qquad \Delta L \leqslant 2 \times 10^{-14} \text{ m}.$$

In practice the lower limit to $\Delta\nu_L$ is set by the temperature fluctuations, which cannot be reduced below about 0.01 K over a period of a few hours. If the coefficient of linear expansion of the material between the end mirrors is that of a typical glass ($\gtrsim 10^{-6}$ K^{-1}), this gives

$$\frac{\Delta\nu}{\nu} \gtrsim 10^{-8}, \qquad \Delta\nu \gtrsim \text{few MHz}. \tag{7.11}$$

To obtain lower values of $\Delta\nu$ (such as the value of a few kHz quoted above) it is necessary to use an elaborate feed-back system in which the frequency of the laser line is compared with that of a narrow atomic or molecular line, the error signal being processed and fed to a piezo-electric crystal which is used to adjust the cavity length of the laser.

7.1.2 Directionality

The directionality of laser beams has already been discussed briefly in Section 6.3, and a sketch of the angular divergence of a beam can be seen in Fig. 6.12. If the end mirrors have the same radius R the beam waist is situated mid-way between the mirrors; for the uniphase mode it has the radius (see Eq. (6.37))

$$w_0 = \left(\frac{\lambda L}{2\pi}\right)^{1/2} \left(\frac{2R - L}{L}\right)^{1/4}.$$

At a large distance from the laser the angular divergence of the beam is given by Eq. (6.36),

$$\alpha = \frac{\lambda}{\pi w_0}. \tag{7.12}$$

A way of decreasing α by increasing the effective value of w_0 is to expand the beam by passing it through a telescope, as illustrated in Fig. 7.2. In the uniphase mode the beam intensity has a Gaussian profile in the transverse

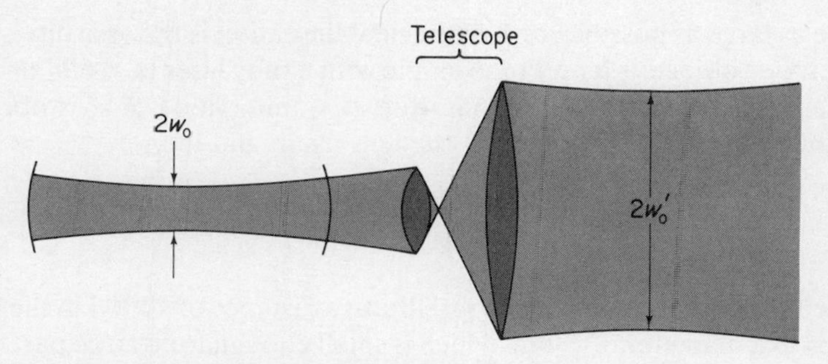

Fig. 7.2. The use of an expanding telescope to increase the beam waist diameter and decrease the angular divergence of the beam. The divergences are shown greatly exaggerated.

direction, and the wavefronts (surfaces of constant phase) are spherical in shape, meeting the edges of the beam at right angles. The telescope converts the diverging wavefront that it receives to a larger area wavefront that is initially diverging, flat, or converging, depending on the setting of the telescope. The Gaussian intensity profile is preserved. In Fig. 7.2 the beam is shown as initially converging after passing through the telescope. Away from the telescope the beam profile is given by the same expression (6.34) that applies inside the laser, namely

$$w(z) = w_0'\left[1+\left(\frac{z}{z_0}\right)^2\right]^{1/2}, \tag{7.13}$$

where w_0' is the new waist radius, the distance z is now measured from the new waist position, and (Eq. (6.35))

$$z_0' = \frac{\pi w_0'^2}{\lambda}.$$

Whatever we do to magnify or focus the beam the product $z_0 w_0^2$ is an invariant, provided that the Gaussian profile of the beam is retained. We see from Eq. (7.13) that the beam is approximately parallel for $|z| \leq \frac{1}{2}z_0$, and so a useful guide to the achievable length and radius of an approximately parallel beam is

$$\frac{\text{length}}{(\text{radius})^2} \leq \frac{\pi}{\lambda}. \tag{7.14}$$

For example the 632.8 nm light from a He–Ne laser can be formed into a beam which has a radius of 1 mm and is approximately parallel for 5 m, or a beam which has a radius of 25 mm and is approximately parallel for 3 km.

For illumination at very large distances the final angular divergence α must be as small as possible. This implies (Eq. (7.12)) that the waist diameter must be as large as possible, and so the final limitation is the aperture size of the telescope objective lens. For example with a ruby laser ($\lambda = 694$ nm) and a lens aperture of diameter 200 mm the minimum value of α is obtained when the beam waist is situated at the lens itself, and then

$$\alpha = \frac{694 \times 10^{-9}}{\pi \times 10^{-1}} = 2.2 \times 10^{-6} \text{ rad.}$$

In travelling to a communications satellite at a distance of 1000 km the beam expands to a diameter of 4.4 m, which is small enough for a large part of the energy in the beam to be intercepted. For transmission to a larger distance (e.g. to the moon, 3.84×10^5 km), an astronomical telescope can be used in reverse to give a large aperture (~10 m diameter) and thus a much smaller half-angle (~4×10^{-8} rad, giving a beam diameter ~30 m at the moon's surface).

7.1.3 Coherence

When a continuous laser operates in a single mode the coherence time (see Section 5.2) depends only on the linewidth $\Delta\nu$, and is given by

$$\tau = \frac{1}{2\pi\,\Delta\nu}.$$

The coherence length is then

$$l = \frac{c}{2\pi\,\Delta\nu}. \tag{7.15}$$

As mentioned in Section 7.1.1, $\Delta\nu$ is typically $\gtrsim$ few MHz, giving a coherence length of up to ~10 m. Use is made of this in holography, as we shall see in Section 7.2.3.

The coherence length is reduced if the laser operates in more than one mode (and if the relative phases of the modes are random). For example, two axial modes separated in frequency by $c/2L$ (Eq. (6.12)) gives a coherence length

$$l = \frac{c}{2\pi(c/2L)} = \frac{L}{\pi}.$$

If several axial modes are excited, covering the range of frequencies in the Doppler-broadened laser transition, the coherence length reduces to that of an incoherent discharge source using the same transition.

The spatial coherence transverse to a laser beam can also be extraordinarily large. Because nearly all the photons in an oscillating cavity mode are coherent with each other, the spatial coherence inside and outside the cavity is almost complete when the laser operates in a single mode, and this coherence extends across the whole width of the beam. When the mode is an axial mode for example, the phase of the electric field is essentially constant across the beam. This nearly complete spatial coherence also exists when more than one axial mode is excited, but obviously it is reduced when there is more than one transverse mode.

The spatial coherence gives rise to the familiar *speckle pattern* produced when a rough surface (such as a wall of the laboratory) is illuminated by laser light. Each part of the illuminated area can be thought of as a point source, and the electric field at a point in space is the vector sum of the light received from these sources. The path differences depend on the position of the point, and so a three-dimensional interference pattern is produced, having a visibility (see Section 5.4) that depends on the degree of coherence between the point sources, and hence on the degree of spatial coherence of the laser beam. An interference pattern is formed also on the retina of the eye, and the details of the pattern depend on the position of the eye and the direction in which it is looking, as well as on the focal length of the lens of the eye. The pattern changes when the eye is moved or refocussed. The formation of the interference field is illustrated in Fig. 7.3.

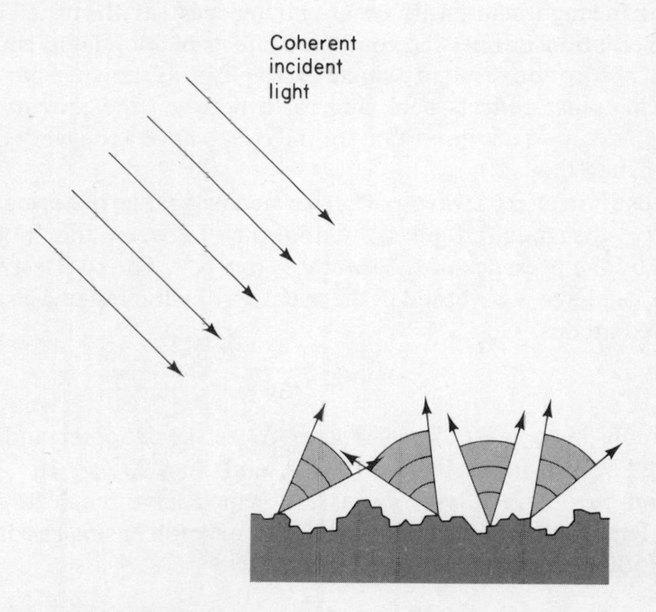

Fig. 7.3. The light scattered by a rough surface produces an interference field having a visibility that depends on the degree of spatial coherence of the illuminating beam. A speckle pattern is seen when this field is detected at the retina of the eye.

7.1.4 The generation of short pulses

The fact that lasers tend to oscillate in more than one mode at the same time is not always a nuisance. By using several modes, and forcing a definite relationship between their phases, it is possible to produce pulses of extremely short duration. The technique is called *mode-locking* or *phase-locking*.

Mode-locking with a saturable dye

One way of obtaining short pulses is to place a cell containing a *saturable dye* in front of one of the end mirrors. For low intensities of light this type of dye absorbs in the normal way over the appropriate wavelength range, but at high intensities a substantial number of the dye molecules are raised from the ground state to an excited state, giving a reduced number in the ground state and hence a reduced absorption coefficient. At very high intensities the dye saturates and becomes nearly transparent.

The laser light initially has a non-uniform intensity when several independent modes are excited; it fluctuates randomly, as shown in Fig. 5.12(*b*) for chaotic light. All the fluctuations help to saturate the dye, but those with the highest intensity are attenuated least, as illustrated in Fig. 7.4. After several transit times the combined effects of stimulated emission and saturated absorption result in a single sharp pulse, as shown in part (*d*) of the figure. This pulse reflects back and forth between the mirrors, with the period $2L/c$, and so it is important for the dye to have a recovery time which is shorter than this.

The pulse is in effect a wave-packet formed by a range of normal modes of the cavity, the required phase relationship between the modes being imposed by the presence of the saturable dye cell. The shortest width that the pulse can have is governed in the usual way by the range $\Delta\nu$ spanned by the excited modes,

$$\Delta t_{\min} \sim \frac{1}{2\pi\ \Delta\nu}.$$

For example in the case of a gas laser $\Delta\nu$ is the Doppler and pressure-broadened width of the line, $\sim 10^8$ Hz, and then $\Delta t_{\min} \sim 10^{-9}$ s. Shorter pulses can be obtained with dye lasers, which have much larger homogeneous line widths (see Section 6.4.3). For example, using rhodamine 6G, with $\lambda \simeq 600$ nm and $\Delta\lambda = 0.2$ nm, gives

$$\Delta\nu = \frac{c}{\lambda} \cdot \frac{\Delta\lambda}{\lambda} = 1.7 \times 10^{11}\ \text{Hz}, \qquad \Delta t_{\min} \sim 1\ \text{ps}.$$

In fact this, and shorter times, have been achieved in practice. The physical length of a 1 ps pulse is only 0.3 mm!

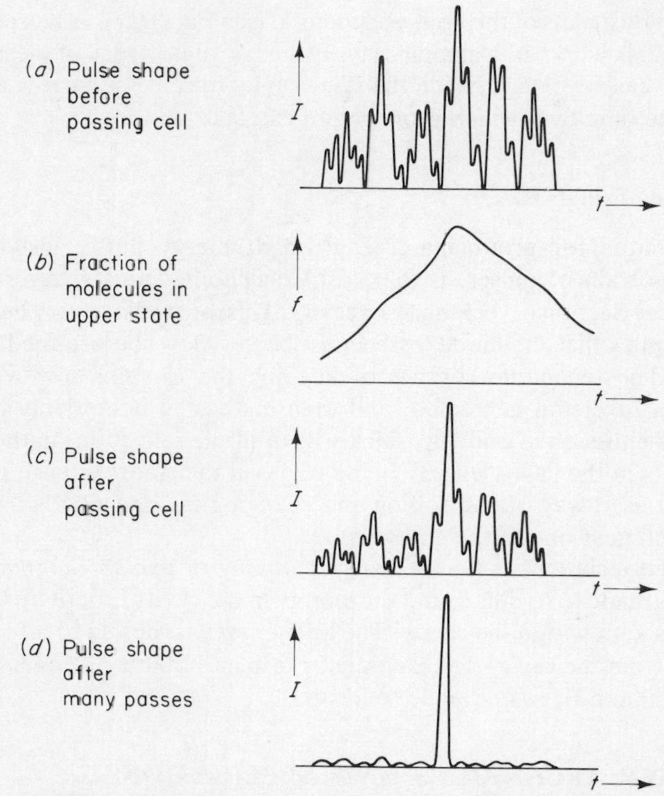

Fig. 7.4. (*a*) Before passing through the dye cell the light intensity fluctuates randomly. (*b*) The dye molecules are excited by photon absorption, causing the cell to be more transparent at high intensities. (*c*) The more intense fluctuations are attenuated least. (*d*) After many passes (and also intensity amplification by stimulated emission between passes) only the largest fluctuations survive.

Mode-locking with a Pockels cell

As an alternative to the saturable dye cell, a Pockels cell can be used. In this the anisotropic propagation characteristics of a crystal are changed by applying a potential difference V across it (see Section 3.2.4). It can be arranged that the refractive index of the crystal is different for two orthogonal directions (x and y, say) of polarization, and that the difference changes linearly with V. When a beam which is linearly polarized at 45° to the x and y directions is passed through the crystal the state of polarization of the emerging beam then depends on V. When the crystal is used in conjunction with linear polarizers it is then possible to control the intensity of the light by means of the applied potential (see Fig. 3.27). A Kerr cell (see Section 3.2.4) works in a similar way. Either of these types of *electro-optic modulator* can be placed inside the laser cavity to select a narrow pulse by limiting the time over which propagation can take place.

Generation of single pulses

A technique for producing a single high-energy pulse, instead of a continuous train of pulses, is that of *Q-switching*. The 'Q' refers to the Q-value (see Section 6.1) of the laser cavity. This must have a very high value (which implies that the mode losses must be very low) before the laser can oscillate. The technique consists of keeping the Q-value low while the population inversion is created, and then increasing it suddenly to allow stimulated emission to build up quickly by multiple reflections in the cavity. This results in the cavity energy being released in a short intense pulse. A frequently used way of controlling and changing the Q-value is to place a Pockels cell near one of the end mirrors.

A related technique is that of *cavity-dumping* or *photon-dumping*. Here the laser is made to oscillate, and the energy in the cavity is built up to a high level but is kept within the cavity. The light energy is then suddenly allowed to escape from the cavity (for example by using a suitable arrangement of a Pockels cell and Brewster angle reflector).

7.2 LASER TECHNIQUES AND APPLICATIONS

Ordinary light sources can be used to give beams having a narrow bandwidth or well defined directionality, but only at the cost of having a very low intensity. Alternatively a high intensity is possible, but then the bandwidth is large and the directionality is usually poor. If this type of beam is focussed to a spot the temperature at the spot cannot exceed that of the original source. The uniqueness of laser beams lies in the fact that they can have at one and the same time the properties of narrow bandwidth, good directionality, and high intensity, and they can be focussed to give a spot of extremely high brightness. They can also have a high degree of temporal and spatial coherence, or they can be formed into pulses of extremely short duration. These exceptional properties have made possible many new types of spectroscopic studies and many new technical applications. A small selection of these is described briefly in the following sections.

7.2.1 Frequency doubling

Because laser operation becomes more and more difficult as the light frequency increases (see Section 6.1) it is important to be able to generate high frequency beams from laser beams of lower frequency. This can be achieved by making use of the non-linear behaviour of media at high light intensities. For example, as we have seen in Section 3.2.5, an intense beam of frequency ν passing through a suitable dielectric medium can give rise to light of frequency 2ν, travelling in the same direction. This process is known as *frequency doubling*. Alternatively beams of frequency ν_1 and ν_2 can be mixed to give the sum frequency $\nu_1 + \nu_2$ (as well as the difference frequency $|\nu_1 - \nu_2|$ and the doubled frequencies $2\nu_1$ and $2\nu_2$). Higher-order effects, giving for example the third harmonic $3\nu_1$, or the mixed frequency $2\nu_1 + \nu_2$, have also been seen. In some applications a laser beam is frequency doubled more than once, by passing it through more than one cell, giving a final frequency of 4ν or more.

The two requirements that must be satisfied before frequency doubling can take place efficiently (see Section 3.2.5) are (i) that the medium must be transparent at the frequencies ν and 2ν, and (ii) that the frequency doubled radiation must be in phase with the incident radiation (see Fig. 3.28), which requires that the *phase-matching condition*

$$n(2\nu) = n(\nu) \tag{7.16}$$

be satisfied (where $n(\nu')$ is the refractive index of the medium at the frequency ν'). To see what happens when the second condition is not satisfied, let us consider a primary beam

$$E = E_0 \cos(\omega_1 t - k_1 z)$$

travelling through a dielectric which extends from $z = 0$ to L. If we regard the regions in which $\cos^2(\omega_1 t - k_1 z)$ is a maximum as the 'generators' of the frequency doubled light (as in Fig. 3.28), then the generator at $z = 0$, $t = 0$ produces a frequency doubled wavelet which reaches the position $z = \frac{1}{2}L$ at the time $t = \frac{1}{2}(L/v_2)$, where $v_2 = c/n(2\nu_1)$. The phase of the primary beam at this position and time is

$$\phi = \omega_1 t - k_1 z = \left(\frac{\omega_1}{v_2} - k_1\right)\frac{L}{2} = [n(2\nu_1) - n(\nu_1)]\frac{\pi L}{\lambda},$$

where λ is the wavelength in free space. If this phase is $\pi/2$ the wavelets generated at $z = 0$ and $\frac{1}{2}L$ are out of phase and cancel each other. Similarly all the wavelets generated from $z = 0$ to L cancel in pairs, and the resultant intensity of the frequency doubled beam is zero. The length for which this happens is therefore

$$L_0 = \frac{\lambda}{2[n(2\nu_1) - n(\nu_1)]}.$$

More generally, it can be shown (see Problem 7.2) that the intensity of the frequency doubled beam is

$$I \propto \frac{\sin^2 (L\pi/L_0)}{[n(2\nu_1) - n(\nu_1)]^2}. \tag{7.17}$$

The maximum value occurs when condition (7.16) is satisfied, and then I is proportional to L^2. The intensity has half the maximum value when

$$|n(2\nu_1) - n(\nu_1)| = \frac{0.44\lambda}{L}. \tag{7.18}$$

We see from Eq. (7.18) that $n(2\nu)$ and $n(\nu)$ must be very close to each other for efficient doubling. For example if $\lambda \sim 500$ nm and $L \sim 10^{-2}$ m, we need $|n(2\nu) - n(\nu)| \leqslant 2 \times 10^{-5}$. A way of achieving this with a birefringent crystal is described in Section 3.2.4. The crystal is tuned to the phase matching condition by adjusting its orientation (see Fig. 3.29). Alternatively the crystal can be tuned by adjusting its temperature, since refractive indices are temperature-dependent. Yet another solution is to use a gas as the frequency doubling medium, since the refractive index is then approximately constant because it is nearly unity at all frequencies. This has the additional advantage that the gas is more likely than a condensed medium to be transparent at the doubled frequency. For high efficiencies it is usually necessary for the gas atoms to be nearly resonant at the frequencies ν and 2ν (i.e. the atom must have excited states with energies near $h\nu$ and $2h\nu$). Alkali metal vapours are often used for this purpose. When considering high-order effects, such as the production of the frequencies $3\nu_1$ or $2\nu_1 + \nu_2$, metallic vapours or gases are usually essential if the final frequency is very high (for example in the vacuum ultraviolet), because other media would absorb too strongly at these frequencies.

7.2.2 Doppler-free spectroscopy

Atomic and molecular spectra consist typically of many groups of closely spaced lines. Because the atoms inevitably have a random thermal motion the absorption lines are Doppler broadened (see Section 6.3), often by an amount that is greater than the spacings of the lines in each group. Worse still, the emission lines from conventional light sources are also Doppler-broadened, and so the study of individual lines often required much hard work and ingenuity in the pre-laser area.

Of course the invention of the laser did not immediately solve all the technical problems of spectroscopy, because although laser light can be produced with a very narrow bandwidth the atoms and molecules to be studied still have a thermal motion and so their absorption lines are still Doppler broadened. What is needed is a means of effectively 'freezing' the motion of the atoms to be studied, and it is the techniques for doing this that we now discuss. The uses to which these techniques have been put are unfortunately outside the scope of this book, although it must be said that they have led to many important and interesting experiments in atomic and molecular spectroscopy.

Crossed beam technique

Atomic absorption lines are Doppler broadened because the atoms have a variable component of velocity in the direction of the absorbed light beam, and so one obvious way of reducing the Doppler width is to reduce the randomness of the thermal motion by ensuring that the atoms move only at right angles to the light beam, as shown in Fig. 7.5. In practice the collimated

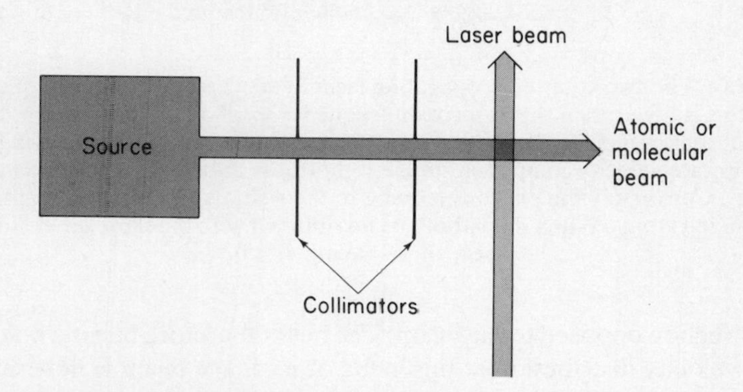

Fig. 7.5. Crossing an atomic beam and a laser beam at right angles to each other reduces the effective Doppler width of the atomic absorption lines.

atomic beam still has a small range of directions, and the laser beam may also be slightly divergent, resulting in an effective width (see Eq. (6.26))

$$\Delta\nu \sim \Delta\nu_{\text{Doppler}} \times \Delta\theta \sim 10^{-6}\nu_0\left(\frac{T}{A}\right)^{1/2} \Delta\theta,$$

where T is the effective temperature of the atomic source, A is the atomic weight and $\Delta\theta$ is the half-angular range of the relative atom–photon direction. This width is often small enough to be of the same order as that of the laser beam itself, but the atomic beam necessarily has only a low density, and the volume of overlap of the two beams is small, giving a rather small rate of absorption events.

Lamb dip spectroscopy

The basis of a better technique is illustrated in Fig. 7.6. If the frequency ν of a single laser beam is chosen to be lower than the absorption frequency ν_0 of a stationary atom, the only atoms that can absorb photons from the laser beam are those having a velocity component (see Eq. (6.24))

$$v_z = \frac{c(\nu - \nu_0)}{\nu_0}$$

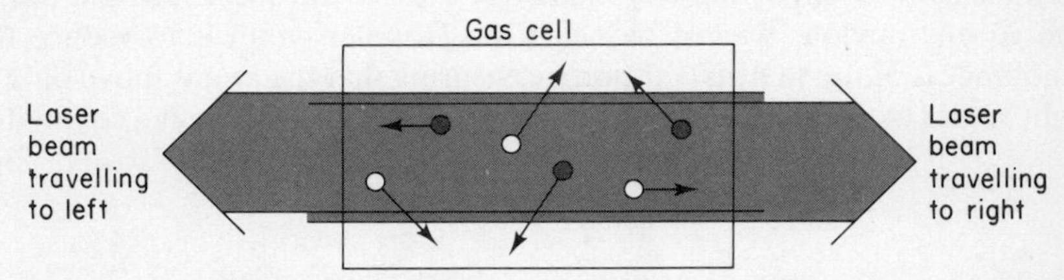

Fig. 7.6. The two counter-propagating laser beams have the same frequency ν. If this is lower than the absorption frequency ν_0 of a stationary atom, then the photons moving to the left can be absorbed only by those atoms having the appropriate velocity component to the right (these atoms are unshaded in the figure). Conversely the photons moving to the right can be absorbed only by the shaded atoms. When $\nu = \nu_0$ both beams interact with the same set of atoms, namely those having $v_z = 0$.

in the direction opposed to the beam. The beam therefore burns a hole in the atomic velocity distribution at this value of v_z. If the beam is now reflected back through the gas cell (or the beam is divided and reflected in some other way to give two counter-propagating beams), two holes are burnt, as shown in Fig. 7.7. The width of the holes is related to the homogeneous natural and pressure-broadened width Γ of the transition (see Section 6.3). The power absorbed from each laser beam is related to the area of the corresponding hole. When the beam is intense the value $P(v_z)$ at the minimum of a hole is small, and the absorption tends to saturate.

An interesting effect occurs when the laser frequency is scanned through the central frequency ν_0 (most lasers can be tuned over their line profiles, and dye lasers can of course be tuned over a much wider range). As long as ν and ν_0 differ by more than Γ the two holes are separate from each other and the absorption of one laser beam is not influenced by the presence of the other. But when ν and ν_0 coincide there is only one hole burnt, at the centre of the velocity distribution, and then both laser beams compete for the same sub-group of atoms, namely those moving at right-angles to the beam directions. If the laser beams are very intense the area of this single hole is not much larger than that of either of the separated holes of Fig. 7.7, and so

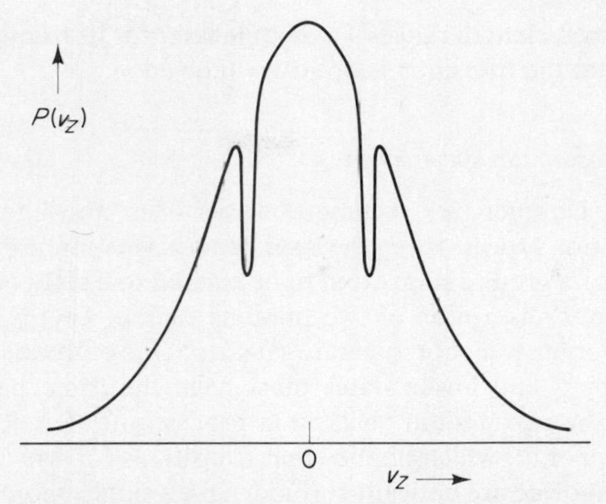

Fig. 7.7. Two holes burnt in the atomic velocity distribution by two counter-propagating laser beams of the same frequency ν, when $\nu \neq \nu_0$.

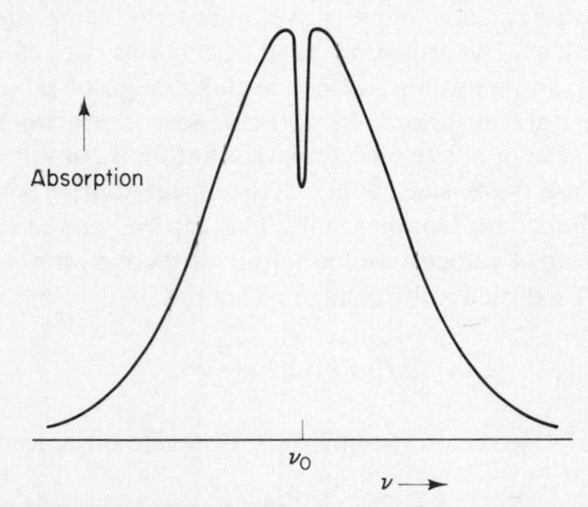

Fig. 7.8. The Lamb dip produced by saturated absorption when the common frequency ν of two counter-propagating laser beams is scanned across the Doppler profile of an absorption line.

the absorption of each beam decreases sharply, as shown in Fig. 7.8. The effect is known as *saturated absorption*, and the dip in the absorption curve is known as a *Lamb dip*.

The importance of the Lamb dip is that its width is the homogeneous width Γ, which is usually very much smaller than the Doppler width. Absorption spectra of atoms and molecules can therefore be studied free from all the complications of thermal motion, provided that the transitions are within the wavelength range of a tunable laser (or its frequency doubled output) and that the transition is optically allowed.

Two-photon absorption spectroscopy

A different Doppler-free technique arises from non-linearities in the atomic absorption process when the laser beam is very intense. In particular it is possible for a ground state atom to be excited to a state of energy E by the simultaneous absorption of two photons each of energy $\frac{1}{2}E$, provided that the appropriate selection rules are satisfied. One of these selection rules is that the upper and lower states must have the same parity (a fuller discussion of the two-photon absorption process and its selection rules is given in Section 4.5), which implies that transitions for which two-photon absorption is allowed are optically forbidden for a single photon absorption. Therefore any Doppler-free method based on two-photon absorption is in this sense complementary to the Lamb dip method for single photon absorption.

The two photons need not be travelling in the same direction before absorption, and so by adjusting their directions we can adjust their combined linear momentum, and hence the change of momentum of the atom when they are absorbed. In particular we can arrange for the two photons to travel in opposite directions, so that their combined momentum is zero. In this case there is no change in the momentum or kinetic energy of the atom, and hence no Doppler shift. This can be seen also from the fact that when an atom of velocity component v_z absorbs a photon of energy $h\nu_0$ moving in the $+z$ direction its change of energy is

$$\Delta E = h\nu_0\left(1-\frac{v_z}{c}\right),$$

while if it absorbs two photons moving in opposite directions its change of energy is

$$\Delta E = h\nu_0\left(1-\frac{v_z}{c}\right)+h\nu_0\left(1+\frac{v_z}{c}\right) = 2h\nu_0,$$

which is independent of v_z.

The technique is therefore simple: two counter-propagating beams are derived from a tunable laser and passed through an absorption cell. The rate of absorption is usually monitored by observing the subsequent single-photon decay of the upper level to some third level of the atom. The technique is known as *two-photon absorption spectroscopy*. Higher laser beam powers are required than for saturated absorption spectroscopy, but there is the compensating advantage that all the atoms of the gas cell can contribute to the absorption signal, whatever their values of v_z.

7.2.3 Holography

The coherence properties of laser light find one of their most spectacular applications in the subject of *holography* or *wavefront reconstruction*. Dennis Gabor first demonstrated the reconstruction of a wavefront in 1948 (*Nature*, **4098**, 777), but high-quality reconstruction became possible only after the invention of the laser.

The physical basis of holography is illustrated in Fig. 7.9. The photographic plate receives (i) a spherical wave from the pinhole aperture, which we shall regard as the object whose wavefront is to be reconstructed, and (ii) a plane wave, which we shall call the reference wave, formed by the prism.

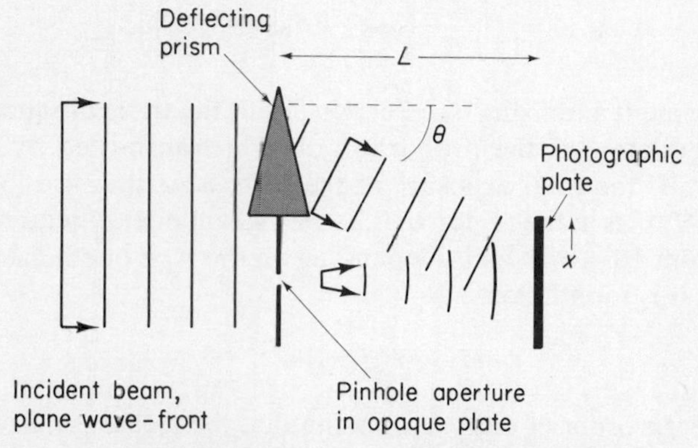

Fig. 7.9. A deflecting prism and a pinhole aperture cause a plane wave and a spherical wave to fall on the photographic plate, where they create an interference pattern.

Because these two wavefronts are derived from the same incident plane wave they are coherent with each other (assuming that the plane wave is produced by a laser, and that it has a high degree of spatial and temporal coherence). The distance from the pinhole to a position x on the photographic plate is

$$l(x) = (L^2 + x^2)^{1/2} = L + \frac{1}{2}\frac{x^2}{L} + \cdots,$$

(where x is measured from the centre of the plate) and therefore the electric field of the spherical wave has the value

$$E_{\mathrm{obj}}(x) = A_0 \exp\left[\mathrm{j}\left(\omega t - \frac{kx^2}{2L} + \phi_0\right)\right]$$

at the plate, where A_0 is approximately constant (for $x \ll L$) and A_0 and ϕ_0

are assumed to be real. Similarly the field of the reference wave at the plate is

$$E_{\text{ref}}(x) = A_r \, e^{j(\omega t + kx \sin\theta + \phi_r)},$$

where A_r and ϕ_r are both real constants. The intensity at the plate is therefore

$$\begin{aligned} I(x) &\propto |E_{\text{ref}} + E_{\text{obj}}|^2 \\ &= A_r^2 + A_0^2 + A_r A_0 \exp\left[j\left(-\frac{kx^2}{2L} + \phi_0 - kx \sin\theta - \phi_r\right)\right] \\ &\quad + A_r A_0 \exp\left[j\left(\frac{kx^2}{2L} - \phi_0 + kx \sin\theta + \phi_r\right)\right]. \end{aligned} \tag{7.19}$$

After development and fixing (see Section 9.2.2) the transmittance H of the photographic plate (i.e. the proportion of light transmitted by the plate) depends on x. If the total exposure of the plate is neither too low nor too high, so that $H(x)$ is in the region of 0.2 (the radiant energy required for this is typically from 10^{-2} to 1 J/m^2, depending on the type of emulsion and the wavelength), it is found that

$$H(x) \propto [I(x)]^{-\gamma/2},$$

where γ is of the order of unity. Assuming that the reference wave is much stronger than the object wave, we find that

$$\begin{aligned} H(x) \propto A_r^{-\gamma-2}\Big\{ & A_r^2 - \frac{\gamma}{2} A_0^2 - \frac{\gamma}{2} A_r A_0 \exp\left[j\left(-\frac{kx^2}{2L} - kx \sin\theta + \phi_0 - \phi_r\right)\right] \\ & - \frac{\gamma}{2} A_r A_0 \exp\left[j\left(\frac{kx^2}{2L} + kx \sin\theta - \phi_0 + \phi_r\right)\right]\Big\}. \end{aligned} \tag{7.20}$$

We see from this that the developed photographic plate is unlike an ordinary photograph in that is contains information about the phase of the object wave, as well as its amplitude. Gabor called it a *hologram*, from the Greek word 'holos' meaning 'the whole'.

To extract the phase information from the hologram it is necessary to illuminate it with a second suitable reference wave. Figure 7.10 shows an example of this, the reference wave being simply

$$E'_{\text{ref}} = A'_r \, e^{j(\omega t + \phi'_r)}$$

at the position of the hologram. The electric field transmitted by the

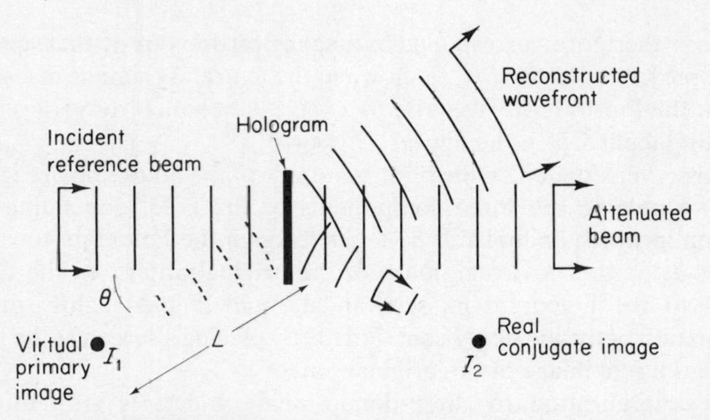

Fig. 7.10. Reconstruction of the wavefront of the original point object by illumination of the hologram with a plane wave reference beam. A conjugate image is also produced.

hologram is

$$E = E'_{\text{ref}} \times H(x) \propto A'_r A_r^2 \exp[j(\omega t + \phi'_r)] - \frac{\gamma}{2} A'_r A_0^2 \exp[j(\omega t + \phi'_r)]$$
$$- \frac{\gamma}{2} A'_r A_r A_0 \exp\left[j\left(\omega t - \frac{kx^2}{2L} - kx \sin\theta + \phi\right)\right]$$
$$- \frac{\gamma}{2} A'_r A_r A_0 \exp\left[j\left(\omega t + \frac{kx^2}{2L} + kx \sin\theta - \phi\right)\right]. \qquad (7.21)$$

Each of the four terms of this equation gives rise to a simple form of propagating wave. The first term is an attenuated continuation of the incident beam, as shown in the figure. The second term is the smallest (since $A_0 \ll A_r$) and it also corresponds to a continuation of the incident beam, with some diffraction caused by the x-dependence of A_0.

To see the significance of the third term let us consider the form of the wavefront from a point object placed at the position labelled I_1 in the figure. The distance from I_1 to the position x on the hologram is

$$l'(x) = [(L\cos\theta)^2 + (L\sin\theta + x)^2]^{1/2} = L + x\sin\theta + \frac{x^2}{2L} + \cdots,$$

and therefore the field at the hologram is

$$E'(I_1) = A'_0 \exp\left[j\left(\omega t - \frac{kx^2}{2L} - kx\sin\theta + \phi'\right)\right]. \qquad (7.22)$$

We see that this field is the same as that given by the third term of Eq. (7.21).

This term therefore corresponds to a spherical wavefront that appears to have come from the point I_1, as shown in the figure. By similar reasoning we see that the fourth term gives rise to a wavefront which converges towards the point labelled I_2 in the figure.

An observer situated on the right-hand side of the hologram of Fig. 7.10 is therefore able to see three components of the field, depending on the direction in which he looks. If he looks through the hologram towards the position I_1 he sees a virtual image of the original object. In the direction normal to the hologram he sees an attenuated and slightly diffracted continuation of the incident beam. Finally by placing a screen at the position I_2 he sees a real image of the original object.

The generalization to three-dimensional objects is straightforward. Assuming that the surfaces of the objects diffuse, rather than reflect, the light falling on them, each surface point can be regarded as equivalent to the pinhole object of Fig. 7.9. A part of the laser beam used to illuminate the objects must be suitably reflected or refracted to form a reference beam which can recombine with the scattered light at the photographic plate. Figure 7.11(a) shows an arrangement used in the undergraduate teaching laboratories at Manchester.

In order for the scattered light to be coherent with the reference beam the object must move by much less than one wavelength during the exposure, and the coherence length of the laser light has to be at least twice the depth of the field being viewed. The same coherence length is required in the reconstruction process as well.

Usually each part of a hologram receives light from all points on the object surfaces (apart from those points that are hidden from the hologram by other objects), and in this sense there is a redundant amount of recorded information, and so the hologram is called a *redundant hologram*. This gives rise to the interesting property that any small part of the hologram can be broken off (or masked off) and used as a separate hologram to reconstruct the wavefront of the whole object. Another interesting property is that the wavelength λ_2 used in the construction process need not be the same as the wavelength λ_1 used for making the hologram. A magnification of the order of λ_2/λ_1 can be achieved with a suitable reconstruction arrangement.

A different type of hologram is the *phase hologram*. This differs from the ordinary hologram in the way that it is developed after being exposed to the scattered and reference waves. It is developed in such a way that the resultant hologram is transparent over all its area, but with the variable intensity $I(\mathbf{r})$ being transformed into a variable optical thickness of the developed plate. In the reconstruction stage the phase hologram alters the phase of the reference beam which passes through it, giving a reconstructed image. The intensity of this image is higher than for an ordinary hologram because of the transparency of the phase hologram.

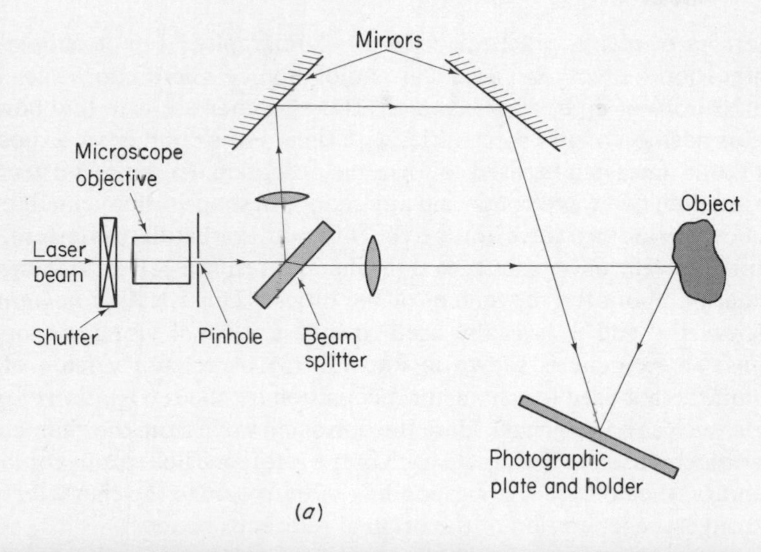

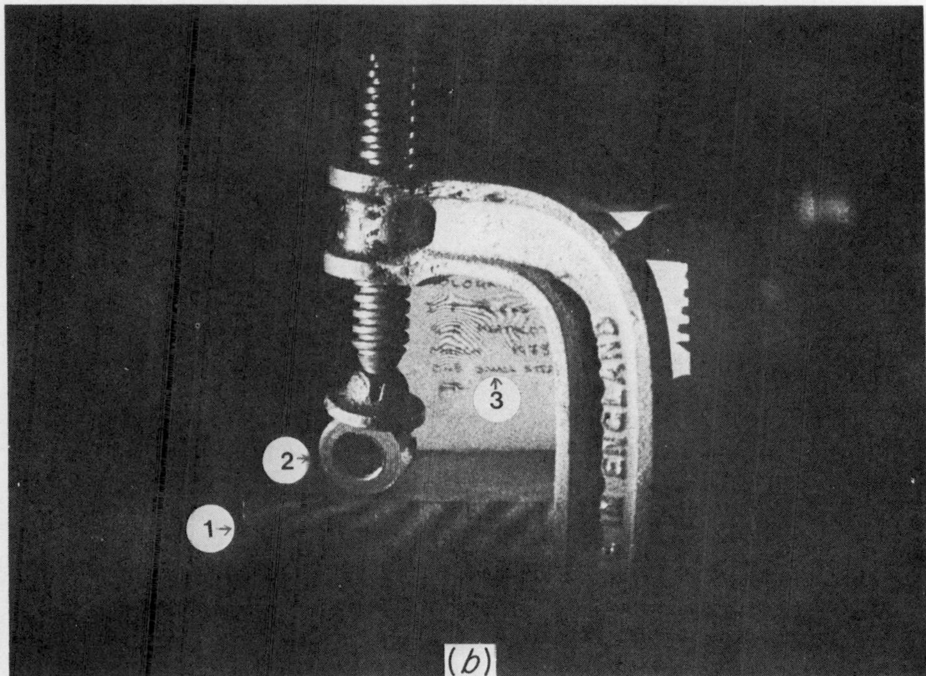

Fig. 7.11. (*a*) An arrangement used in the undergraduate teaching laboratories at Manchester to produce holograms. (*b*) Reconstruction of a double-exposure interference hologram, made by two students (S. R. Heathcote and D. P. Burke). Between the two exposures the G-clamp was tightened slightly. The fringe patterns at 1 and 2 show the stresses in the metal plate and ring, while those at 3 show the movement of a piece of paper.

There are many practical uses of holography. For example the magnification effect is used in holographic microscopes to give magnifications of up to approximately 100. Another use is to find how the shape or position of an object varies with time. For example two exposures at different times can be used to form the hologram. In the reconstruction there are then two wavefronts, and any change in shape or movement of the object causes the two wavefronts to be different, so that they interfere. The pattern of interference (similar to thin-film interference patterns) then gives information about the movement of the object. This is called *holographic interferometry*, and is typically used to measure small vibrations or distortions: an example is shown in Fig. 7.11(*b*). A related version of this technique is that used for character recognition (or the recognition of other objects, such as fingerprints). Here the diffracted wave from the character or other object is used to illuminate each of the reference holograms contained in a library; the hologram corresponding most closely to the character gives the strongest regeneration of the original reference beam.

Another potentially important use of holography is in the computer storage of data. Arrays of tiny holograms (diameter <1 mm) can be built up, and it can be arranged that as each hologram is illuminated by a narrow laser beam it produces a characteristic pattern of beams that are detected by an array of photodetectors. In this way a large amount of data can be stored in a small space.

7.2.4 Communications

Information is usually transmitted by (i) modulating the amplitude of a sinusoidal carrier wave, or (ii) modulating the frequency of the carrier wave, or (ii) selectively removing some of the pulses from a beam consisting of a regular train of narrow pulses. These three techniques are called *amplitude modulation*, *frequency modulation*, and *pulse-code modulation* respectively. The rate at which information can be transmitted depends on the frequency bandwidth available. For example with a television frequency of 300 MHz and a bandwidth of 8 MHz it is possible to transmit the necessary information to give a high-resolution flicker-free picture. Because the bandwidth is an appreciable fraction of the frequency of the carrier wave, only a limited number of television channels can be transmitted in a given locality. The bandwidth required for radio communication is much smaller, and that for telephone links is smaller still, approximately 5 kHz, but because the carrier wave is also of lower frequency the number of channels is again limited.

The advantage of using laser light for transmitting information is that the frequency of the carrier wave ($\simeq 5\times10^{14}$ Hz for visible light) is several orders of magnitude higher than that used for television, radio, and telephonic communications. For example if the available bandwidth of laser light were to be 0.1% of the carrier frequency it would be possible in principle to carry of the order of 6×10^4 television channels or 10^8 telephone channels, which is probably more than enough to cater for the whole of present world-wide needs! The bandwidths so far achieved are much smaller than this, but are still large enough to make the laser a potentially very important item of communications technology. The basic components of a simple type of laser communication system are shown in Fig. 7.12.

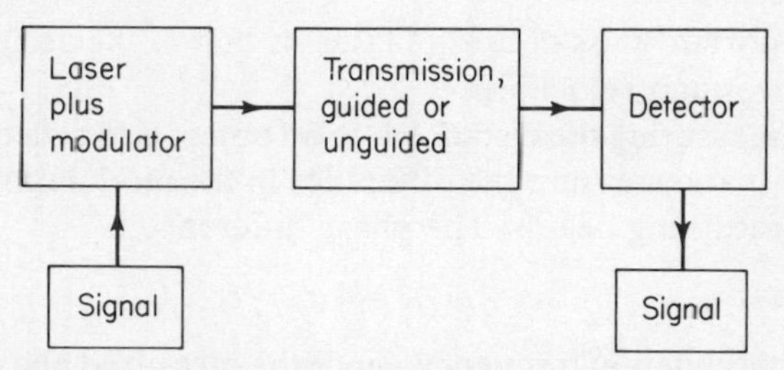

Fig. 7.12. The basic components of a simple laser communications system.

One obvious way of modulating the amplitude of a continuous laser beam is by modulating the pumping power supplied to the laser. The frequency that can be achieved is limited by the rate at which the pumping power level can be changed, by the speed of the pumping cycle, and by the lifetime of the upper laser level when the laser is operating. As we saw in Section 6.4.4, these factors are especially favourable for the semi-conductor diode lasers, which can be amplitude modulated at up to approximately 10^{10} Hz.

Another way of achieving amplitude modulation is to periodically interrupt or attenuate the laser beam by means of an electro-optic modulator (e.g. a Pockels cell, see Section 3.2.4). The modulator can be placed either inside or outside the laser cavity. When pulse code modulation is required, a continuous train of narrow pulses is produced by one of the techniques described in Section 7.1.4, and the pulses are externally interrupted by passing the train through a switched electro-optic cell.

Transmission of the modulated laser beam can be either through the earth's atmosphere (unguided transmission) or through a light-pipe or optical fibre (guided transmission see Section 2.5). The unguided method is the simpler and is often used for short distances ($\sim$few km), but it suffers from the vagaries of the weather (rain, fog, and snow give large attenuations). Even in clear weather water vapour and carbon dioxide cause absorption, and thermal inhomogeneities cause distortions and changes in direction of the beam. For reliable communication it is necessary therefore to use guided transmission, and to this end there has been much development work on fibre wave guides, particularly in connection with telephonic communication. These guides are described briefly in Section 2.5. They consist typically of cylindrical quartz fibres of diameter approximately 50 μm, and are normally used in bundles. The attenuation of the fibres can be <1 dB/km, allowing transmission over tens of km, and even further if amplification stages are used.

7.2.5 Measurement of distance and velocity

Another important class of uses of lasers is that of the accurate measurement of distances and velocities.

One way of measuring the distance L to an object is to reflect a modulated laser beam off it and measure the difference in the modulation phase of the outgoing and returning beams. The phase difference is

$$\Delta\phi = \omega\,\Delta t = 4\pi L\nu/c,$$

where ν is the modulation frequency. Since the measured phase difference is the angle remaining after an unknown integral multiple of 2π has been subtracted, several different modulation frequencies must be used in order to determine the distance unambiguously. Commercial geodolites based on this method can give accuracies of approximately 1 mm per km.

An alternative method is to use pulsed transmission and to measure the time-of-flight to the object and back. This is known as the *pulse echo technique.* In the case of the earth–moon distance the flight time is 2.6 s, and by measuring this to an accuracy of 0.5 ns the distance has been established to within ±15 cm.

For the measurement of the velocity of a distant object the Doppler shift of the returning reflected beam can be utilized. The beat frequency between the outgoing and returning beams gives the value of the shift. This method is sometimes called *optical Doppler radar.* For an object moving with the velocity component v in the direction of observation, the Doppler shift, and hence the beat frequency, is

$$\Delta\nu = 2\frac{v}{c}\nu = \frac{2v}{\lambda}.$$

For a wavelength of 1 μm and a velocity of 1 mm/s the resultant frequency is 2 kHz, which can easily be measured. This provides a powerful means of studying fluid flow and particle diffusion. It can be used for example to measure the flow of blood in arteries in the eye.

7.2.6 Heating and cutting

As a final example of the use of laser light we consider the transfer of electromagnetic energy to produce localized heating in an object. The unique importance of the laser for this purpose can be gauged from the fact that the most powerful laser systems produce peak pulse powers ($\gtrsim 1$ TW) that are comparable with the mean world-wide consumption of electrical power (but only for very brief time intervals of course!).

Figure 7.13 shows a laser beam being focussed to a small spot of radius r by a lens of focal length f. If the laser produces a uniphase beam with a Gaussian intensity profile (see Section 6.3) then its divergence semi-angle θ is related to the radius w_0 at its waist by (see Eq. (6.36))

$$\theta = \frac{\lambda}{\pi w_0} \approx \frac{\lambda}{\pi R}.$$

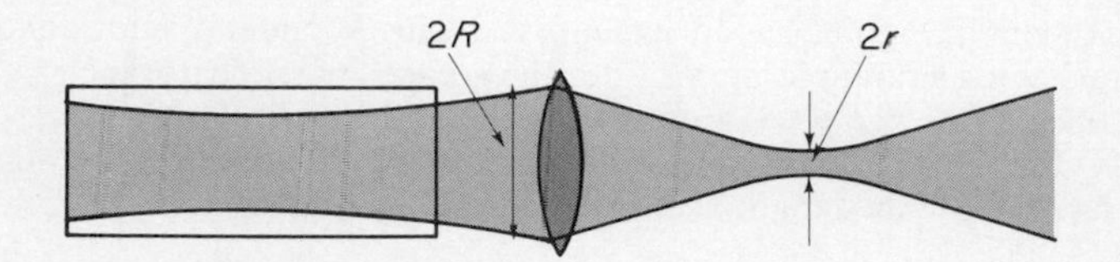

Fig. 7.13. A laser beam of radius R is focussed by a lens to a small spot of radius r.

The lens therefore produces (neglecting aberrations) a spot of radius

$$r = f\theta \approx \frac{f\lambda}{\pi R}. \tag{7.23}$$

This radius is very small, of the order of the laser wavelength. for a laser operating in transverse modes we can use the equation

$$\theta = \frac{0.61\lambda}{R}$$

for the diffraction limited divergence of a beam, together with the relationship $r = f\theta$, to obtain

$$r = \frac{0.61\lambda f}{R}, \tag{7.24}$$

a somewhat larger value than that given by Eq. (7.23). Using this larger value we find that the power density at the focussed spot is

$$E = \frac{P}{\pi r^2} \sim \frac{PR^2}{\lambda^2 f^2}\ \mathrm{W/m^2},$$

where P is the power produced by the laser. For example a CO_2 laser operating at a power level of 500 W, with $f/R \sim 5$, produces a spot of radius approximately 50 μm (allowing for lens aberrations and laser multi-moding) and gives a power density $\sim 100\,\text{kW/mm}^2$, sufficient for most welding, cutting, and drilling operations.

Even higher power densities, perhaps sufficient for the initiation of nuclear fusion reactions, can be obtained by passing the laser beam through laser amplifying stages (i.e. stages containing population inverted media in which the beam is amplified by stimulated emission), and by focussing several such beams at the same spot.

PROBLEMS: CHAPTER 7

(Answers to the problems are given in Appendix E.)

7.1 The exit lens of Fig. 7.2 has an aperture of radius R, and it is required to produce a beam having a radius less than R for the greatest possible distance z. Show that the radius at the beam waist must be made equal to $R/\sqrt{2}$.

7.2 Verify Eq. (7.17) by vectorially summing the frequency-doubled wavelets generated along the length L of the medium.

CHAPTER 8

The scattering and absorption of electromagnetic radiation

In the discussion of Chapter 3 all the scattering and absorption processes that result in the loss of photons from a beam passing through a medium were lumped together and described as 'absorption', and we saw that the strength and frequency dependence of this absorption essentially determine the refractive index of the medium. At that stage it was not necessary to enquire into what happens to the scattered or absorbed energy. In Chapter 6 we went one stage further and discussed the modes of de-excitation of atoms which have absorbed a photon, since this is relevant to the pumping mechanisms of lasers. In fact there are many other aspects to the scattering and absorption of photons by matter, and the more important of these are the subject of the present chapter. We shall concentrate on the essential physical mechanisms of the scattering and absorption processes, making use of the results obtained in previous chapters, where these are helpful. An understanding of most of the processes to be discussed is necessary for the next chapter, on the detection of radiation.

We start in the first main section (8.1) of the chapter with the simplest type of scatterer, the free electron. When the photon energy $h\nu$ is sufficiently small (much smaller than the electron rest energy m_0c^2) the only scattering process that can occur for this scatterer is elastic scattering, giving changes in the direction, but not the frequency, of the scattered photons. We shall see that the physical mechanisms and the cross-section for this scattering process, called Thomson scattering, can be obtained from a straightforward classical treatment. At higher energies the scattering is usually called Compton scattering. Here we are able to gain physical insight by an entirely different approach, treating the photon–electron interaction as a billiard-ball type of collision between point particles. This yields the Compton scattering formula for the change of wavelength of the scattered photon. We finish the section by discussing the more general approach to photon–electron scattering, of which Thomson and Compton scattering are particular examples.

In the middle part of the chapter (Section 8.2) we deal with the scattering and absorption of photons by atoms and molecules. When the scattering is elastic, and hence the internal energy of the atom is not altered, the process is called Rayleigh scattering. Although the atom is not excited we shall see that the existence of excited states is nevertheless important in determining the scattering mechanism and scattering cross-section. Three essentially different physical mechanisms apply when the photon energy is respectively much smaller than, approximately equal to, and much greater than an internal excitation energy of the atom. We shall see also that there is a strong connection between the cross-section for Rayleigh scattering and the cross-section for absorption (which has already been briefly discussed in Chapters 3 and 4). In the region of an isolated atomic state these cross-sections are proportional to each other: they have the same Lorentzian dependence on energy, and their absolute magnitudes are characterized by quantities called the partial widths of the scattering and absorption processes. This connection enables us to unify the expressions for the cross-sections of the various types of scattering and absorption processes.

Section 8.2 continues (Subsections 8.2.2 and 8.2.3) with a discussion of two processes in which the internal energy of the atom or molecule is altered by the interaction with a photon. The first of these is Raman scattering, in which the vibrational energy of a molecule is increased or decreased by collision with a photon. In the second process, the photo-electric effect, the energy given to the atom or molecule is sufficient to cause the ejection of an electron. We finish the section (Subsection 8.2.4) with a discussion of the process in which the energy of the photon is used to create an electron–positron pair.

In Section 8.3 the treatment of scattering and absorption processes is completed by considering the physical mechanisms that can occur only in condensed media. We start by discussing the way in which low frequency radiation is scattered by the density fluctuations in a medium, created for example when an acoustic wave travels through a liquid. This scattering process is known as Brillouin scattering. At higher frequencies energy can be given to the individual electrons of the medium, giving rise to the photoconductivity and photoemission processes. We finish the section (and the chapter) by discussing the physical mechanisms of these processes, in conductors and semi-conductors.

8.1 SCATTERING OF RADIATION BY FREE ELECTRONS

As mentioned above, we start by separately considering the two cases $h\nu \ll m_0c^2$ and $h\nu \not\ll m_0c^2$ (where m_0c^2 is the electron rest mass energy). For each of these a simple physical model of the scattering process is possible. The first model gives the scattering cross-section as a function of the photon energy and the angle of scattering, while the second gives the change in energy of the scattered photon. After that we consider the exact, general treatment, which gives the scattering cross-section at all photon energies.

8.1.1 Thomson scattering

If we treat the radiation passing through a medium of free electrons classically, and express the electric field as

$$\mathbf{E} = \mathbf{i}_x E_0 \cos(\omega t - kz),$$

we see that an electron with z-coordinate z_0, mass m_0 and charge $-e$ has the equation of motion

$$m_0 \frac{\mathrm{d}^2 x}{\mathrm{d}t^2} = -eE_0 \cos(\omega t - kz_0), \tag{8.1}$$

in the absence of any damping forces. The solution of this equation is

$$x = x_0 \cos(\omega t - kz_0),$$

where

$$x_0 = \frac{eE_0}{m_0\omega^2}.$$

The electron therefore acts as a Hertzian dipole, and it re-radiates at the same angular frequency ω, with a mean power given by the Larmor formula (3.38),

$$P_{\text{rad}} = \frac{e^2}{6\pi\varepsilon_0 c^3}\overline{\left(\frac{\mathrm{d}^2 x}{\mathrm{d}t^2}\right)} = \frac{e^4 E_0^2}{12\pi\varepsilon_0 m_0^2 c^3}. \tag{8.2}$$

This power can come only from the incident radiation, and so the intensity of the incident beam is reduced in passing through the medium. (Another way of finding Eq. (8.2) is to include in Eq. (8.1) the damping term given by the radiative reaction (3.48), and to calculate the rate at which work is done on the electron by the incident radiation.)

The net effect is therefore that some part of the incident radiation (initially travelling in the z direction) reappears as radiation of the same frequency, but travelling in a different direction. This type of elastic scattering is called *Thomson scattering*. The power per unit area in the incident beam is (Eq. (3.41))

$$P_{\text{beam}} = \tfrac{1}{2}\varepsilon_0 c E_0^2,$$

and therefore the cross-section per electron for the re-radiation process (see Sections 3.2.2 and 4.2 for a definition of cross-section) is

$$\sigma_T = \frac{P_{rad}}{P_{beam}} = \frac{8\pi r_0^2}{3} = 6.652 \times 10^{-29}\ \text{m}^2, \tag{8.3}$$

where

$$r_0 = \frac{e^2}{4\pi\varepsilon_0 m_0 c^2} = 2.818 \times 10^{-15}\ \text{m}. \tag{8.4}$$

The radius r_0 is called the classical electron radius, and σ_T is called the *Thomson cross-section* for elastic scattering.

To obtain information about the distribution in direction of the scattered radiation, we consider the direction Θ shown in Fig. 8.1. The Hertzian dipole

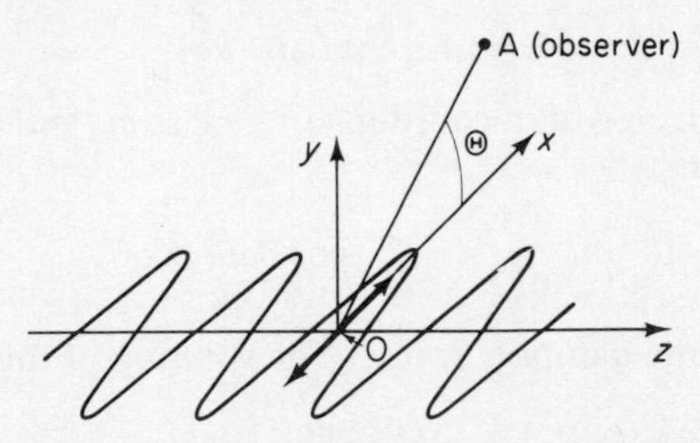

Fig. 8.1. The plane wave, linearly polarized in the x-direction, acts on the charged particle and causes it to oscillate as a Hertzian dipole. Θ is the angle between the direction of the dipole and the direction of an observer.

formed by the oscillating electron lies in the direction of polarization of the incident beam, which we take to be the x direction, and Θ is the angle between the direction of the dipole and the direction of an observer. It was shown in Section 3.1.2 that the plane of polarization of the radiation is the plane containing the dipole direction, and that the mean power radiated at the angle Θ is proportional to $\sin^2\Theta$. To be quantitative, the mean power radiated into a solid angle $d\Omega$ at the angle Θ is

$$P_{rad}(\Theta, d\Omega) = \frac{e^4 E_0^2}{32\pi^2 \varepsilon_0 m_0^2 c^3} \sin^2\Theta\, d\Omega.$$

By dividing this by P_{beam} we obtain the partial cross-section for scattering into the solid angle $d\Omega$ only,

$$\sigma(\Theta, d\Omega) = r_0^2 \sin^2\Theta\, d\Omega.$$

By dividing further by $\mathrm{d}\Omega$ we now obtain the partial cross-section per unit solid angle in the direction Θ. This is called the *differential cross-section* for the scattering process, and is represented by

$$\frac{\mathrm{d}\sigma(\Theta)}{\mathrm{d}\Omega} = r_0^2 \sin^2 \Theta. \tag{8.5}$$

To express the differential cross-section in terms of the spherical polar angles (θ, ϕ) of the direction OA, we note that

$$\cos \Theta = \sin \theta \cos \phi,$$

which gives

$$\frac{\mathrm{d}\sigma(\theta, \phi)}{\mathrm{d}\Omega} = r_0^2 (1 - \sin^2 \theta \cos^2 \phi). \tag{8.6}$$

When the incident beam is unpolarized the differential cross-section is effectively averaged over the direction ϕ, giving

$$\frac{\mathrm{d}\sigma(\theta)}{\mathrm{d}\Omega} = \frac{1}{2\pi} \int_0^{2\pi} r_0^2 (1 - \sin^2 \theta \cos^2 \phi)\, \mathrm{d}\phi = \tfrac{1}{2} r_0^2 (1 + \cos^2 \theta), \tag{8.7}$$

We see from Eq. (8.3) that the Thomson cross-section is extremely small, being about eight orders of magnitude smaller than the area of an atom ($\sim 10^{-20}$ m^2). There is nevertheless one well-known instance in which this type of scattering can be seen with the naked eye. This occurs during a total eclipse of the sun. The light from the sun then reaches the earth only after being Thomson scattered by the electrons in the sun's corona, a region of extremely high temperature ($\sim 10^6$ K) extending far beyond the sun's surface. The corona then appears as a bright ring surrounding the eclipsed sun.

For the classical treatment of Thomson scattering to be valid the length x_0 of the oscillating dipole must be much smaller than the wavelength λ of the scattered radiation, or else the dipole cannot be considered as a Hertzian radiator. This condition can be expressed as

$$\frac{x_0}{\lambda} = \frac{eE_0}{m_0 \omega^2 \lambda} \ll 1. \tag{8.8}$$

Another condition is that the energy of the dipole must be much larger than the energy $\hbar\omega$ of the photons that comprise the field, thus allowing us to treat the re-radiation as a continuous process. This second condition is

$$\frac{\tfrac{1}{2} m_0 \left(\frac{\mathrm{d}x}{\mathrm{d}t}\right)^2_{\max}}{\hbar\omega} = \frac{e^2 E_0^2}{2 m_0 \hbar \omega^3} \gg 1. \tag{8.9}$$

Combining the two conditions, we find that

$$2m_0\hbar\omega^3 \ll e^2E_0^2 \ll (m_0\omega^2\lambda)^2,$$

or equivalently,

$$\hbar\omega \ll \frac{e^2E_0^2}{2m_0\omega^2} \ll 2\pi^2 m_0c^2. \tag{8.10}$$

We see that if

$$\hbar\omega \ll m_0c^2 \tag{8.11}$$

a value of E_0, say E_0', can be found to satisfy (8.10). Since the cross-section is independent of E_0 the value deduced from the classical treatment for the field E_0' is valid for all fields. Therefore the only important condition for the validity of Eqs. (8.6) and (8.7) is that E_0' should exist, i.e. that condition (8.11) should hold, as we assumed at the beginning of the section.

8.1.2 Compton scattering

When condition (8.11) is not satisfied it is more appropriate to consider the scattering process in terms of individual photon–electron collisions. This is illustrated in Fig. 8.2. The electron is initially at rest, with a rest energy m_0c^2, but as a result of the collision it acquires energy and momentum from the photon, and moves away with velocity v, mass

$$m = \frac{m_0}{(1-v^2/c^2)^{1/2}},$$

and total energy

$$E = mc^2.$$

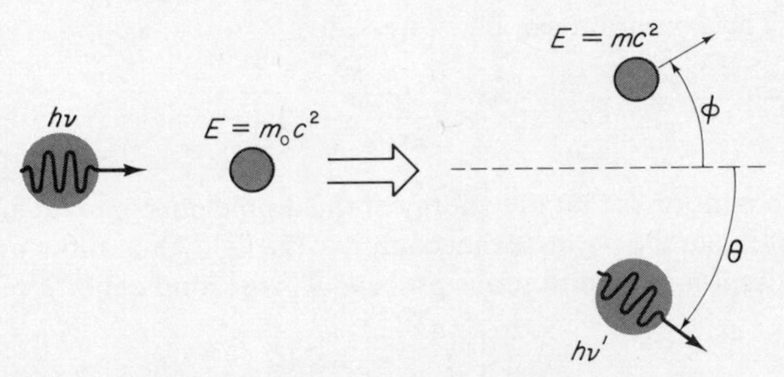

Fig. 8.2. Compton scattering of a photon by an electron. After scattering the photon has a reduced frequency ν'.

Its momentum is p_e, where

$$E^2 = m_0^2c^4 + p_e^2c^2, \tag{8.12}$$

Since the electron is free and initially at rest, conservation of total energy requires that the recoil kinetic energy T_e of the electron be given by

$$T_e = (m - m_0)c^2 = h\nu - h\nu', \tag{8.13}$$

where $h\nu$ and $h\nu'$ are respectively the initial and final energies of the photon. Conservation of linear momentum in the direction of the incident photon gives the relationship

$$\frac{h\nu}{c} = \frac{h\nu'}{c}\cos\theta + p_e\cos\phi, \tag{8.14}$$

while conservation of linear momentum in the transverse direction requires that the outgoing electron and photon move in the same plane, and that

$$\frac{h\nu'}{c}\sin\theta = p_e\sin\phi, \tag{8.15}$$

By combining Eqs. (8.12) and (8.13) we find that

$$p_e^2 = \left(\frac{h\nu}{c}\right)^2 - 2\left(\frac{h\nu}{c}\right)\left(\frac{h\nu'}{c}\right) + \left(\frac{h\nu'}{c}\right)^2 + 2m_0c\frac{h(\nu-\nu')}{c}.$$

Similarly from Eqs. (8.14) and (8.15) we obtain

$$p_e^2 = \left(\frac{h\nu}{c}\right)^2 + \left(\frac{h\nu'}{c}\right)^2 - 2\left(\frac{h\nu}{c}\right)\left(\frac{h\nu'}{c}\right)\cos\theta.$$

These two equations now yield

$$\frac{\nu'}{\nu} = \frac{1}{1+\gamma(1-\cos\theta)}, \tag{8.16}$$

where

$$\gamma = \frac{h\nu}{m_0c^2}. \tag{8.17}$$

Equation (8.16) can also be expressed in terms of wavelengths, when it becomes the *Compton scattering formula*

$$\lambda' - \lambda = \frac{h}{m_0c}(1-\cos\theta). \tag{8.18}$$

The quantity

$$\lambda_c = \frac{h}{m_0c} = 2.426 \times 10^{-12}\ \text{m} \tag{8.19}$$

is known as the *Compton wavelength* of the electron.

The dependence on θ of the kinetic energy of the recoil electron follows from Eqs. (8.13) and (8.16) namely

$$T_e = h\nu \frac{\gamma(1-\cos\theta)}{1+\gamma(1-\cos\theta)}. \tag{8.20}$$

This dependence is shown in Fig. 8.3 for a few values of $h\nu$. We see that T_e has nearly its maximum value

$$(T_e)_{max} = h\nu \frac{2\gamma}{1+2\gamma} \tag{8.21}$$

over a wide range of angles in the backward direction, especially for high values of $h\nu/m_0c^2$. This is of special significance for the detection of γ-rays (see Section 9.3).

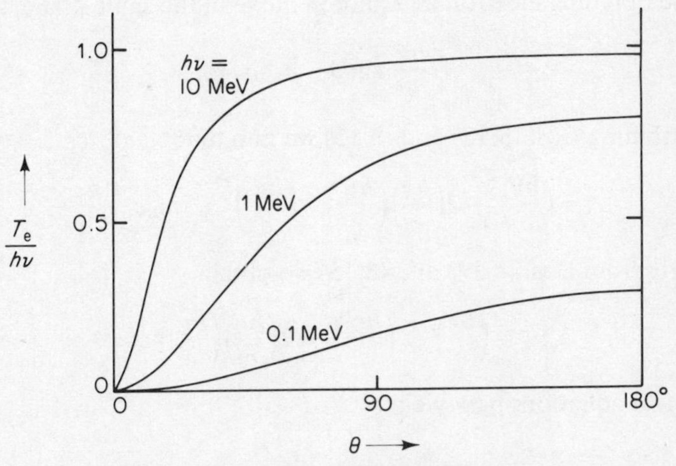

Fig. 8.3. Dependence of the electron recoil energy T_e on the angle of scattering of the photon, for three values of the initial photon energy $h\nu$.

The early experiments

Changes in the wavelength of scattered X-rays had first been noticed experimentally by Sadler and Mesham in 1912. In this and later work it was found that when an approximately monochromatic beam of X-rays is passed through a solid or gas some of the scattered X-rays have the *same* wavelength as the incident beam (and so have been elastically scattered by the atoms of the medium, see Section 8.2) while others have an *increased* wavelength, approximately as given by Eq. (8.18). In 1922 Compton, who had already carried out many experiments and proposed many different interpretations, made the new suggestion that some of the electrons of the scattering medium are essentially free as far as the scattering of the incident X-rays is concerned, and that energy and momentum are conserved in a photon–electron scattering process, so giving the correct interpretation for the scattered X-rays of increased wavelength. This was an important landmark in the development of the photon concept (see Chapter 1).

An experiment which demonstrated vividly the reality of the photon was performed three years later by Compton and Simon. They passed a beam of X-rays through a Wilson cloud-chamber and were able to see electron tracks originating from two types of events, namely Compton scattering events in which recoil electrons are produced by scattering on the nearly-free outermost electrons of the molecules in the chamber, and photo-electric events in which X-rays eject inner electrons. An individual X-ray may produce both types of event within the cloud-chamber, first being Compton scattered and then travelling a further distance before initiating a photoelectric event, as illustrated in Fig. 8.4. By identifying such double events Compton and Simon were able to measure θ and ϕ and thus verify that energy and momentum are conserved.

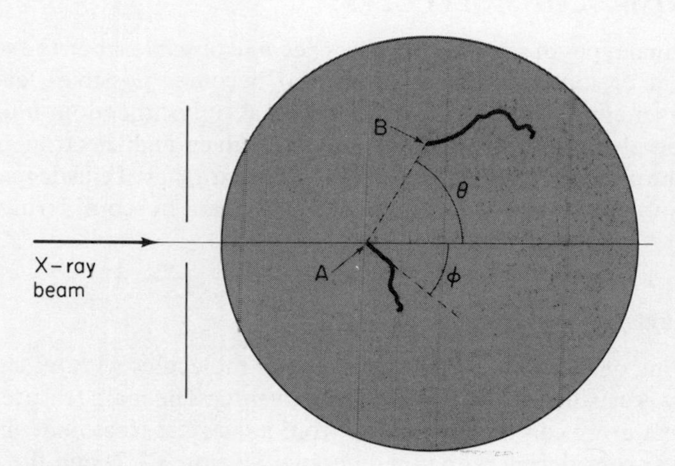

Fig. 8.4. The experiment of Compton and Simon. An X-ray photon that has entered the cloud chamber from the left undergoes Compton scattering at the point A, producing an electron that initially travels at the angle ϕ, before being multiply scattered and slowed down by collisions with the molecules in the chamber. The scattered photon travels to the position B where it ejects a photo-electron. By connecting the points A and B on the cloud-chamber photograph the angle of scattering θ can be deduced.

The Klein–Nishina formula

Although Eqs. (8.16) to (8.20) give the dependence of $h\nu'$ and T_e on the scattering angle, the method we have used gives no information about the total or differential cross-sections. To do this requires a full quantum-mechanical treatment of the process, which is beyond the scope of this book. The result is that the differential cross-section for the scattering of unpolarized light is

$$\frac{d\sigma(\theta)}{d\Omega} = \frac{1}{2} r_0^2 \frac{\nu'^2}{\nu^2}\left(\frac{\nu}{\nu'} + \frac{\nu'}{\nu} - \sin^2\theta\right). \tag{8.22}$$

This is known as the *Klein–Nishina formula*. In the limit of small photon energies ($\gamma \ll 1$) it gives, as it must, the Thomson formula (8.7). When γ is not small the differential cross-section still has the Thomson value in the forward direction, but at larger angles it falls away increasingly quickly as γ increases, as can be seen in Fig. 8.5(a). The total cross-section therefore also decreases as γ increases, as shown in part (b) of the figure.

8.2 SCATTERING AND ABSORPTION OF RADIATION BY ATOMS AND MOLECULES

Additional types of scattering process become possible when the scatterer is an atom or molecule. Absorption also becomes possible, leading to internal excitation of the atom, or to the break-up of the atom into an ion and a photoelectron. If the photon energy is high enough electron–positron pairs can be created. These processes, together with that of elastic scattering, are the subject of the present section. We start by considering elastic scattering.

8.2.1 Rayleigh scattering

Scattering of radiation by neutral atoms or molecules without change of frequency is usually known as *Rayleigh scattering*. The main features of this type of scattering can be understood from a classical treatment similar to that used in considering radiation damping (Section 3.1.7) and the absorption and refraction of radiation (Section 3.2.2).

For simplicity we start by considering a model in which the atom or molecule contains a single electron which can oscillate at the natural frequency ω_i. In the presence of the electric field $\mathbf{i}_x E_0 \exp(j\omega t)$ the equation of motion of the electron is given by Eq. (3.67), and its displacement by Eq. (3.68), namely

$$x = \frac{eE_0}{m_0} \frac{e^{j\omega t}}{(\omega - \omega_i)^2 - j\omega\gamma}. \tag{8.23}$$

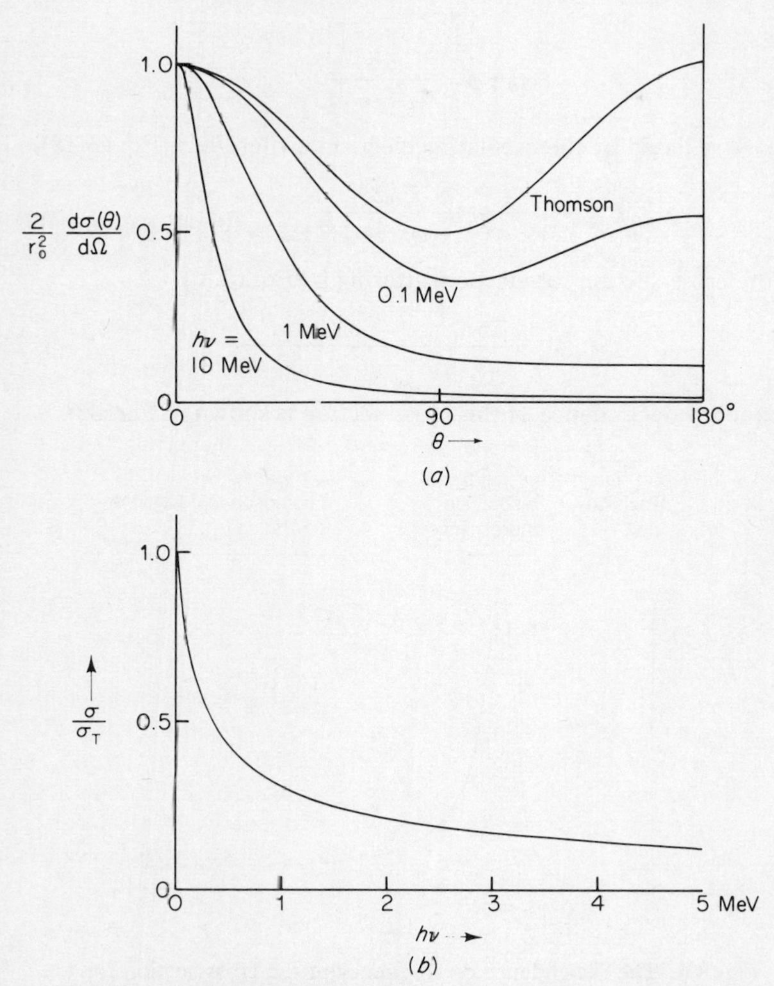

Fig. 8.5. (*a*) Differential cross-section and (*b*) total cross-section for Compton scattering of X-rays and γ-rays by free electrons. σ_T is the Thomson cross-section (8.3).

Here γ is the decay constant of the electron motion. It is the inverse (see Eq. (3.53)) of the lifetime τ of the atomic state which corresponds to the frequency ω_i (and which therefore has the excitation energy $\hbar\omega_i$),

$$\gamma = \frac{1}{\tau}. \tag{8.24}$$

The real part of x is

$$x_r = \frac{eE_0}{m_0}\frac{\cos(\omega t + \phi)}{[(\omega^2 - \omega_i^2)^2 + \omega^2\gamma^2]^{1/2}},$$

where

$$\tan\phi = \frac{\omega\gamma}{(\omega^2-\omega_i^2)}.$$

The power radiated by the oscillating electron is therefore (Eq. (3.38))

$$P = \frac{e^4 E_0^2 \omega^2}{12\pi\varepsilon_0 m_0^2 c^3[(\omega^2-\omega_i^2)^2+\omega^2\gamma^2]},$$

and so the cross-section for elastic scattering is (Eq. (8.3))

$$\sigma_{sc}(\omega) = \frac{8\pi r_0^2}{3}\frac{\omega^4}{(\omega^2-\omega_i^2)^2+\omega^2\gamma^2}. \quad (8.25)$$

The frequency dependence of this cross-section is shown in Fig. 8.6.

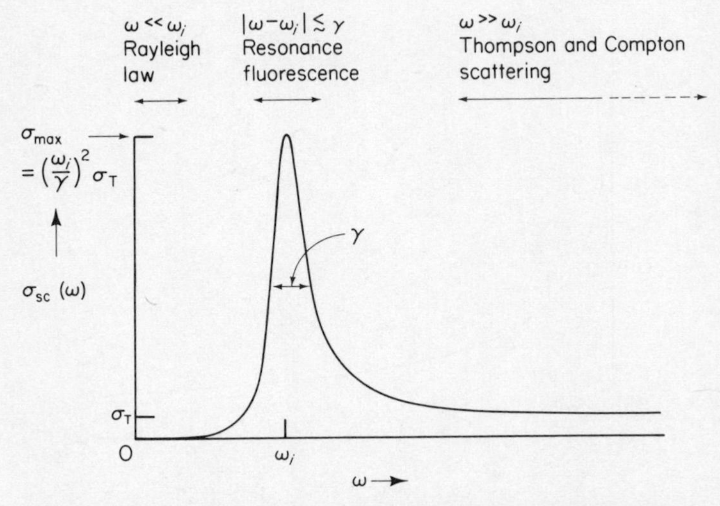

Fig. 8.6. The dependence on frequency of the cross-section for elastic scattering of radiation by a one-electron atom having the single resonance frequency ω_i. For clarity the ratio γ/ω_i has been greatly exaggerated.

The Rayleigh law of scattering

In the region of low frequency, for which $\omega \ll \omega_i$, the terms in ω^4 and $\omega^2\omega_i^2$ in the denominator of Eq. (8.25) can be ignored in comparison with the term ω_i^4, as can the term $\omega^2\gamma^2$ (because $\gamma \ll \omega_i$), and so the cross-section becomes

$$\sigma_{sc}(\omega) = \frac{8\pi r_0^2}{3}\left(\frac{\omega}{\omega_i}\right)^4 = \frac{8\pi r_0^2}{3}\left(\frac{\lambda_i}{\lambda}\right)^4, \qquad (\lambda \gg \lambda_i) \quad (8.26)$$

This is the *Rayleigh law of scattering*. It applies for example to the scattering of visible light by the earth's atmosphere, since the relevant frequencies ω_i of the N_2 and O_2 molecules lie in the ultra-violet region. Blue light has a wavelength approximately half that of red light, and therefore its scattering cross-section is approximately 16 times larger, causing the sky to be blue. The sky becomes more and more violet (and also darker) as the observer's height increases, because then only the shortest wavelength visible photons have a high enough probability of being scattered towards the observer.

Resonance fluorescence

The second region indicated on Fig. 8.6 is the resonance region, for which $|\omega - \omega_i| \lesssim \gamma$. Still assuming that $\gamma \ll \omega_i$, we can put

$$(\omega^2 - \omega_i^2)^2 \simeq (\omega - \omega_i)^2 (2\omega_i)^2,$$

and hence the scattering cross-section in this region can be expressed as

$$\sigma_{sc}(\omega) = \frac{8\pi r_0^2}{3} \frac{(\omega_i/2)^2}{(\omega - \omega_i)^2 + (\gamma/2)^2}. \tag{8.27}$$

The peak value is

$$\sigma_{sc,max} = \frac{8\pi r_0^2}{3} \left(\frac{\omega_i}{\gamma}\right)^2 = \sigma_T \left(\frac{\omega_i}{\gamma}\right)^2.$$

The magnitude of this can be estimated by using the classical value of γ (Eq. (3.50)), which leads to

$$\sigma_{sc,max} \simeq \frac{6\pi c^2}{\omega^2} = 6\pi \lambdabar^2. \tag{8.28}$$

In the visible region this gives a cross-section of about 4×10^{-14} m^2, which is 15 orders of magnitude greater than the Thomson cross-section, and is much larger even than the physical area of the scattering atom itself ($\sim 10^{-20}$ m^2). The cross-section has a Lorentzian shape about the maximum value, with the width γ (i.e. the full width at half maximum).

The essential physical reason for the large value of $\sigma_{sc,max}$ is that the atom is in resonance with the incident radiation at this frequency. Real atoms have many excitation energies $\hbar\omega_i$ for which the elastic scattering cross-section has this resonant form. The quantum-mechanical description (which we shall discuss in more detail later) of the resonant elastic scattering process is that a photon of energy $\hbar\omega_i$ is absorbed by the atom, and that the atom then fluoresces, re-emitting a photon of the same energy. For this reason the scattering is usually called *resonance fluorescence*.

Compton scattering

When ω is much greater than ω_i the electron of the atom is essentially free as far as the scattering process is concerned, and so the incident photon is Compton scattered by the electron and the cross-section approaches the value given by the Klein–Nishina formula (8.22). For a real atom $\hbar\omega$ must be much greater than the highest excitation energy of the atom (which ranges from 13.6 eV for hydrogen to over 100 keV for the heaviest atoms).

8.2.2 Inelastic scattering and absorption

A process which can compete with elastic scattering is inelastic scattering. Near a resonance frequency ω_i a photon can be absorbed by the atom and the atom can then decay back to the ground state, giving elastic scattering, or it can decay to other excited states (provided that $\hbar\omega_i$ is not the lowest excitation energy of the atom), giving a photon of lower energy. This inelastic process is called *reactive absorption*.

The transition probability for the decay of the state i to a lower state j with the emission of a photon of energy $\hbar\omega_{ji}$ is the Einstein coefficient for spontaneous emission, A_{ij} (Eq. (4.19)). The probability for the initial absorption from the ground state (which we now label 1) to the state i is proportional (at the resonance frequency) to the Einstein coefficient for absorption, B_{1i} (Eq. (4.18)), which is in turn proportional to A_{i1} (Eqs. (4.29), (4.30)). Therefore the reactive absorption cross-section for absorption at the resonance frequency ω_{1i}, followed by emission at the frequency ω_{ji}, is proportional to the product $A_{i1}A_{ij}$,

$$\sigma_r(\omega_{1i} \to \omega_{ji}) \propto A_{i1}A_{ij}.$$

Similarly, the elastic resonance cross-section is proportional to A_{i1}^2,

$$\sigma_{sc}(\omega_{1i} \to \omega_{1i}) \propto A_{i1}^2.$$

The cross-section for *all* events in which a photon is removed from the incident beam, regardless of the outcome of the absorption event, is therefore

$$\sigma_{abs}(\omega_{1i}) = \sigma_{sc}(\omega_{1i} \to \omega_{1i}) + \sum_{j=2}^{i-1} \sigma_r(\omega_{1i} \to \omega_{ji}) \propto A_{i1} \sum_{j=1}^{i-1} A_{ij}.$$

Now the transition probabilities A_{ij} are related to the decay constant γ that appears in the equations of the previous sub-section, because we see from Eqs. (4.30) and (8.24) that

$$\sum_{j=1}^{i-1} A_{ij} = \tau^{-1} = \gamma.$$

Because of this it is usual to represent the transition probabilities by different symbols: A_{ij} becomes γ_j, the *partial width* for the reactive absorption $(\omega_{1i} \to \omega_{ji})$; A_{i1} becomes γ_{sc}, the *elastic scattering width*; and the sum of the A_{ij} becomes γ_t, the *total width* for the resonance absorption. Then

$$\gamma_t = \gamma_{sc} + \sum_{j=2}^{i-1} \gamma_j, \tag{8.29}$$

and

$$\begin{aligned} \sigma_{rj} &\propto \gamma_{sc}\gamma_j, \\ \sigma_{sc} &\propto \gamma_{sc}^2, \\ \sigma_{abs} &\propto \gamma_{sc}\gamma_t. \end{aligned} \tag{8.30}$$

All these cross-sections have the same Lorentzian frequency dependence in the region of the resonance frequency, with the same half-width γ_t.

The various scattering processes can now be regarded as *transitions* from the initial state of the system (photon plus ground state atom in the present example), usually called the *entrance channel*, to the relevant final state, called the *exit channel*. The cross-section of a particular process is proportional to the product of the width of the entrance channel and the width of the exit channel. This is illustrated in Fig. 8.7.

The magnitudes of the cross-sections could be estimated from the classical result of the previous sub-section, but at this stage it is more informative to give the result of the quantum-mechanical calculations. For resonance fluorescence this is

$$\sigma_{sc}(\omega) = 2\pi r_0^2 f_{1i}^2 \frac{\omega_{1i}^2}{(\omega - \omega_{1i})^2 + (\gamma_t/2)^2}, \tag{8.31}$$

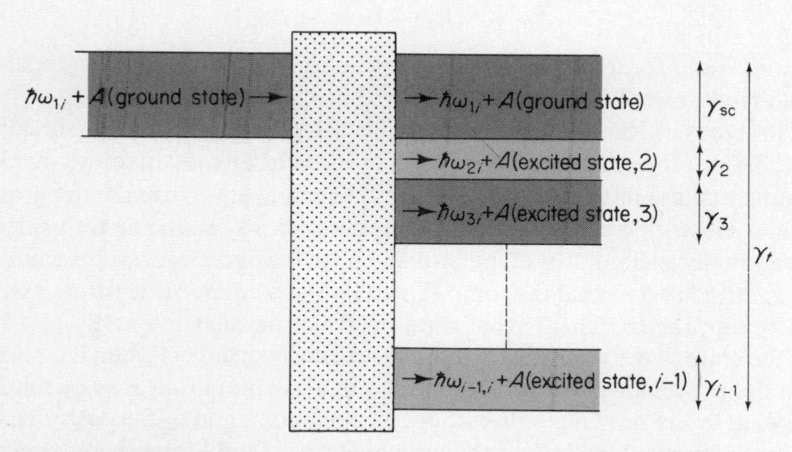

Fig. 8.7. Representation of a resonant or nearly-resonant scattering or reaction event as a transition between two channels, each characterized by a width γ. The cross-section is proportional to the product of the widths of the two channels.

where f_{1i} is the oscillator strength for the transition $1 \to i$. When γ_t is dominated by the scattering width γ_{sc} (for example, when i is the lowest optically accessible state of the atom), we find from Eq. (4.35) that

$$\gamma_{it} \simeq \gamma_{sc} = A_{i1} = \frac{e^2 \omega_{1i}^2 f_{1i}}{2\pi m \varepsilon_0 c^3},$$

and that the maximum value of the cross-section is then

$$\sigma_{sc,max} = 2\pi \lambda\!\!\!^{-2},$$

in approximate agreement with the classical estimate (8.28).

The result that the cross-section for a nearly resonant process is proportional to the product of the widths of the entrance and exit channels is completely general, and is independent of the nature of the projectiles in the entrance and exit channels. It applies, for example, when the entrance channel corresponds to photon absorption and the exit channel corresponds to a non-radiative decay, such as ionization or molecular dissociation. Each exit channel of this type has an associated partial width (when the channel is energetically allowed), and the summation in Eq. (8.29) must extend over these channels.

8.2.3 Raman scattering

In the previous sub-section we were concerned with nearly resonant inelastic scattering. Although the cross-sections for non-resonant inelastic processes are very much smaller they are still of interest. When the scattering involves photons in both the entrance and exit channels,

$$\hbar\omega + A(E_1) \to A(E_2) + \hbar\omega'$$

where E_1 and E_2 are different excitation energies of an atom or molecule, it is referred to as *Raman scattering*.

Two types of Raman scattering are illustrated in Fig. 8.8. In the first type (Fig. 8.8(a)) the scattered photon has a lower frequency than the incident photon, and the molecule is excited from its initial state (usually the ground state of electronic and vibrational excitation) to an excited state (usually a vibrationally excited state). The broken line in the figure represents what can be regarded as a virtual (i.e. non-existent) intermediate state formed in the scattering process. This virtual state need not be near in energy to a real excited state of the molecule, but the scattering is strongest when it is so, and then the process is more aptly described as absorption followed by fluorescence, as in the previous sub-section. The phenomenon of fluorescence was of course known long before the time of Raman, and Stokes had explained its origin and pointed out that in all the known examples the fluorescent light has a frequency *lower* than that of the exciting light.

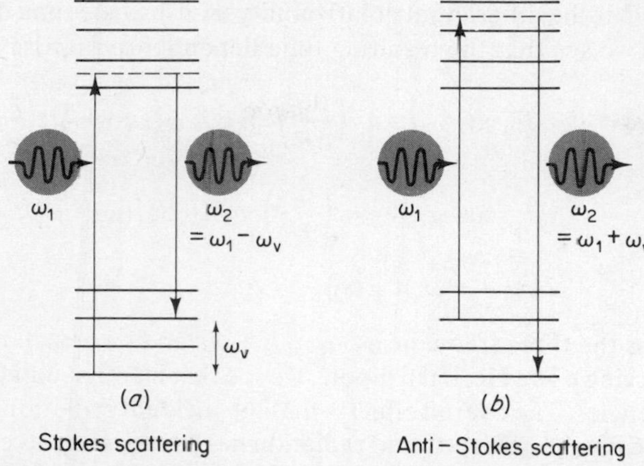

Fig. 8.8. Inelastic scattering involving the excitation or de-excitation of a molecular vibrational level of energy $\hbar\omega_v$.

Figure 8.8(*b*) shows the reverse process in which the molecule is initially in an excited state and the photon energy *increases* while the molecular excitation energy decreases. These two scattering processes can therefore be described as inelastic and super-elastic scattering respectively, although it is more usual to distinguish between them by calling the two cases *Stokes scattering* and *anti-Stokes scattering*, as shown in the figure. The probability that a molecule is initially in a vibrationally excited state is usually very small for a system of molecules in thermal equilibrium at temperature T (because the vibrational energy is typically much larger than kT), and therefore anti-Stokes scattering is usually much weaker than Stokes scattering.

To see how the cross-section for Raman scattering depends on the properties of the molecule let us consider the polarization $\mathbf{p}$ induced in a diatomic molecule by an electric field $\mathbf{E}$ (Fig. 2.9),

$$\mathbf{p} = \alpha\varepsilon_o\mathbf{E}.$$

Here α is the electric polarizability (see Section 2.1) of the molecule. This polarizability changes if the size of the molecule changes. Now when the molecule is vibrating at the frequency ω_v the inter-nuclear separation R has the time dependence

$$R(t) = R_0 + a \cos(\omega_v t + \phi),$$

where a is the amplitude of vibration (assumed to be small) about the equilibrium separation R_0, and ϕ is an unimportant phase factor. The time dependence of α is then

$$\alpha(t) = \alpha_0 + \left(\frac{d\alpha}{dR}\right)_0 a \cos(\omega_v t + \phi),$$

where $\mathrm{d}\alpha/\mathrm{d}R$ is the differential polarizability. If E has the time dependence $E_0 \cos(\omega_1 t)$ we see that the resulting time dependence of p is

$$p(t) = \varepsilon_0 \alpha_0 E_0 \cos \omega_1 t + \varepsilon_0 \left(\frac{\mathrm{d}\alpha}{\mathrm{d}R}\right)_0 a \cos \omega_1 t \cos(\omega_v t + \phi)$$

$$= \varepsilon_0 \alpha_0 E_0 \cos \omega_1 t + \tfrac{1}{2}\varepsilon_0 \left(\frac{\mathrm{d}\alpha}{\mathrm{d}R}\right)_0 a E_0 \{\cos[(\omega_1 + \omega_v)t + \phi]$$

$$+ \cos[(\omega_1 - \omega_v)t + \phi]\}.$$

This contains the three frequencies ω_1, $(\omega_1 + \omega_v)$ and $(\omega_1 - \omega_v)$. If we regard the time-varying **p** as a Hertzian dipole, we see that it emits radiation at these three frequencies. The overall effect is that the incident radiation is scattered either elastically (i.e. the scattered radiation has a frequency ω_2 equal to ω_1), or inelastically ($\omega_2 = \omega_1 - \omega_v$, Stokes scattering), or super-elastically ($\omega_2 = \omega_1 + \omega_v$, anti-Stokes scattering). We see also that Raman scattering can take place only if the differential polarizability of the molecule is non-zero.

The subject of Raman scattering received a considerable impetus when Woodbury and Ng discovered in 1964 that an exceptionally high intensity of Raman scattering (relative to the intensity of the incident light) can be produced when a laser beam is used as the incident light source. This results from the stimulated emission of the inelastic or super-elastic light, and the process is known as *stimulated Raman scattering*. The Stokes and anti-Stokes versions are illustrated in Fig. 8.9.

The Stokes and anti-Stokes scattering processes are stimulated by the presence in the exciting beam of photons having the frequency of the

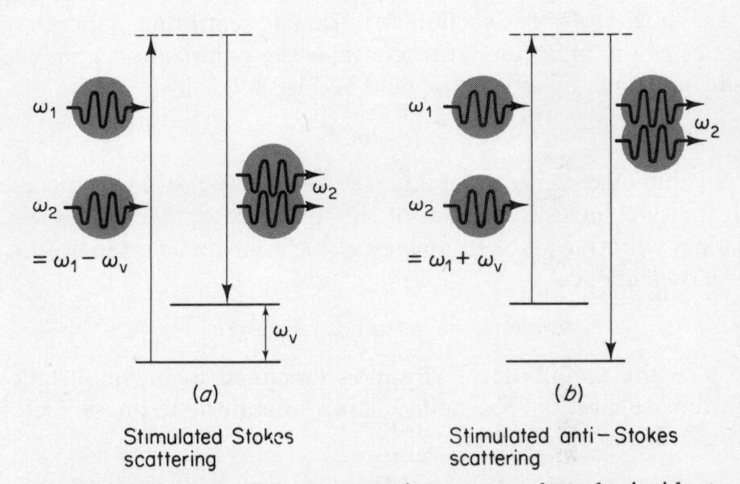

Fig. 8.9. Stimulated Raman scattering occurs when the incident light already contains the scattered frequency.

scattered light. As in the case of the ordinary stimulated emission process (see Section 4.2), the number of photons scattered into a normal mode of frequency ω_s is proportional to the number n_s of photons already in that mode. Unstimulated Raman scattering also occurs, giving a total rate proportional to (n_s+1). The incident beam initially contains only the one frequency ω_1, and the initial Raman scattering events are un-stimulated, but when the beam is very intense these initial events quickly lead to a coherent set of photons at the frequency ω_2, in analogy to the way in which laser action is triggered by spontaneous emission events. The process can be aided by multiple reflections of the scattered beam inside a cavity, as in laser amplification. The Raman scattered light has almost the same directionality, monochromaticity, and polarization as the incident laser beam, which makes the technique eminently suitable for the spectroscopic study of vibrational and rotational levels of molecules. It is particularly convenient to be able to study these levels using visible light, thus avoiding the difficulties involved in direct absorption and emission studies at the infra-red frequencies ω_v.

8.2.4 The photoelectric effect

When the photon energy exceeds the ionization energy of an electron in a free atom or molecule the photoelectric process

$$h\nu + A \rightarrow A^+ + e$$

can occur. This process is also called *photoionization*. The kinetic energy of the photoelectron is

$$T_e = h\nu - I \tag{8.32}$$

where I is the ionization energy. The values of I range from a few eV for atomic valence electrons to 116 keV for the K shell electrons of uranium. Some experimentally determined values are shown in Fig. 8.10.

The cross-section for photoionization from a particular shell or sub-shell is a maximum at the threshold energy (I) and decreases quickly at higher energies. This behaviour is illustrated in Fig. 8.11, which shows the total photoionization cross-section for argon. We see that at the K edge the cross-section increases by a factor of approximately 10, which is the typical order-of-magnitude rise at other edges. If this rise were determined only by the approximate physical sizes of the K and L shells (the K orbitals have an area about one tenth that of the L orbitals) and by the numbers of electrons in the shells (2 in the K shell and 8 in the L shell), the rise would be very much smaller.

We can understand the reason for the large increase in the cross-section at an ionization threshold by considering conservation of linear momentum in the photoejection process. The photoelectron tends to be ejected along the

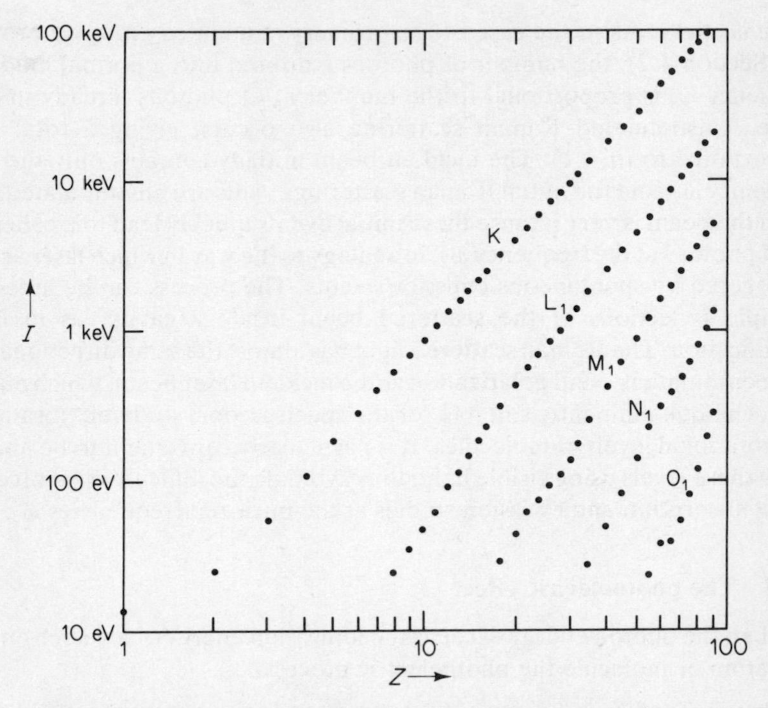

Fig. 8.10. Experimentally determined ionization energies for atomic electrons in the K, L_1, M_1, N_1, and O_1 shells (corresponding to the orbitals 1*s*, 2*s*, 3*s*, 4*s*, and 5*s* respectively).

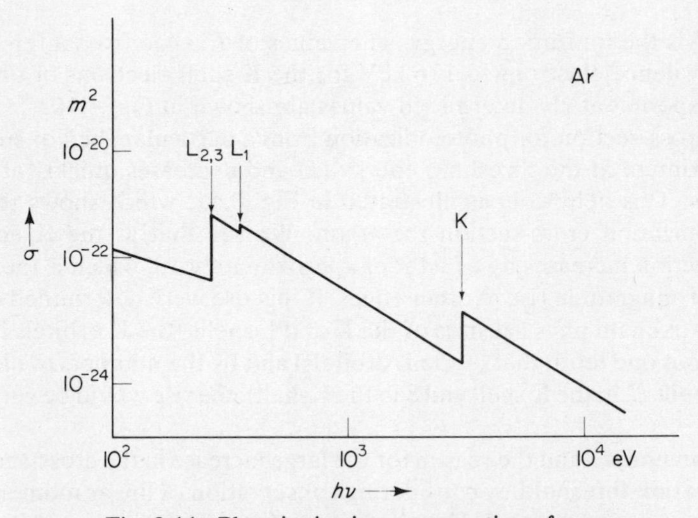

Fig. 8.11. Photoionization cross-section of argon.

polarization direction of the light (using the co-ordinates defined in Fig. 8.1, the differential cross-section is proportional to $\cos^2 \Theta$ at non-relativistic energies), with a momentum which tends to be much larger than that of the photon. The momentum difference is taken up by the residual ion, which moves in approximately the opposite direction to the ejected electron. Just above the K edge in argon this momentum difference is much smaller for K electron ejection than for L electron ejection. Furthermore, the K-shell electrons are initially nearer to the nucleus than the L-shell electrons, and so interact more strongly with it, thus facilitating the momentum exchange between the ejected electron and residual ion. The need to conserve momentum is also the essential physical reason for the decrease in the cross-sections above the ionization edges: the momentum that must be exchanged between the electron and ion increases continuously as the photon energy increases.

Experimental measurements (and quantum-mechanical calculations) show that for non-relativistic energies the total cross-section for ejection of a K-shell electron is given by

$$\sigma_K \simeq 3\times10^{-27} Z^2 \left(\frac{h\nu}{1\ \mathrm{keV}}\right)^{-7/2} \mathrm{m}^2, \tag{8.33}$$

where Z is the atomic number of the atom. The total cross-section for all the other atomic shells is approximately 5 times smaller than this, when $h\nu > I_K$.

Decay of the ion

The atomic ion created by a photoelectric event in which an inner-shell electron has been ejected can initially decay in one of two ways. The first of these is by the emission of a characteristic X-ray. For example, in an argon atom from which a K-shell electron has been ejected, the K-shell hole (which we denote by K^{-1}) can be filled by an electron from the L_1 sub-shell (or from one of the other sub-shells) with the emission of an X-ray,

$$\mathrm{Ar}(\mathrm{K}^{-1}) \rightarrow \mathrm{Ar}(\mathrm{L}_1^{-1}) + h\nu.$$

This process is called *X-ray fluorescence*. The transition rate is approximately proportional to $(h\nu)^2$ (see Eq. (4.35)). The other decay mode is by electron ejection. For example the K-shell hole of argon can be filled by an L_1 electron, with the simultaneous emission of another L_1 electron,

$$\mathrm{Ar}(\mathrm{K}^{-1}) \rightarrow \mathrm{Ar}(\mathrm{L}_1^{-2}) + e.$$

This is known as *Auger decay*. The two holes in the final ion need not be in the same sub-shell, nor even in the same shell. The transition rate for Auger decay depends only weakly on the energies of the initial and final hole states.

The total decay rate of the ion is of course the sum of the rates for the fluorescence and Auger processes. Figure 8.12 shows the probability that the decay occurs by fluorescence when the initial hole is in the K-shell. Each X-ray fluorescence or Auger decay is followed by one or more further decays, the entire process finishing only when the residual ion is in its ground state. The final charge state of the ion may be quite high: for example a K-shell hole in a Pb atom, which has 6 shells ($n = 1$ to 6), can lead to a final charge state greater than 10.

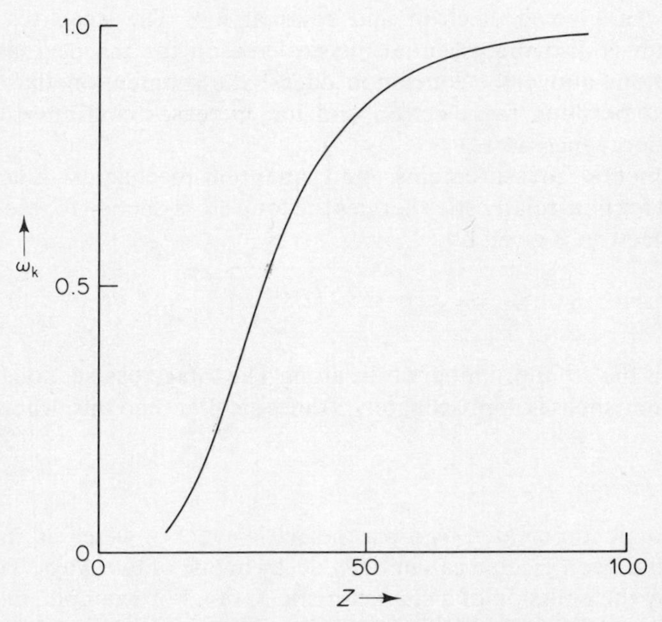

Fig. 8.12. The dependence on Z of the probability of X-ray fluorescence when the atom has a K-shell hole.

8.2.5 Pair production

The scattering and absorption processes that we have discussed have involved changes in the energy and direction of the incident photon, and the conversion of some or the whole of the photon energy to internal excitation energy of an atom and/or kinetic energy of an ejected electron. Another possibility is that the photon is annihilated and its energy used to create new particles.

Two necessary conditions for the creation of one or more particles by a photon are that the total charge of the created particles must be zero, and that the sum of their rest energies must be less than the photon energy. The created particles are invariably charged particles, since only then is the electromagnetic interaction large enough for the creation process. The lightest known charged particles are the electron (e^-) and its anti-particle the positron (e^+), each having a rest energy (m_0c^2) of 0.511 MeV. Therefore the particle creation process requiring the least energy is that in which a photon (more aptly called a γ-ray, because of its high energy) creates an electron and a positron,

$$\gamma \rightarrow e^- + e^+. \tag{8.34}$$

This process is called electron–positron *pair production.* The excess energy ($E_\gamma - 2m_0c^2$) appears as kinetic energy of the electron and positron.

Because the momentum of an electron or positron of total energy E (rest energy plus kinetic energy) is always less than E/c, the total momentum of the electron–positron pair is necessarily less than the momentum (E_γ/c) of the photon. It is therefore not possible to conserve energy *and* momentum unless another particle, or system of particles, is involved in the pair-production process, and takes up the difference in momentum between the γ-ray and the electron–positron pair. This function is usually performed by the nucleus of an atom, or by the whole atom.

The kinetic energy of the recoil nucleus is negligible compared with E_γ, and so energy conservation requires that

$$T_- + T_+ = E_\gamma - 2m_0c^2, \tag{8.35}$$

where T_- and T_+ are the kinetic energies of the electron and positron respectively. The threshold energy for pair-production is therefore

$$E_{\gamma,\text{thresh}} = 2m_0c^2 = 1.022\ \text{MeV}. \tag{8.36}$$

The particle production process having the next smallest threshold energy is electron–positron pair production in the field of a free electron,

$$\gamma + e^- \rightarrow e^- + e^- + e^+. \tag{8.37}$$

This is sometimes called the triplet production process (because three new tracks appear in bubble chamber photographs). Its threshold is $4m_0c^2 =$ 2.044 MeV (see Problem 8.2). The creation of particles heavier than the electron and positron becomes possible above a γ-ray energy of a few 100 MeV. (We shall not however discuss these higher energy processes.)

Cross-section for pair production

The techniques of quantum electrodynamics (QED), i.e. the full quantum-mechanical treatment of systems involving charged particles and photons, show that the total cross-section for electron–positron pair production in the field of a bare nucleus of charge Ze is proportional to Z^2, and that it increases as the γ-ray energy increases. The calculated value is shown in Fig. 8.13. When pair production occurs in the field of an atom the electrons of the atom contribute to the yield at energies above 2.044 MeV, although this contribution is always less than $1/Z$ of the nuclear pair production cross section. The atomic electrons also have the opposite effect of slightly reducing the nuclear pair-production cross section by partially shielding the charge of the nucleus.

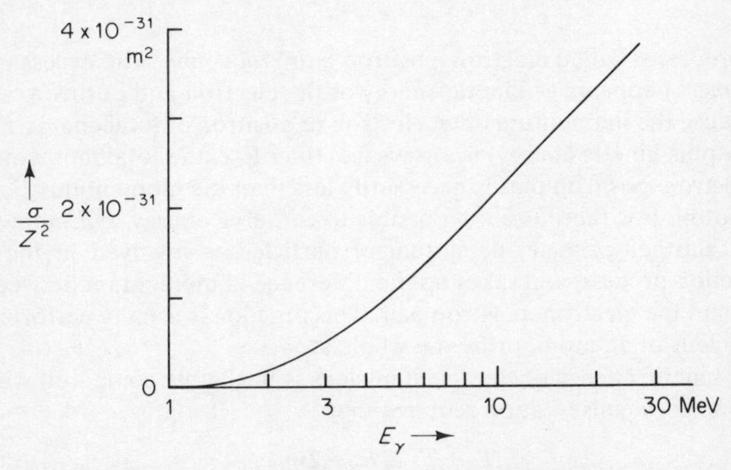

Fig. 8.13. Energy dependence of the cross section for pair production in the field of an unscreened nucleus of charge Z. (Adapted from Davisson and Evans, *Rev. Mod. Phys.*, **24**, 79 (1952).)

The differential cross section for the production of an electron with kinetic energy between T_- and $T_- + dT_-$ can be shown by QED to be

$$d\sigma = r_0^2 Z^2 P \frac{dT_-}{(h\nu - 2m_0c^2)}, \tag{8.38}$$

where the function P is shown in Fig. 8.14. The distribution is symmetric about the mid-point, and it applies also to the initial distribution of positron energy T_+. The electrons and positrons interact with the Coulomb field of the nucleus, and after leaving the atom do not in general have the same kinetic energy.

Positron annihilation

What happens to the positron after its creation is also of some interest to us because it affects the way in which γ-rays are detected (see the next chapter).

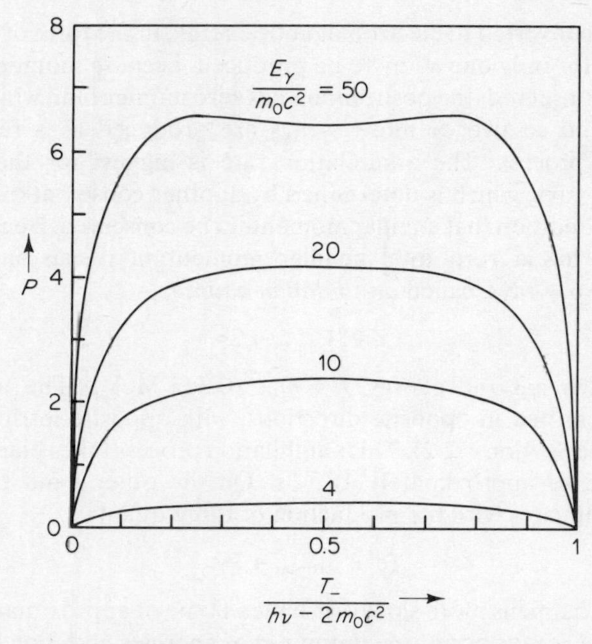

Fig. 8.14. The function P appearing in Eq. (8.38). This shows the distribution of the available energy ($E_\gamma - 2m_0c^2$) between the electron and positron. (Adapted from Davisson and Evans, *Rev. Mod. Phys.*, **24**, 79 (1952).)

The usual course of events is that a positron created by pair production in a solid or liquid loses kinetic energy by exciting or ionizing the atoms that it collides with or passes near to as it travels through the medium. When it has been slowed down sufficiently it becomes captured in the Coulomb field of a free or bound electron of the medium. The positron and electron then revolve around their common centre of mass, forming a system known as *positronium*. This is a system analogous to the hydrogen atom, with the positron taking the place of the proton. As in the case of hydrogen, positronium has discrete energy levels. Once it has been formed, usually in a highly excited level, the positronium decays to lower levels by emitting photons, until it arrives at the ground state.

When in the ground state the electron and positron of the positronium have a zero orbital angular momentum, and their intrinsic spins ($\frac{1}{2}\hbar$ in each case) are aligned either in opposite directions (giving a total spin of 0), or in the same direction (giving a total spin of $\hbar$). These two forms are called singlet and triplet positronium respectively. In both forms the electron and positron are sufficiently close to each other that they are able to interact and mutually *annihilate* in a short time interval. When this happens their rest energies are converted to electromagnetic energy, in the form of γ-rays. It is not possible for only one γ-ray to be produced, because momentum would then not be conserved (the positronium has zero momentum, while the γ-ray does not), and so two or more γ-rays are produced as a result of the annihilation process. The annihilation rate is highest for the minimum number of γ-rays, which is determined by another conservation condition, namely the condition that angular momentum be conserved. Because singlet positronium has a zero total angular momentum it can annihilate by producing two γ-rays, called *annihilation quanta*,

$$(e^-e^+)_{\text{singlet}} \rightarrow 2\gamma, \tag{8.39}$$

each of which has the energy $E = m_0c^2(0.511\text{ MeV})$. The annihilation quanta are emitted in opposite directions, with opposite intrinsic angular momenta (see Section 4.2.2). This annihilation process takes place in a mean time interval of approximately 10^{-10} s. On the other hand triplet positronium annihilates with the production of three quanta

$$(\text{e}^-\text{e}^+)_{\text{triplet}} \rightarrow 3\gamma. \tag{8.40}$$

This process happens more slowly (in a mean time of approximately 10^{-7} s), and the three γ-rays occupy a continuum of energies and directions.

A process which competes with positronium formation (and subsequent decay) is the annihilation of the positron in free flight as it passes near to an electron. The probability of this happening is small compared with the probability of positronium formation when the positron is created in a condensed medium, as is usually the case. Another much less probable mode of decay of a positron occurs when the positron and an inner-shell electron are absorbed together by a nucleus,

$$e^+ + e^- + \text{N} \rightarrow \text{N} + \gamma,$$

with the emission of a single γ-ray.

8.3 SCATTERING AND ABSORPTION OF RADIATION BY SOLIDS AND LIQUIDS

All the processes discussed in Section 8.2 can occur also in condensed matter, and essentially the same theoretical models apply. In addition there are other scattering and absorption processes that depend on specific features of the condensed material, and it is these that are discussed in the present section.

8.3.1 Brillouin scattering

One essentially non-atomic feature of condensed matter that can cause the scattering of electromagnetic radiation is the dielectric non-uniformity caused by acoustic waves. This type of scattering is called *Brillouin scattering*. The acoustic waves may be externally produced, or they may exist as a random wave motion of the medium caused by the thermal motion of the molecules of the medium.

Figure 8.15 shows electromagnetic waves being scattered by an acoustic plane wave travelling in a liquid. The pressure variations in the acoustic wave cause variations in the density and hence in the dielectric constant ε and refractive index n of the liquid, forming a travelling periodic structure. The condition for the reflected electromagnetic waves to be in phase, and hence observable, is the Bragg condition that applies to diffraction of X-rays by crystal planes, namely

$$N\lambda_1 = 2d \sin\theta = 2\lambda_s \sin\theta, \tag{8.41}$$

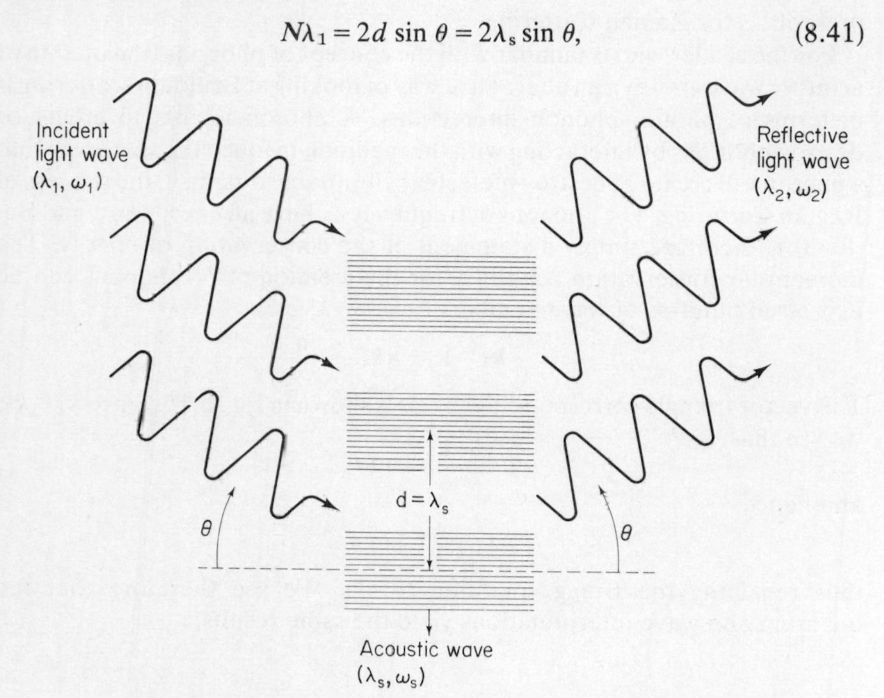

Fig. 8.15. Brillouin scattering by a travelling acoustic wave.

where λ_1, λ_s, and θ are defined in the figure, and N is the order of diffraction. The fact that the acoustic waves are travelling with the velocity v_s means that there is a Doppler shift in frequency, the reflected light having a lower frequency than the incident light (for the situation shown in the figure). The frequency difference is proportional to the component of v_s in the direction of the light, and is given by (Eq. (6.24))

$$\frac{\omega_1 - \omega_2}{\omega_1} = 2\frac{v_s \sin\theta}{c}, \tag{8.42}$$

assuming that $v_s \ll c$ (as is certainly the case). This assumption also implies that the fractional change in frequency is small, so that Eq. (8.41) is not significantly affected by the fact that λ_1 and λ_2 are different.

Combining Eqs. (8.41) and (8.42) we now find that

$$\omega_1 - \omega_2 = \omega_1 \frac{2v_s}{c} \frac{N\lambda_1}{2\lambda_s} = N \frac{2\pi v_s}{\lambda_s} = N\omega_s. \tag{8.43}$$

This shows that the change of frequency can only be an integral multiple of the acoustic frequency. When the acoustic wave travels in the direction opposite to that shown in the figure the frequency is increased and the integer N is negative. Thus inelastic and superelastic scattering are both present, as for Raman scattering.

For the reader who is familiar with the concept of phonons (the quanta of acoustic wave energy), an alternative way of looking at Brillouin scattering is in terms of photon–phonon interactions. A photon is able to create or destroy phonons by interacting with the medium, in much the same way that a photon can create or destroy molecular vibrational quanta in the process of Raman scattering. The phonons of frequency ω_s have an energy $\hbar\omega_s$, and Eq. (8.43) is therefore simply a statement of the conservation of energy. The momentum conservation condition for the creation of N phonons can be expressed in terms of wave numbers $k(=2\pi/\lambda)$ as

$$\mathbf{k}_1 = \mathbf{k}_2 + n\mathbf{k}_s.$$

The vector triangle corresponding to this is shown in Fig. 8.16. Since $k_s \ll k_1$, we see that

$$nk_s = 2k_1 \sin\theta,$$

and hence

$$n\lambda_1 = 2\lambda_s \sin\theta,$$

thus regaining the Bragg condition (8.41). We see therefore that the quantum and wave interpretations yield the same results.

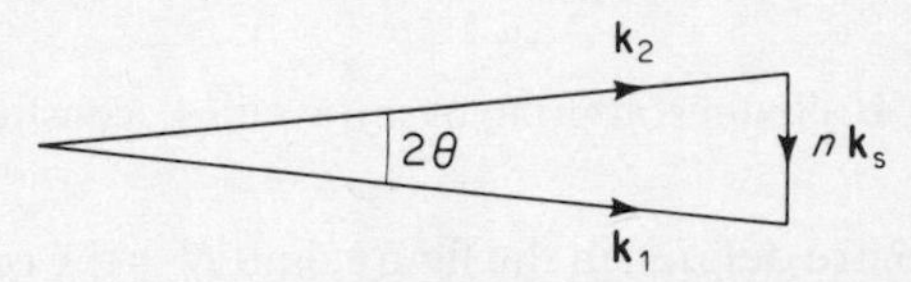

Fig. 8.16. Momentum conservation condition for Brillouin scattering.

An important use of Brillouin scattering is in the determination of the dependence on frequency of the velocity of acoustic waves in a medium, since this dependence can then be used to derive information about the structure of the medium (although what this information is and how it is derived are beyond the scope of the present book). The acoustic waves that are studied are usually those excited by the thermal motion of the molecules of the medium. A laser beam is used as the light source. The angle at which the scattered light is observed gives N/λ_s (Eq. (8.41)), and the change in frequency of the scattered light gives $N\omega_s$ (Eq. (8.43)). The acoustic velocity then follows from

$$v_s = \omega_s \Big/ \left(\frac{2\pi}{\lambda_s}\right).$$

The acoustic waves are attenuated and damped by the medium and therefore (as in the analogous case of damped electromagnetic radiation) the frequency ω_s at a particular angle of scattering is not discrete, but has instead a range of values. The frequency width of the range can be measured by the photon correlation technique described in Section 5.5.2, and can then be used to deduce the attenuation coefficient of the wave. In a typical experiment the acoustic velocity and attenuation coefficient of synthetic or natural polymers are measured as a function of temperature, in order to obtain information on quantities such as the temperature at which a rubber-like polymer changes to a glass-like state.

8.3.2 Photoconduction and photoemission

The photoelectric effect for isolated atoms and molecules has already been discussed in Section 8.2.4. In this section we discuss the new aspects which arise when the photoelectric effect occurs in a solid medium consisting of either a conductor or a semi-conductor.

Some of the processes that can occur when a photon is absorbed in a conductor or semi-conductor are illustrated in Fig. 8.17. Process (*a*) is that of reflection at a metal surface; this has already been discussed in Chapter 2. Process (*b*) is the rather trivial process in which the photon energy is used to heat the metal; this will be referred to again in the next chapter in connection with the detection of thermal radiation. We start therefore with (*c*), the photoelectric process in a conductor.

Photoemission in a conductor

In order to consider the dependence of process (*c*) on the photon energy we need to know the density and range of energy of the electrons in the conduction band. For a free-electron model of a metal the number of conduction electrons having energies between E and $E + \mathrm{d}E$ (measured from the bottom of the conduction band) is given by the Fermi–Dirac distribution

$$\mathrm{d}N(E) \propto \frac{E^{1/2}\,\mathrm{d}E}{\mathrm{e}^{(E-E_f)/kT}+1}. \tag{8.44}$$

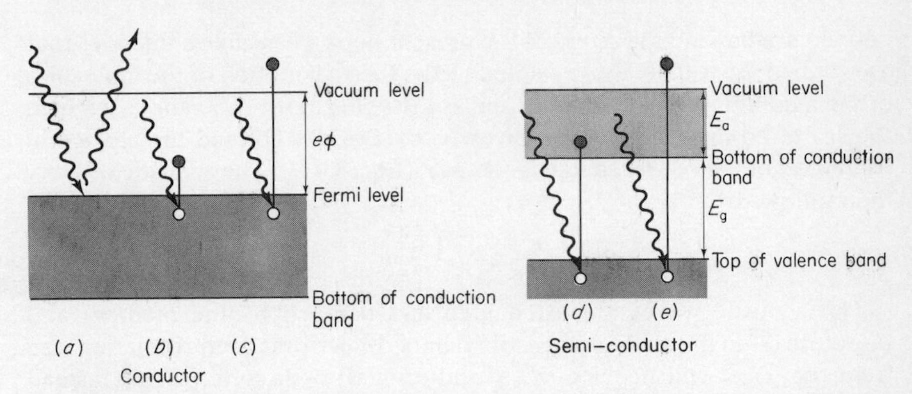

Fig. 8.17. Schematic illustration of some of the results of photon absorption in a conductor or semi-conductor. ϕ is the work function of the conductor, and E_a and E_g are respectively the electron affinity and band gap of the semi-conductor. (*a*) Reflection of the incident photon at a metal surface (this is more probable for a conductor than a semi-conductor); (*b*) absorption of thermal radiation to heat the electrons; (*c*) ejection of an electron from the conductor by the photoelectric effect, with $h\nu > e\phi$; (*d*) photoelectric effect, with $h\nu > E_g$, giving an electron and a hole, either of which may increase the conductivity of the semi-conductor; (*e*) photo-ejection of an electron, with $h\nu > E_g + E_a$.

Here T is the temperature and E_f the Fermi energy. The shape of the distribution is shown in Fig. 8.18(*a*) for zero temperature and a finite temperature.

The dependence of the yield of photoelectrons on the frequency of the incident light is shown in Fig. 8.18(*b*). As the photon energy is increased from zero the yield starts to become appreciable above the cut-off value

$$h\nu_c = e\phi$$

where ϕ (see Fig. 8.17) is the work function of the metal. The electrons then have sufficient energy to cross the gap between the Fermi level and the vacuum level.

As the frequency increases above ν_c more and more conduction electrons have a high enough initial energy to allow them to be ejected from the metal. Those at the Fermi level have an ejection energy given by the well-known *Einstein's photoelectric equation*

$$E_{\max} = h\nu - e\phi. \tag{8.45}$$

Having the necessary initial energy does not ensure ejection however, since (i) the initial direction of the photoelectron might be away from the surface, or (ii) the photoelectron might initially travel towards the surface but then be scattered before reaching it, or (iii) the photoelectron might reach the surface but with a velocity component $v_\perp$ normal to the surface which is

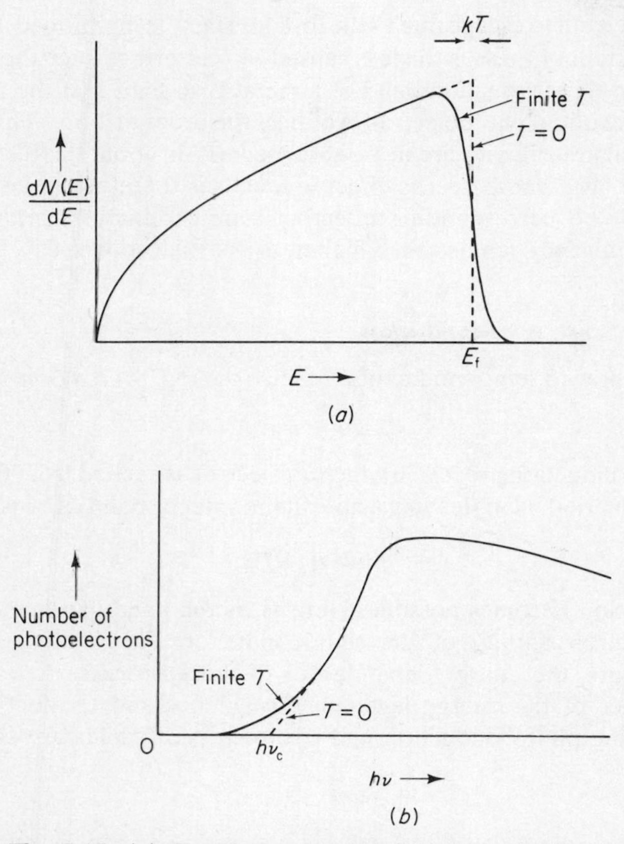

Fig. 8.18. (*a*) The Fermi–Dirac distribution of electron energies for zero temperature and a finite temperature. (*b*) Dependence of the photoemissive yield on the light frequency.

insufficient to allow it to escape (the requirement is that $\frac{1}{2}\, m\, v_{\perp}^2 > E_f + e\phi$), or (iv) it might be reflected back into the metal by the potential discontinuity at the surface. These four factors combine to give a photoemission yield which is typically as shown in Fig. 8.18(*b*).

The absolute value of the photoelectric yield at a given frequency is expressed in terms of the ratio

$$\varepsilon = \frac{\text{number of ejected electrons}}{\text{number of incident photons}}. \tag{8.46}$$

This is known as the *quantum-efficiency* of the material. Its value tends to be small (typically $\sim 10^{-2}$ to 10^{-6}) for conductors because most of the incident light is reflected at the surface, and also because the photoelectrons that are formed often fail to escape due to the four loss factors mentioned above. The most important of these is the loss caused by scattering, since the density of electrons in the conduction band of a metal is so high that the mean path length of the initial photoelectron is only of the order of 1 nm. Thus the only photons that are effective are those absorbed within about this distance from the surface. The photoelectric effect in metals is therefore of less practical value than the corresponding effect in semi-conductors, for which the quantum efficiency can be much higher, as we shall soon see.

Photoprocesses in semi-conductors

Turning now to semi-conductors we see from Fig. 8.17(*d*) and (*e*) that for

$$h\nu > E_g$$

the photoconduction process in which an electron is excited from the valence to the conduction band (leaving a hole in the valence band) is possible, while for

$$h\nu > E_g + E_a$$

photoemission becomes possible. Here E_g is the band gap energy and E_a is the electron affinity of the semi-conductor. As in photoabsorption in conductors, the energy dependences of these processes are related to the densities of the energy levels in the valence and conduction bands. The resulting photoconduction and photoemission yields are sketched in Fig. 8.19.

Photoemission in semi-conductors

The most important difference between the photoemission process in semi-conductors and conductors is that the photoelectrons can travel much

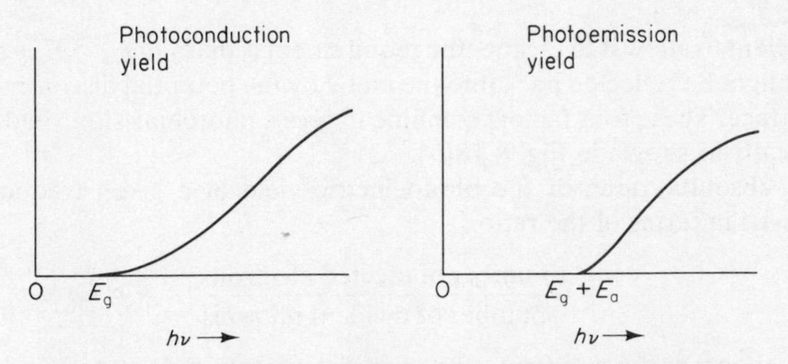

Fig. 8.19. Schematic illustration of the photoconduction and photoemission yields of a semi-conductor as a function of the photon energy.

more freely through a semi-conductor, provided that their kinetic energy is less than E_g. This is because a photoelectron of this energy is unable to interact with valence electrons to produce an electron-hole pair (see Problem 8.4), and it has only a small probability of interacting with the low density of conduction electrons. The mean free path of the photoelectron is typically a few tens of nm. Because of this, and the fact that semi-conductors have lower reflection coefficients for the incident light, the quantum efficiency can be much higher (up to 0.4) than for metals. The quantum efficiencies of two materials frequently used in the detection of visible light are shown in Fig. 8.20. These will be referred to again in the next chapter (Section 9.2.1).

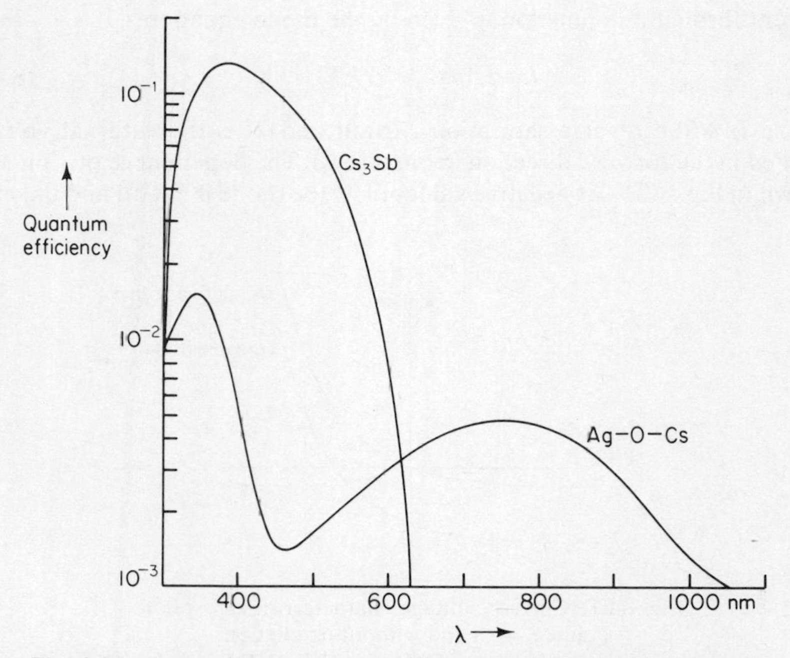

Fig. 8.20. Approximate quantum efficiencies of photocathodes made from semi-transparent coatings of Cs_3Sb and Ag–O–Cs on lime glass (adapted from data in the RCA Photomultiplier Manual).

Photoconduction in semi-conductors

Next we come to the photoconductivity process in semi-conductors. The electrons and holes created by photon absorption diffuse from (or in the case of an applied field are accelerated from) their initial positions and eventually recombine with other charge carriers. The increase in conductivity depends on the rates of generation and recombination of free electrons and holes. For low levels of illumination the increase is proportional to the illumination.

It is usually preferable to use semi-conductors of the extrinsic type, in which small amounts of impurity atoms (dopants) are added, rather than the intrinsic type. The impurities are chosen to give extra energy levels (the donor and acceptor levels) within the gap between the valence and conduction bands (see Fig. 6.22). Photon absorption is then possible at energies lower than E_g, since electrons in the donor levels can be excited into the conduction band, or electrons from the valence band can be excited into the acceptor levels, leading in both cases to an increase in conductivity.

It is also usual for the semi-conductor to be in the form of a *p–n* junction. The device is then called a *photodiode*. In the absence of radiation the current through the junction is given by the diode equation,

$$I = I_s[\exp(eV/kT) - 1], \tag{8.47}$$

where I_s is the reverse saturation current and V is the external voltage applied in the forward direction (from p to n). The dependence of I on V is shown in Fig. 8.21. At negative values of V the diode is cut off and only the

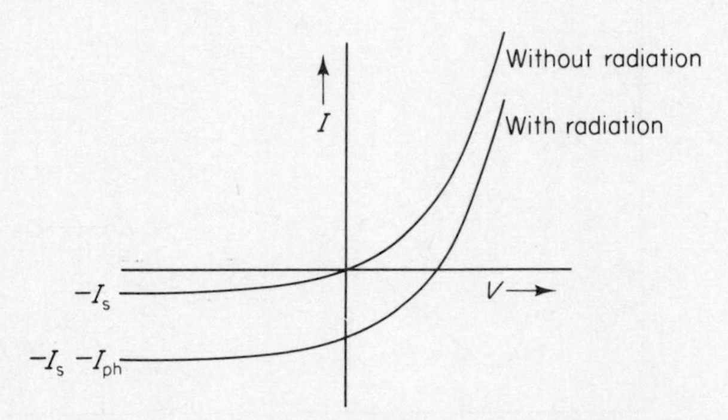

Fig. 8.21. Current–voltage characteristic of a photodiode, with and without irradiation.

reverse saturation current flows, while at positive values there is a large forward current. Irradiation of the junction by photons gives rise to an additional source of electrons and holes, thus causing an additional reverse current I_{ph} to flow across the junction. The current–voltage characteristic of the junction then becomes

$$I = I_s[\exp(eV/kT) - 1] - I_{ph}. \tag{8.48}$$

This characteristic is also shown in Fig. 8.21.

Solar cell

A particular type of photodiode of great practical importance is the *solar cell*. This has a *p–n* junction diffused close to a flat surface, so that as much light as possible can reach the junction. The surface area is typically about 1 cm^2, and often large arrays of several thousand cells are used.

The way in which the photogenerated current I_{ph} is converted into usable power in an external load (R_L) is shown schematically in Fig. 8.22. This is called the *photovoltaic* mode of operation. If the current I flows in the direction shown in Fig. 8.22(a) then the forward bias across the junction is

$$V = -I(R_L + R_{int}),$$

where R_{int} is the internal resistance of the cell and contacts. Since Eq. (8.48) must also be satisfied, the operating point of the cell is at the intersection A shown in Fig. 8.22(b). The power produced in the external load is then

$$P = I_m^2 R_L = \frac{V_m^2 R_L}{(R_L + R_{int})^2},$$

which can be maximized by choosing R_L suitably.

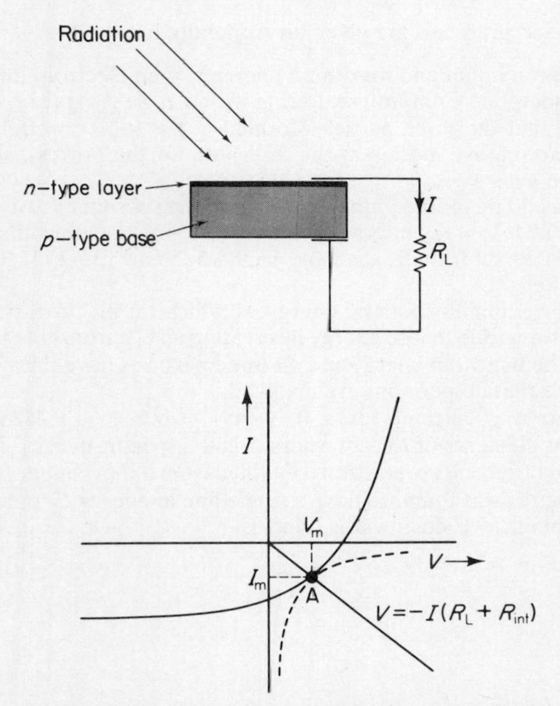

Fig. 8.22. Photovoltaic mode of operation of a solar cell.

The overall efficiency ε of the cell is the ratio of P to the radiation power W falling on the cell,

$$\varepsilon = P/W.$$

To obtain an estimate of the upper limit of ε we can suppose that the incident photons have a mean energy $h\bar{\nu}$ and that every photon creates one electron-hole pair which contributes to I_{ph}. The power in the incident beam is then $(I_{ph}/e)h\bar{\nu}$, and

$$\varepsilon_{max} \sim \frac{I_m V_m}{(I_{ph}/e)h\bar{\nu}} \sim \frac{eV_m}{h\bar{\nu}}.$$

For solar radiation at the earth's surface, $h\bar{\nu} \simeq 2.3$ eV, and for Si solar cells $V_m \simeq 0.4$ V, giving $\varepsilon_{max} \simeq 20\%$. Solar cells have been made from various semi-conductors, including Si, GaAs, and CdS, and the highest efficiencies achieved have been in the range from approximately 8 to 15%, approximately half the theoretical maxima.

PROBLEMS: CHAPTER 8

(Answers to selected problems are given an Appendix E.)

8.1 What are the minimum and maximum energies of an electron, initially at rest, which has undergone Compton scattering with a 1 MeV γ-ray?

8.2 By assuming that the three particles formed in the triplet production process (8.37) have no relative motion at the threshold for the process, show that the threshold energy is 4 m_0c^2.

8.3 What energies do the initial photoelectron and the subsequent Auger electrons have when a 1 MeV γ-ray interacts with a lead atom, the ionization energies of which are 88.0 keV for the K shell and 15.9, 15.2, and 13.0 keV for the L_1, L_2, and L_3 sub-shells?

8.4 Show that the minimum kinetic energy at which an electron travelling in a semi-conductor is able to lose energy by creating an electron-hole pair is 1.5 E_g, where E_g is the band gap energy and all three particles have the same effective mass. Assume that no phonons are involved.

8.5 The attenuation coefficient for soft γ-rays (~0.1 to 1.0 MeV) is always $\sim 10\rho$ m^{-1} for elements of low Z, where ρ is the specific density. Explain this.

8.6 The spectrum of secondary electrons obtained from a thin copper foil irradiated by a γ-ray source was found to have a single line at energy E and a continuum extending from 0.75 E downwards. Find E.

CHAPTER 9

The detection of electromagnetic radiation

The human observer can detect some forms of electromagnetic radiation directly, provided that the power levels are high enough (for example, visible light can be seen by eye, infrared radiation can be felt on the skin, and microwave radiation can be experienced as a deeper-seated heating), but otherwise these and other forms can be detected only by a usually complex chain of events in which the radiant energy is converted into various other forms. In the present chapter we concentrate on the physics of the primary energy conversion processes and on the first steps in the detection chain. The groundwork for this has been laid in the previous chapter, and so we shall now discuss which process are the most suitable for each frequency range, and consider which additional features are important for the accurate measurement of intensities or frequencies.

In the microwave and far infrared regions (see Fig. 1.1) there is little choice about the detection process, since the radiation is able only to heat up the material on which it falls (here we exclude a few special detectors of limited general usefulness, such as those based on amplification by a maser). We start therefore in Section 9.1 with those detectors in which a temperature rise is measured. At shorter wavelengths ($\lesssim 10\ \mu m$) the radiation is able to increase the conductivity of some materials, giving a new means of detection, which is described next. At even shorter wavelengths ($\lambda \lesssim 1\ \mu m$, $h\nu \gtrsim 1$ eV) photoemission of electrons can occur from some solids, giving the possibility of detecting photons through the photoelectrons that they produce. These photoelectrons can be made to produce a large number of secondary electrons for example, as in the photomultiplier. Alternatively the photoelectrons can be used to give a permanent record of the photon flux, as in photography and xerography. These detection techniques are all discussed in Section 9.2. In the last section (9.3) of the chapter we start by discussing the techniques used in the energy range ($h\nu \gtrsim 10$ eV) at which gaseous ionization becomes possible. Here each initial ionization event can be made to produce a large number of secondary ions and electrons, making possible the detection of single photons. We finish with the highest energy range ($h\nu \gtrsim 100$ eV), where it is possible to arrange that the number of secondary particles produced (directly and indirectly) by an incident photon is approximately proportional to the energy of the photon, thus allowing the direct measurement of the photon energy.

9.1 DETECTION OF INFRARED RADIATION

9.1.1 Bolometers and thermocouples

In the range of wavelengths from approximately 100 μm to 1 mm (the far infrared) the usual means of detecting radiation is through its heating effect. The measurement of the radiant power therefore becomes a measurement of temperature, which in turn becomes a measurement of some associated physical or electrical property. It is usual for the absorber to be in the form of a solid and for the measurement to be made either of a change in resistance, when the detector is called a *bolometer*, or of the change in a thermoelectric potential, when it is called a *thermocouple* detector. We start with the bolometer detector.

Bolometer detector

Suppose that radiation of power P watts falls on a body of thermal capacity C joules per degree which loses heat to its surroundings (by conduction and re-radiation) at the rate $G\,\Delta T$ watts when the temperature of the body is ΔT above that of the surroundings. In the steady state, when P is constant and all the radiation is absorbed, the temperature increase is given by

$$\Delta T_0 = P/G. \tag{9.1}$$

On the other hand if the radiant power starts abruptly at the time $t = 0$ and is constant thereafter, the subsequent temperature change is given by the equation

$$C\frac{\mathrm{d}(\Delta T)}{\mathrm{d}t} = P - G\,\Delta T. \tag{9.2}$$

This has the solution

$$\Delta T = \Delta T_0(1 - e^{-t/\tau}),$$

where the time constant τ is

$$\tau = C/G. \tag{9.3}$$

We see that for a high sensitivity the bolometer must absorb the radiation efficiently and must have a low value of G; for a short time constant it must also have a low value of C. To obtain an idea of the orders of magnitude involved, a typical metal bolometer might have an element consisting of a platinum strip of area 2 mm^2 and thickness 0.1 μm, mounted on a thermal insulator; the $C \simeq 6 \times 10^{-7}$ J K^{-1}, $G \sim 10^{-4}$ W K^{-1}, and $\tau \sim 6$ ms.

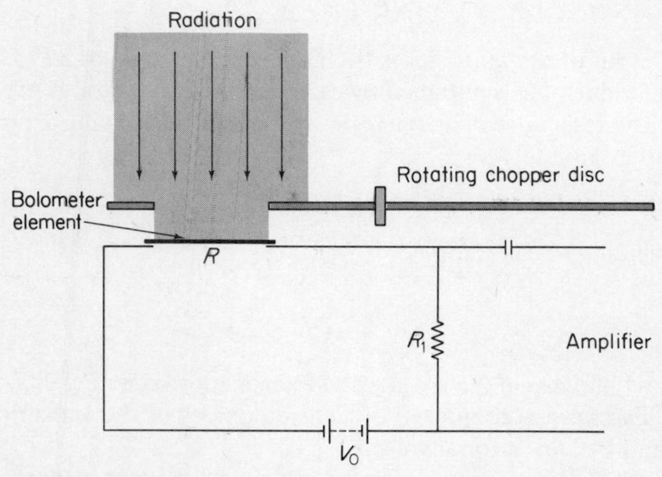

Fig. 9.1. Use of a bolometer to detect infrared radiation.

The way in which the bolometer is used is illustrated in Fig. 9.1. The bolometer element is usually covered with a very thin layer of 'carbon black' or 'gold black', to increase the absorption of the incident radiation. The presence of the rotating disc in front of the element is important in helping to discriminate against the various sources of 'noise' that are present even when there is no incident radiation. The openings in the disc give a square-wave shape to the power P falling on the element. A modulation frequency of approximately 10 Hz is typically used, to ensure that the modulation period is much larger than the time constant τ. The amplifier has a high gain only over a narrow band of frequencies centred at the modulation frequency (this is achieved by phase-locking the amplifier to the modulation signal).

The resistance R of the platinum bolometer element is approximately inversely proportional to the temperature T,

$$R \simeq \text{const}/T.$$

This is true also of other pure metals at room temperature. The coefficient

$$\alpha = \frac{1}{R}\frac{\mathrm{d}R}{\mathrm{d}T} \tag{9.4}$$

therefore has the value

$$\alpha \simeq \frac{1}{T} = 0.0034\ \mathrm{K}^{-1}. \tag{9.5}$$

The change in resistance caused by a temperature change ΔT is

$$\Delta R = \alpha R\ \Delta T,$$

and the output of the amplifier is therefore proportional to $\alpha P/G$.

A larger value of α is obtained by using a semi-conductor as the sensitive element. The resistance of an intrinsic semi-conductor having a band gap E_g is given approximately by

$$R \propto \exp(-E_g/kT).$$

The coefficient α is therefore

$$\alpha \simeq \frac{E_g}{kT^2}, \tag{9.6}$$

which for a band gap of 0.5 eV gives a value of approximately $0.06\ \mathrm{K}^{-1}$. The so-called *thermistors*, composed of sintered oxides of the transition metals Ni, Mn, and Co, are normally used.

Yet another improvement can be made by using a superconducting material to detect the radiation, giving a bolometer which is more sensitive but also more difficult to use. The resistivity of a superconductor changes very quickly with temperature in the region in which the material is on the verge of being superconducting. For example, the resistivity of tin changes from zero to normal values in an interval $\Delta T \sim 0.01$ K in the region of the transition temperature 3.7 K, which means that $\alpha \sim (\Delta T)^{-1} \sim 100\ \mathrm{K}^{-1}$. The minimum detectable power of this type for bolometer is less than 10^{-13} W. At a wavelength of 100 μm this corresponds to less than 5×10^7 photons per second.

Thermocouple detectors

The other type of detector mentioned above for use in the infrared range is the thermocouple detector. One design of this is illustrated in Fig. 9.2. The

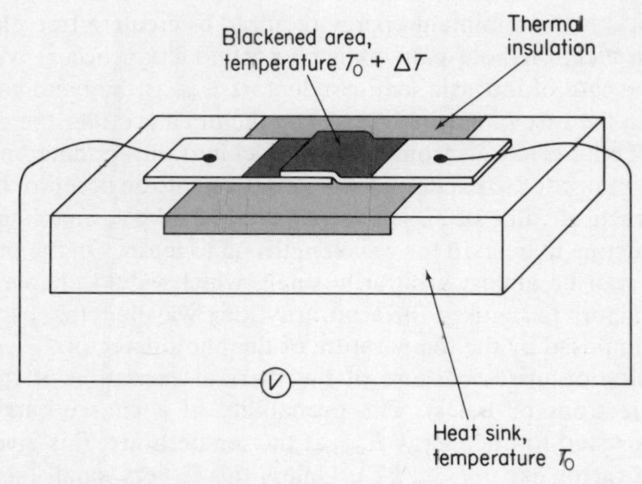

Fig. 9.2. Schematic illustration of a thermocouple detector.

sensitive element has an area which is typically 1 mm × 1 mm, and is made by evaporating thin overlapping films of two different metals onto an insulating backing. The region of overlap is blackened, and so radiation falling on this region causes the temperature of the junction to rise by an amount ΔT above that of the other junctions in the circuit. This causes the thermoelectric EMF

$$E = P_{ab}\,\Delta T$$

to be set up in the circuit, where P_{ab} is the *thermoelectric power* for the two metals of the junction. The metals bismuth and antimony are often used, since this combination has the highest value of P, 102 μV/K. As in the case of the bolometer detectors the intensity of the radiation is usually modulated to give greater sensitivity, and the amplitude of the resulting thermoelectric EMF is measured with a narrow-bandwidth AC amplifier. The minimum detectable power is usually $\geqslant 10^{-10}$ W.

If a sensitive area of more than about 1 mm^2 is required then several thermocouples have to be connected together as a *thermopile*. In this the thermocouples are connected in series, as for example Bi–Sb–Bi–Sb–Bi etc., and arranged in a closely packed array, with every second junction exposed to the radiation and the intermediate ones connected to a heat sink. The EMFs of the junctions add together, each being proportional to the radiation falling on it.

9.1.2 Photoconduction detectors

The mechanism by which the conductivity of semi-conductors is increased by photoabsorption was discussed in Section 8.3.2 (see for example Fig.

8.17). If E_{min} is the minimum energy required to create a free electron or hole, or an electron–hole pair, then photoconduction occurs when $h\nu > E_{min}$. In the case of intrinsic semi-conductors E_{min} is the band gap energy E_g, while in the case of extrinsic semi-conductors it is either the energy $\mathscr{E}_d$ needed to excite electrons from a donor level into the conduction band, or the energy $\mathscr{E}_a$ needed to excite a valence electron into an acceptor level. The smallest practical value of E_g is approximately 0.15 eV, enabling intrinsic semi-conductors to be used for wavelengths up to 8 μm. On the other hand $\mathscr{E}_d$ and $\mathscr{E}_a$ can be almost arbitrarily small, which would enable extrinsic semi-conductors to be used for arbitrarily long wavelengths, except for a limitation imposed by the temperature of the photodetector.

This limitation arises because of the thermal excitation of the charge carriers (electrons or holes). The probability of a charge carrier being thermally excited to the energy E_{min} at the temperature T is given by the Boltzmann factor $\exp(-E_{min}/kT)$. Unless this is very small ($\lesssim 10^{-7}$) the number of thermally excited carriers is much larger than the number excited by the incident radiation. In practice the operating temperature and maximum detected wavelength are connected by

$$T\lambda_{max} = T\frac{hc}{E_{min}} \lesssim 8\times 10^{-4}\ \text{K m}. \tag{9.7}$$

Therefore a detector operated at the temperature of liquid helium, 4.2 K, would be capable of detecting wavelengths up to approximately 190 μm, if it were to possess the appropriate value of $\mathscr{E}_d$ or $\mathscr{E}_a$ (approximately 0.006 eV). This represents the maximum convenient wavelength limit for this type of detector. Figure 9.3 shows the wavelength dependencies of four different materials at their typical operating temperatures.

As in the case of bolometer detectors the sensitivity is improved by modulating the intensity of the incident radiation. The minimum detectable power depends very much on the material and the operating temperature, but tends to lie in the range 10^{-12} to 10^{-10} W for detectors of area 1 mm^2.

9.2 DETECTION OF VISIBLE RADIATION

9.2.1 Photoemission detectors

As discussed in Section 8.3.2, photoemission of electrons from solid materials becomes possible at wavelengths shorter than about 1.2 μm ($h\nu \gtrsim 1$ eV). By using techniques in which single photoelectrons can be detected it then becomes possible to detect single photons, giving a sensitivity several orders of magnitude higher than that available at longer wavelengths. The single electrons are detected by a process of multiplication, as we shall see, and the combination of a photoemission stage

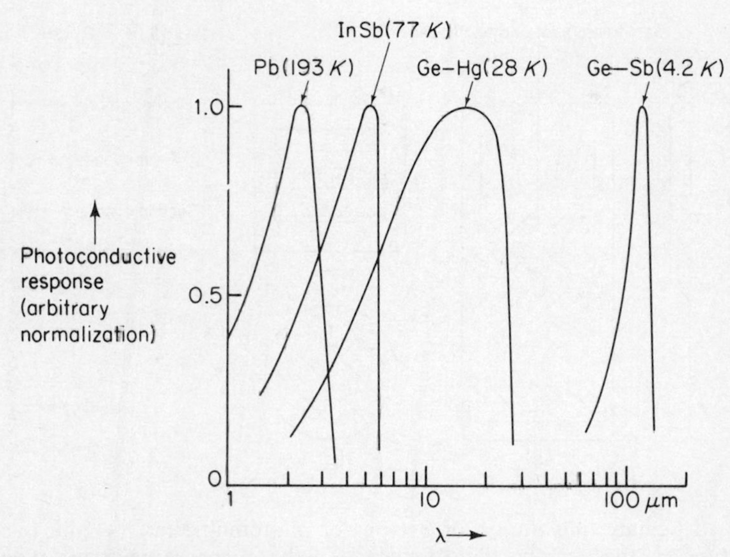

Fig. 9.3. Approximate wavelength dependence of the photoconductive response of 2 intrinsic (PbS and InSb) and 2 extrinsic (Ge–Hg and Ge–Sb) semiconductors (taken from the data given by R. A. Smith, F. F. Jones and R. P. Chasmar, The detection and measurement of infrared radiation, and M. Posner and W. Raith, *Methods of Experimental Physics*, **4A**, edited by V. W. Hughes and H. L. Schultz).

followed by an electron multiplier stage is usually known as a *photomultiplier.*

Three types of photomultiplier are illustrated in Fig. 9.4. In all three types a semi-transparent photocathode is deposited on the inside of the front face of an evacuated tube. Photoelectrons are ejected from the photocathode with an efficiency that depends on the wavelength (see Fig. 8.20), and are accelerated away from the photocathode by an electric field. The area of the photocathode is typically from 1 to 20 cm^2, but can be as large as several hundred cm^2.

In the first type shown the electrons are focussed onto a curved dynode (i.e. a metal plate held at a positive potential with respect to the cathode). At this point their energy (given by the potential difference between the photocathode and dynode) is typically a few 100 eV. On hitting the dynode each electron produces several secondary electrons of much lower energy. The mechanism by which this happens is similar to that of photoejection itself (see Fig. 8.17(c)). The average number of secondary electrons from a single incident electron is known as the *secondary emission ratio* δ. The value of δ depends on the dynode material and on the incident electron energy, but for the dynode material commonly used, copper beryllium, it is typically from 3

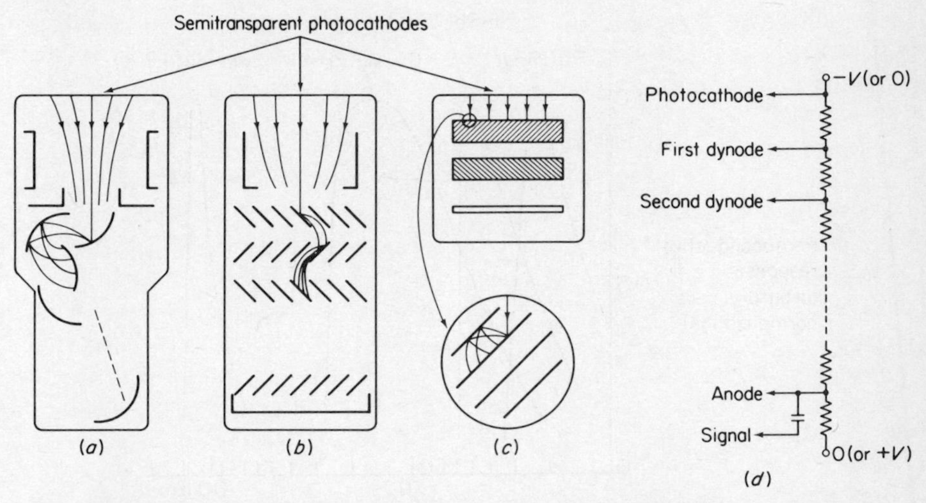

Fig. 9.4. Schematic illustration of 3 types of photomultiplier; (*a*) the focussing dynode type, (*b*) the 'venetian blind' type, and (*c*) the 'microchannel array' type (with the inset showing an enlargement of the entrance to a microchannel array). Part (*d*) shows a resistor chain used to apply the appropriate voltages to the dynodes of the focussing and venetian-blind photomultipliers (the potential difference V is typically a few kV, and either the cathode or anode may be at earth potential).

to 5 at the incident energy used in the photomultiplier (~100 eV). The secondary electrons are accelerated and focussed by the electrostatic field towards a second dynode, as shown in the figure, and on hitting this each of them again produces an average of δ further secondary electrons. This process continues for a total of n dynodes, where n is usually from 6 to 16, by which time the average total number of secondary electrons per initial photoelectron is

$$G = \delta^n. \tag{9.8}$$

With $\delta = 4$ and $n = 10$, the effective gain is then

$$G = 4^{10} \simeq 10^6.$$

This cloud of secondary electrons is finally collected on the last electrode, the anode, where it produces a voltage pulse of magnitude

$$v = \frac{Ge}{C}, \tag{9.9}$$

where C is the effective capacity of the anode. With $C \sim 10$ pF the value of v is ~0.02 V, thus giving a pulse that can easily be further amplified and detected by conventional circuitry. The gain G increases sharply as the overall voltage V across the photomultiplier tube is increased (because δ increases with increasing electron energy, for the energies used in the photomultiplier), but in practice the gain is usually arranged to be of the order of 10^7 when it is required to detect single photons.

In the second type of photomultiplier, shown in Fig. 9.4(b), the dynodes are of the 'venetian blind' type. There is then no need to focus the secondary electrons during the transits between the dynodes. Nor is there any need to focus the initial photoelectrons onto the first dynode, and so the photocathodes can be flat instead of curved. The operation of the photomultiplier is otherwise as for the first type. There are also several other versions of these two types of photomultipliers. The first dynode is sometimes made from a material having a very low threshold energy for photoejection, since this then also has a high value of δ. For example, p-type gallium phosphide covered with a thin layer of caesium (GaP–Cs) is like this, with a value of δ of approximately 50 for an incident electron energy of 1 keV.

A different type of photomultiplier is shown in Fig. 9.4(c). In this the electron multiplication is performed inside narrow-bore open channels arranged together in a close-packed array known as a *microchannel array*. The array is made from a glass having a high value of δ, and a potential difference of the order of 2 kV is applied between the entrance and exit faces. A photoelectron hitting the glass surface just inside the entrance to a channel, as shown in the enlargement in Fig. 9.4(c), starts by producing secondary electrons. These are accelerated by the electric field that exists inside the channel, and hit the channel wall to produce further secondary electrons; this process continues until the end of the channel is reached. The overall gain is typically $\sim 10^4$. A second microchannel array is frequently used, as shown in the figure, to increase the gain. The channels in each array are usually inclined as shown in order to prevent the feed-back of positive ions (which arise as a by-product of the secondary emission process, or from collisions with the background gas, and which travel in the opposite direction to the electrons). At the exit of the second microchannel array the electrons can be collected at an anode, as in the case of the other two types shown. Alternatively these electrons can be further accelerated onto a screen, as in a television tube, in order to give a greatly enhanced image of the light that initially fell on the photocathode. The device is then known as an *image intensifier*.

The detection efficiency of a photomultiplier is determined essentially by the quantum efficiency ε (see Section 8.3.2) of the primary photoemission event. For photocathodes which respond in the visible region, such as those made of a semi-transparent coating of Cs_3Sb, ε is usually approximately 0.1 (see Fig. 8.21), while in the near infrared region it is an order of magnitude smaller. The major source of noise is usually provided by thermionic emission of electrons from the photocathode, giving a *dark current* equivalent to $\gtrsim 10^3$ photons per second. With a quantum efficiency of 10^{-1}, we see that the minimum detectable constant intensity is $\gtrsim 10^4$ photons per second, or $\gtrsim 2.10^{-15}$ W at 500 nm; this can of course be considerably improved by modulating the incident light and using phase sensitive detection, as in the case of the infrared detectors. Single pulses of light can be detected if the output signal is above the dark noise output, which requires that the flash should contain more than about 20 photons. It is interesting to note that without modulation the typical photomultiplier detector is less sensitive at detecting a constant intensity than is the human eye, which under the most favourable circumstances can see a constant object, such as a faint star, if a few hundred photons per second are received. The eye is able also to see a weak single flash of light if it contains more than a few hundred photons, so in this case the photomultiplier detector is the more sensitive.

9.2.2 Photography and xerography

Photography is a familiar method for detection at visible and X-ray wavelengths; it requires high intensities but gives a permanent record. The photographic emulsion contains $\sim 10^7$ to 10^{10} crystallites or grains of silver bromide (sometimes with a small admixture of silver iodide) per mm^2, held in suspension by the gelatine of the emulsion.

The primary process is photoelectric ejection of an electron from a bromine or iodine atom. Since the minimum energy for this to happen is approximately 1 eV, emulsions usually cannot be made sensitive to wavelengths longer than approximately 1.2 or 1.3 μm. The ejected photoelectrons move about in the conduction levels of the crystal until they become trapped at trapping centres. They then attract some of the interstitial Ag^+ ions which are always present, to form a speck of silver which acts as a latent image. In the subsequent process of 'development' the developer gives electrons to the latent image, so attracting further silver atoms. The final process of 'fixing' results in a permanent image.

The radiant energy required to produce a silver grain depends on the wavelength of the light and the sensitivity of the emulsion, and also on the development process. The minimum energy density required for a visible image is $\sim 3 \times 10^{-12}$ J/mm^2 ($\sim 10^7$ photons per mm^2).

Xerography

A more recently developed method of recording permanent images is that of *xerography*, a name derived from the Greek words xeros and graphos, which together mean 'dry writing'.

The xerographic process starts with a layer of positive charge being deposited on one surface of a film of semi-conductor material, sometimes referred to as the *photoreceptor*. The charge deposition is usually achieved by means of a corona discharge device which is placed near to the surface of the photoreceptor and is traversed across its surface. The second, key, stage is illustrated in Fig. 9.5. An image of the object (usually a flat opaque object such as a page of typescript) is formed on the charged surface of the photoreceptor, and the incident photons create electron–hole pairs inside the photoreceptor. The electrons then drift towards the upper surface under the influence of the electric field in the photoreceptor, and neutralize the surface charge in those regions in which photons have been received. The un-neutralized charges remain on the surface. In the third stage coloured (usually black) powder particles are placed near to the surface so that they are attracted to the charged areas and adhere there to produce a visible image. The final stages depend on whether or not multiple copies are required, but they consist essentially of transferring the powder particles to the final sheet of paper by electrostatic attraction, and fixing them there by a suitable heat treatment.

The properties required of the photoreceptor are (i) that it should give free electron–hole pairs with high efficiency on irradiation, (ii) that the electrons and holes should be able to drift through the material without significant trapping or recombination, and (iii) that an un-neutralized surface charge should not leak away too quickly. Amorphous selenium possesses all these properties, and is widely used for commercial xerography.

9.3 THE DETECTION OF ULTRAVIOLET, X, AND γ-RADIATION

These types of radiation can of course be detected by any of the methods so far described, but the higher energy per photon gives the further

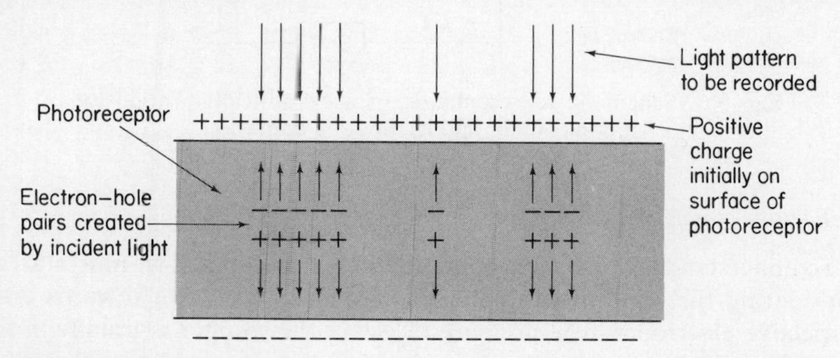

Fig. 9.5. Key stage in the xerographic process. The charge layer initially deposited on the upper surface of the photoreceptor is partially neutralized by the electron-hole pairs created by the incident light.

possibility of using gaseous ionization as the primary detection event. The higher energy also means that photoelectrons can be produced in solids, liquids, or gases with an energy high enough to be directly measurable, thus giving the possibility of a direct measurement of the photon energy. These two new features are discussed in the present section.

9.3.1 Gas ionization counters

The geometry and mode of operation of a typical gas filled ionization counter are represented schematically in Fig. 9.6. The incident radiation passes through a window and ionizes the gas contained in the chamber. An electrostatic field exists inside the chamber, and under its influence the photoelectrons and residual positive ions drift in opposite directions and are collected at the anode and cathode respectively. This gives rise to a voltage signal which is externally amplified and recorded by conventional electronic circuitry. The electric field is usually high enough for the initial photoelectrons to be able to create further electron–ion pairs when they collide with neutral gas atoms on their way to the central anode, thus giving an amplified anode signal, but for convenience we start by discussing the simpler case in which there is *no* such amplification.

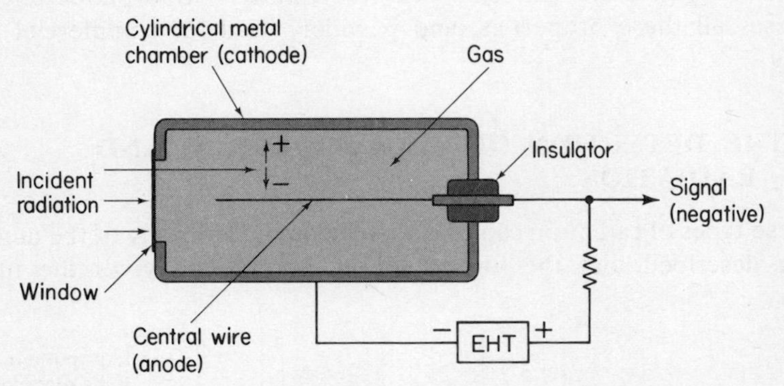

Fig. 9.6. Schematic representation of a cylindrically shaped ionization counter and its mode of operation.

Collection of electrons and ions

To understand the operation of the counter in more detail we must start by considering the way in which the charged particles move towards their respective electrodes. The pressure of gas in the counter is usually in the range from 0.1 to 10 atmospheres, and so the electrons and ions suffer many collisions with the gas atoms (or molecules) before they reach the electrodes. The easier case to consider is that of the positive ions. The time between collisions is usually so short that the energy gained from the electric field is small compared with the mean thermal energy. After each collision an ion therefore starts with a random direction and with a velocity corresponding approximately to the Maxwellian distribution of velocities of the neutral gas atoms. Before the next collision, after a time t_c, the ion increases its velocity component in the direction of the applied electric field E by the amount at_c, where a is the acceleration of the ion. The mean velocity in the direction of the field, called the *drift velocity* of the ion, is therefore

$$\bar{u} = \tfrac{1}{2}a\bar{t}_c = \frac{1}{2}\left(\frac{eE}{M}\right)\left(\frac{\lambda}{\bar{c}}\right), \tag{9.10}$$

where M is the mass of the ion, λ its mean free path, and $\bar{c}$ its mean thermal velocity. For example, in the case of argon ions in 1 atmosphere of argon gas and a field of 50 V/mm, $\lambda = 0.07$ μm, $\bar{c} = 400$ m/s, and $\bar{u} = 11$ m/s. We see that the drift velocity is much smaller than the thermal velocity, as assumed in deriving Eq. (9.10). The mean time taken for a positive ion to reach the chamber wall (of radius typically ~10 mm) and be collected is usually of the order of 1 ms.

Equation (9.10) is inaccurate for electrons because the drift velocity is often not small compared with the mean velocity of the gas atoms. Also if the gas is composed of atoms, rather than molecules, the electrons can scatter only elastically at energies below the first excited state energy of the atom, which means that their mean velocity $\bar{c}$ becomes larger than the mean velocity that they would have in thermal equilibrium. However, if we use Eq. (9.10) as a rough guide, and note that $\bar{c} \propto M^{-1/2}$, we find that

$$\frac{\bar{u}_e}{\bar{u}_{ion}} \sim \left(\frac{M_{ion}}{m_e}\right)^{1/2},$$

which for argon has the value 270. In fact $\bar{u}_e$ for the conditions given above is approximately 400 m/s, a factor 36 greater than u_{ion}. The electron drift velocity can be made larger by adding a small quantity of a polyatomic molecule (for example, 5% of CO_2) to argon. An electron can then lose energy by exciting the low lying rotational and vibrational states of the molecules giving the electron a lower value of $\bar{c}$ and hence increasing $\bar{u}_e$. The lower electron energy also causes the mean cross section for electron–argon atom collisions to decrease considerably, because of the existence of a deep minimum in this cross section at low energies (this is caused by the way in which the electrons are diffracted by the atom, and is known as the Ramsauer effect). The net effect of adding the polyatomic molecules is therefore an increased value of λ, as well as a decreased value of $\bar{c}$, giving an order of magnitude increase in $\bar{u}_e$ (to 4500 ms^{-1} for the conditions stated above).

When the electron and ion from a single photoionization event have both been collected the additional charges $\pm e$ at the electrodes cause a change in the voltage between the electrodes, of magnitude

$$v = -\frac{e}{C}, \tag{9.11}$$

where C is the effective capacitance of the two electrodes. This voltage appears as a signal at the anode, as shown in Fig. 9.6. The build-up to this maximum value of the signal is illustrated in Fig. 9.7. There is an initial fast

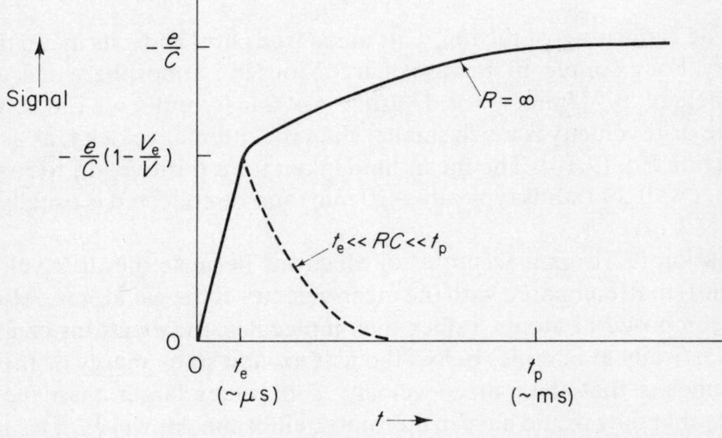

Fig. 9.7. Form of the anode signal for a single photoionization event. The full curve shows the voltage signal induced by the collection of a single electron-ion pair. The broken curve show the anode signal obtained when the time constant (RC) of the anode circuit is between the electron and ion collection times (t_e and t_p respectively).

rise caused by the motion and collection of the electron, followed by a slower rise due to the positive ion. The height of the fast portion can be deduced by considering the energy changes caused by the collection of the electron. If the point at which the electron and ion are created has a potential V_e with respect to the chamber wall, and the potential difference between the central anode and the wall is V, then the kinetic energy of the electron when it strikes the anode is $e(V-V_e)$. The dissipation and loss of this energy is reflected in the decrease of the energy of the capacitor from $\frac{1}{2}CV^2$ to $\frac{1}{2}C(V+v_e)^2$, where v_e is the anode signal at the time of arrival of the electron at the anode. Assuming that the positive ion has not moved appreciably at this time, we find that

$$e(V-V_e) = \tfrac{1}{2}CV^2 - \tfrac{1}{2}C(V+v_e)^2 \simeq -CVv_e,$$

and therefore

$$v_e = -\frac{e}{C}\left(1 - \frac{V_e}{V}\right), \tag{9.12}$$

as shown in the figure. The later arrival of the ion gives the further energy loss eV_e and so we regain the expression given by Eq. (9.11).

In the case of a cylindrically shaped ionization counter, as shown in Fig. 9.6, the potential V_e increases quickly near the central anode and is approximately equal to the potential of the chamber wall over most of the volume of the chamber. Therefore the signal v_e has approximately its maximum magnitude of $-e/C$ for most of the photoionization events. For a single electron this signal is too small (1.6×10^{-8} V for $C = 10$ pF) to be detectable by a counter operated in this way. The reason for this is that the anode signal would have to be shaped (as shown by the broken curve in Fig. 9.7) by allowing the slower components to leak away through the load resistance R, with the time constant RC, and the anode signal would also have to be amplified by an amplifier having a bandwidth $\gtrsim 1$ MHz. But the noise level of the load resistor and amplifier would then be far greater (typically $\gtrsim 10^{-4}$ V, see Problem 9.2) than the signal for the electron–ion pair.

When used in the continuous mode the mean current flowing through the detector is

$$\bar{I} = n\alpha e,$$

where n is the rate of arrival of photons and α is the probability that a photon produces an electron–ion pair (α is approximately unity if all the photons are stopped in the gas of the counter). The mean DC signal at the anode is then

$$v = n\alpha eR. \tag{9.13}$$

The noise level can be made very small by using an amplifier having a small bandwidth (see Problem 9.2). The minimum detectable rate n is then comparable with that of the photomultiplier.

Detection of X-rays and γ-rays

When the photon energy is high the initial photoelectron is able to create several further electron–ion pairs by ionizing collisions with the gas atoms. The average energy W required to produce an electron–ion pair by electron collision is typically approximately 30 eV. For example, a 1 MeV γ-ray would give a photoelectron of nearly the same energy, which would give rise to approximately 3×10^4 pairs, all of which would contribute to the signal. The height of the anode pulse (the broken curve of Fig. 9.7) is then large enough to be detectable above the noise level.

Internal amplification

As mentioned before, it is more usual to operate gas-filled ionization counters with a high voltage difference between the anode and outer wall. The initial photoelectrons (or, in the case of X-rays and γ-rays, the initial electrons created by collisions of the photoelectrons) are then accelerated in the electric field of the counter, giving rise to further electron–ion pairs. This internal amplification causes an increase in the signal pulse height. The dependence of the pulse height on the applied voltage is shown schematically in Fig. 9.8.

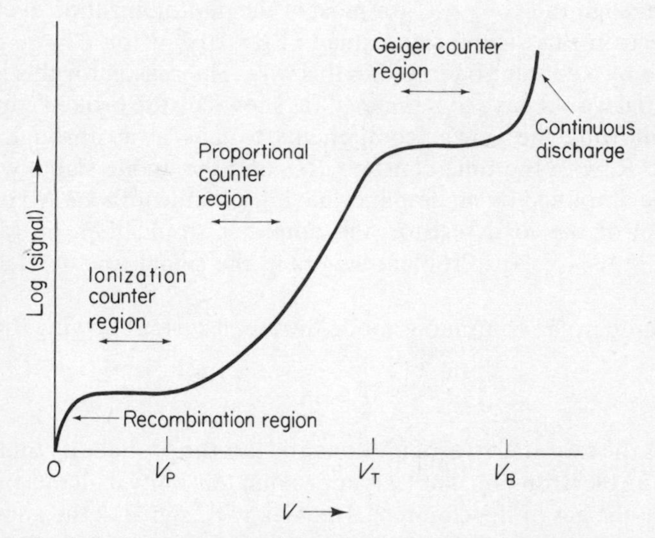

Fig. 9.8. Dependence of the signal height on the voltage applied across a gas ionization counter.

The first region shown in the figure is that in which the applied voltage is so low that some of the electrons and ions recombine before they can be collected. Then follows the region in which all the charged particles are collected, but without internal amplification, as we have discussed so far. In this region the magnitude of the signal is approximately independent of V. Counters operated in this way are known simply as ionization counters.

The next region starts at a voltage V_p which is typically in the range from a few tens of volts to a few hundreds of volts. The electric field is then sufficiently large in the vicinity of the anode ($\gtrsim 10^6$ V/m is needed in the case of argon at 1 atmosphere) that the initially produced electrons sometimes acquire between collisions enough energy to enable them to ionize further gas atoms. The resulting secondary electrons in turn produce further secondaries, giving rise to an *avalanche* of electrons near to the anode. The overall amplification or gain G (i.e. the mean number of electron–ion pairs per initial pair) is dependent on V but is approximately independent of the initial number of pairs or of the position at which they are formed (at least in the case of the usual cylindrical geometry, for which the amplifications occur only when the electrons enter the strong field in the immediate vicinity of the central anode). The signal is therefore proportional to the initial number of pairs, and for this reason the detector is known as a *proportional counter* when it is operated in this region. The value of G is typically between 10^2 and 10^4, which implies that the continuous mode of operation must still usually be used for ultraviolet photons, but single X-ray and γ-ray photons can be detected with the pulsed mode. It is also possible to measure the energies of X-rays and γ-rays, by making use of the fact that W is nearly independent of the initial photoelectron energy; a discussion of how this is done is deferred to the next section on scintillation detectors, for which the method of energy measurement is very similar.

Multi-wire proportional counters

In many experiments (particularly in the field of elementary particle physics) it is necessary to measure the positions and/or directions of incident γ-rays. This can be done by using an array of separate counters, but a more convenient and accurate method is to use a single counter which contains several different anode wires. This type of counter, illustrated in Fig. 9.9, is known as a *multi-wire proportional counter.* There is a strong electrostatic field in the vicinity of each wire, and as in the case of the single-wire counter illustrated in Fig. 9.6, the internal amplification processes occur only in these regions of strong field. The diameters and separations of the wires (typically ~20 μm and ~2 mm respectively) are such that the internal amplification avalanches are each localized at one of the wires. The signal from each wire is separately amplified and recorded, giving the required position sensitivity

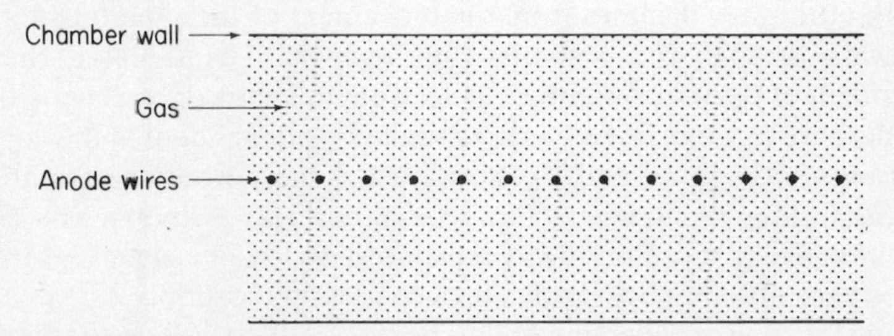

Fig. 9.9. Portion of a multi-wire proportional counter. The diameter of the anode wires is typically about 10 μm, and their spacing is typically about 1 mm. A high potential difference (~4 kV) is applied between the wires and the chamber walls.

in the direction transverse to the wires. It is usual to operate these detectors at quite high electrostatic field levels to give relatively high drift velocities and short electron collection times (~30 ns).

It is possible also to measure the longitudinal position at which an avalanche occurs, by a technique known as 'current sharing'. In this each wire has an amplifier at both of its ends, and the relative magnitude of the signals received at the two ends is used to deduce the position of the avalanche along the wire. Other techniques (e.g. using chamber walls segmented in the longitudinal direction) are also used.

Geiger counters

At higher applied voltages the gain of the gas ionization counter (Fig. 9.6) increases to a value at which the number of slowly moving positive ions created in the avalanche is large enough to electrostatically shield the anode and reduce the strong field in its vicinity. This limits the gain, so that the signal is then no longer proportional to the initial number of electron–ion pairs.

Above the voltage V_T (see Fig. 9.8), which is typically in the range from a few 100 V to a few kV, the avalanche builds up until the positive ion sheath reduces the field near the anode to nearly zero. The avalanche also propagates along the length of the anode. This happens because photons emitted by the excited gas atoms and ions are re-absorbed elsewhere in the gas (or at the cathode) to give further photoionization. The result is that the positive ion sheath is complete, so that no further amplification is possible. The total number of ions in the sheath is then very large and independent of the initial number of pairs. This result can also be achieved by reducing the gas pressure, rather than increasing V, since this increases the mean free path and hence increases the mean energy gained between collisions, giving a higher probability of ionization.

The positive ion sheath has nearly the same potential as the anode, and so the bulk of the signal comes from the movement of the sheath towards the outer wall, through the potential difference V. The signal is therefore independent of the initial number of electron–ion pairs or of the positions at which they are formed. It is also approximately independent of the voltage V in the range between V_T and V_B, above which there would be a continuous discharge inside the chamber. This type of detector is known as a *Geiger–Müller counter*, or more briefly as a *Geiger counter*. The signal is large and requires very little, if any, further (external) amplification.

During the time of collection of the sheath, ~100 to 500 μs, there is only a weak electric field in the region between the sheath and the anode, and the Geiger counter cannot respond to further incident photons. This is known as the *dead time* of the counter. Unless suitable precautions are taken the arrival of the sheath at the outer chamber wall can re-trigger the counter by the ejection of electrons from the wall, leading perhaps to a continuous series of such discharges. The usual way of preventing this is to use suitable external circuitry to reduce the applied voltage below the value V_T after a pulse has started, and to keep the voltage there for a time slightly longer than the ion collection time. The voltage is then returned to its working value and the detector is again able to respond to new ionization events. This method gives a slightly increased dead time for the detector, but ensures that regeneration does not occur.

Detection efficiencies and energy ranges

The detection efficiencies and photon energy ranges of the ionization counter, proportional counter and Geiger counter depend on the transmission characteristics of the window and the photoionization cross section of the gas. In the case of ultraviolet wavelengths the energy range is restricted by the fact that the photons must have an energy greater than the ionization energy of the gas (NO is often used, having the unusually low ionization energy 9.26 eV) but lower than the energy at which the window starts to absorb strongly (LiF has the highest such energy, 11.8 eV). The detection efficiency is very high for ultraviolet photons, however, since they are nearly all absorbed by the gas, and the probability that an absorption results in photoionization is high (typically $>80\%$).

For the detection of X-rays, the window of the detector must have a low mean atomic weight to give a low absorption cross-section, and it must be thin. On the other hand for high efficiency the gas must have a high atomic weight and is often used at high pressure. It is also possible to obtain some crude energy selectivity by an appropriate choice of the gas. For example, krypton ($Z = 36$) has its K edge at 14.3 keV (see Fig. 8.10), and its absorption cross section is such that if a proportional counter is filled to a pressure of 1 atmosphere half the incident radiation would be absorbed in a length of approximately 0.01 m at energies immediately above the edge, but only 0.1% is absorbed in the same length at energies immediately below the edge.

Finally, in the case of γ-rays the gas of the counter is nearly transparent. For example 1 MeV γ-rays are attenuated by a factor e in the distance

$$l = \alpha/\rho,$$

where ρ is the density of the gas and α depends only weakly on Z and has the value $120 \rightarrow 180\ \mathrm{kg/m^2}$. Xenon at a pressure of 100 atmospheres has $\rho = 580\ \mathrm{kg/m^3}$ and $l \simeq 0.3$ m. On the other hand the wall of the chamber has a density an order of magnitude larger than this, and hence the primary photoionization events that trigger the detector tend to take place in the wall rather than the gas. The detection efficiency is nevertheless rather low, of the order of 1% at 1 MeV.

9.3.2 Scintillation detectors

As an alternative to detecting X-rays and γ-rays by collecting the charged particles that they produce, one may make use of the fact that these charged particles excite by collision the atoms or molecules of the medium through which they travel, and that the excited systems often de-excite by emitting photons. In this way part of the energy of the incident photon is transformed into the energy of several lower energy photons. The advantages of doing this are that the lower energy photons (which are usually in the visible and near-ultraviolet range) are easily detectable with a photomultiplier, and that the number of these photons can be used as a measure of the energy of the incident photon. Another advantage is that solid and liquid materials can be used, so giving a higher detection efficiency for γ-rays. These ideas form the basis of the scintillation detector.

Principles of operation

The way in which the scintillation detector works is illustrated schematically in Fig. 9.10. The incident photon (which we shall refer to as a γ-ray, since the scintillation detector is usually used only for energies above about 50 keV) passes through an internally reflecting container and enters the scintillator, shown in the figure as a solid. It may then pass through the scintillator without interacting, or be elastically scattered before passing out of the scintillator; in either case the γ-ray escapes detection. Alternatively the γ-ray may interact with the atoms of the scintillator through the photoelectric process, or by undergoing Compton scattering, or by pair production. These three types of interaction are discussed in Sections 8.2.4, 8.1.2, and 8.2.5 respectively.

The initial interaction shown in the figure is a photoelectric event in which an atomic inner electron is ejected with kinetic energy T_e, leaving behind an excited ion. As the electron travels through the scintillator it loses energy by interacting and colliding with the atoms, and is brought to rest in a short distance (the electron range is approximately $0.4/\rho$ mm for an electron of energy 0.1 MeV and approximately $10/\rho$ mm at 2 MeV, where ρ is the ratio of the density of the scintillator to that of water). Some of the excited and ionized atoms along the track of the electron then decay by emitting photons as illustrated in the figure. The details of the excitation and decay processes depend on the nature of the scintillator, and will be discussed later, but we may note now that the mean number of photons is proportional to the initial electron energy. When the emitted photons arrive at the photomultiplier photocathode, after travelling directly from the electron track or being reflected at the internal surface of the container, they give rise to photoelectrons which are then multiplied and detected, as described in Section 9.2.1.

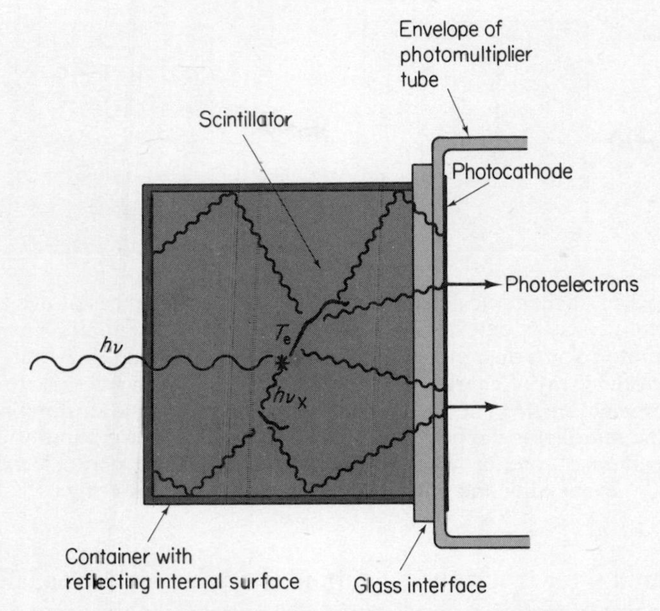

Fig. 9.10. Schematic illustration of the detection of a γ-ray by a scintillator and photomultiplier tube. The initial photoelectric event results in an electron of kinetic energy T_e and an excited ion. As the electron travels through the scintillator it gives rise to photons, some of which produce photoelectrons at the photocathode. The excited ion decays by emitting a characteristic X-ray, which is absorbed in the scintillator to produce a photoelectron and hence further photons. Other sequences of events are also possible.

Some of the energy of the incident γ-ray is initially locked-up in the excitation energy of the absorbing atom, but this is quickly released when this atom emits characteristic X-rays or Auger electrons (see Section 8.2.4). In the example shown in Fig. 9.10 a characteristic X-ray is emitted, which is soon re-absorbed to give a photoelectron and hence further low energy photons. In this way the whole of the energy of the γ-ray contributes to the production of low energy visible (or near u/v) photons. Furthermore, the mean number of these photons is proportional to $h\nu$, and so the mean charge (or equivalently, current) inside the photomultiplier is also proportional to $h\nu$.

Other possible sequences of events are illustrated in Fig. 9.11. In the first example the incident γ-ray is Compton scattered by an atomic electron, giving the electron a recoil energy T_e and leaving the scattered γ-ray with the energy

$$h\nu' = h\nu - T_e.$$

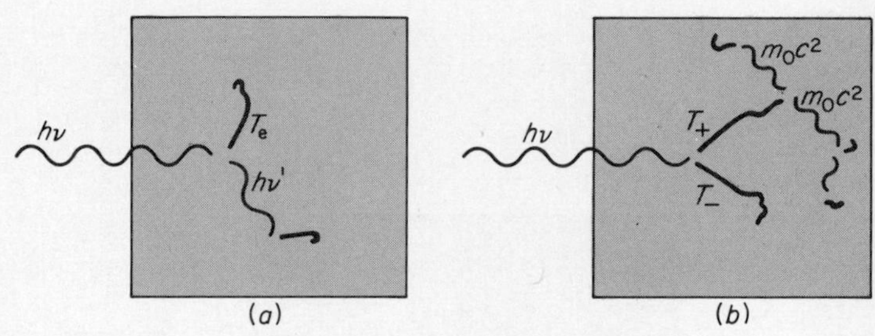

Fig. 9.11. Schematic illustrations of two other sequences of events when a γ-ray is detected in a scintillator. (a) The initial event is Compton scattering, giving an electron of kinetic energy T_e and a scattered γ-ray of energy $h\nu'$; the scattered γ-ray then gives rise to a photoelectron. (b) The initial event is pair production; after stopping in the scintillator the positron is annihilated to give two photons of energy m_0c^2, one of which (in this example) gives a photoelectric event while the other undergoes Compton scattering.

The scattered γ-ray is absorbed in turn to give a photoelectron, although a further Compton scattering is of course also possible. In the second example the incident γ-ray interacts with the field of a nucleus to give an electron and positron which have the combined kinetic energy

$$T_+ + T_- = h\nu = 2m_0c^2.$$

The positron is stopped by the scintillator in the same way as the electron, giving rise to visible and near ultraviolet photons. It then combines with an electron to give positronium (see Section 8.2.5), which quickly decays by being transformed into two annihilation quanta, each of energy $h\nu = m_0c^2$. These are in turn absorbed in the scintillator and converted into electron kinetic energy. In the example shown one of them produces a photoelectron while the other undergoes Compton scattering, followed by the production of a photoelectron.

Attenuation length in the scintillator

The probabilities of the photoelectric, Compton scattering and pair-production processes are related to their cross sections σ_{PE}, σ_C, and σ_{PP} respectively, as discussed in Chapter 8. For a monoenergetic γ-ray beam passing through a scintillator, the intensity after traversing a distance z is given by

$$I(z) = I(0)\,e^{-N_a\sigma z},$$

where

$$\sigma = \sigma_{PE} + \sigma_C + \sigma_{PP}, \tag{9.14}$$

and N_a is the number of scattering or absorbing atoms per unit volume. The cross-section for elastic (Rayleigh) scattering is omitted here since we are interested only in absorption and scattering events which result in a transfer of energy to the scintillator. A more convenient quantity for our present purposes is the attenuation coefficient

$$\mu = \mu_{PE} + \mu_C + \mu_{PP} = N_a(\sigma_{PE} + \sigma_C + \sigma_{PP}). \quad (9.15)$$

The inverse of this is the attenuation length

$$\lambda = \mu^{-1}.$$

This is the distance in which the beam is reduced in intensity by the factor e.

The attenuation coefficients and attenuation lengths of a material of low Z (aluminium, $Z = 13$) and high Z (lead, $Z = 82$) are shown in Fig. 9.12 for photon energies from 10 keV to 100 MeV. The general features of these curves follow from the energy dependences and Z dependences discussed in Chapter 8 for the three processes. For example, the Z dependences of the cross sections per atom (at constant $h\nu$) are approximately Z^5 for the photoelectric effect, approximately Z for Compton scattering, and approximately Z^2 for pair production. It is clear from these Z dependences, and from the data in the figure, that a scintillator must have a high mean value of Z (scintillators are nearly always compounds, not pure elements) if energetic γ-rays are to be detected efficiently.

Incomplete energy deposition

In the examples shown in Figs. 9.10 and 9.11 all the energy of the incident γ-ray is deposited in the scintillator, but, as the reader will have realized, examples also exist in which one or more of the secondary products or scattered γ-rays escapes from the scintillator. The most likely products to escape are the Compton scattering γ-rays and the annihilation quanta. The photoelectrons, Compton recoil electrons, Auger electrons and characteristic X-rays all have a much smaller range or attenuation length, and so are much less likely to escape.

When an annihilation quantum escapes from the scintillator the discrete amount of energy 0.511 MeV is lost to the detection process. On the other hand when a Compton scattered γ-ray is lost the energy carried away occupies a continuum of values since it depends on the angle of scattering θ. The energy which contributes to the detection process by being retained in the scintillator is the kinetic energy T_e of the recoil electron. This energy lies between 0, which occurs when $\theta = 0$, and (see Eq. (8.20))

$$(T_e)_{max} = h\nu \frac{2h\nu}{2h\nu + m_0c^2}, \quad (9.16)$$

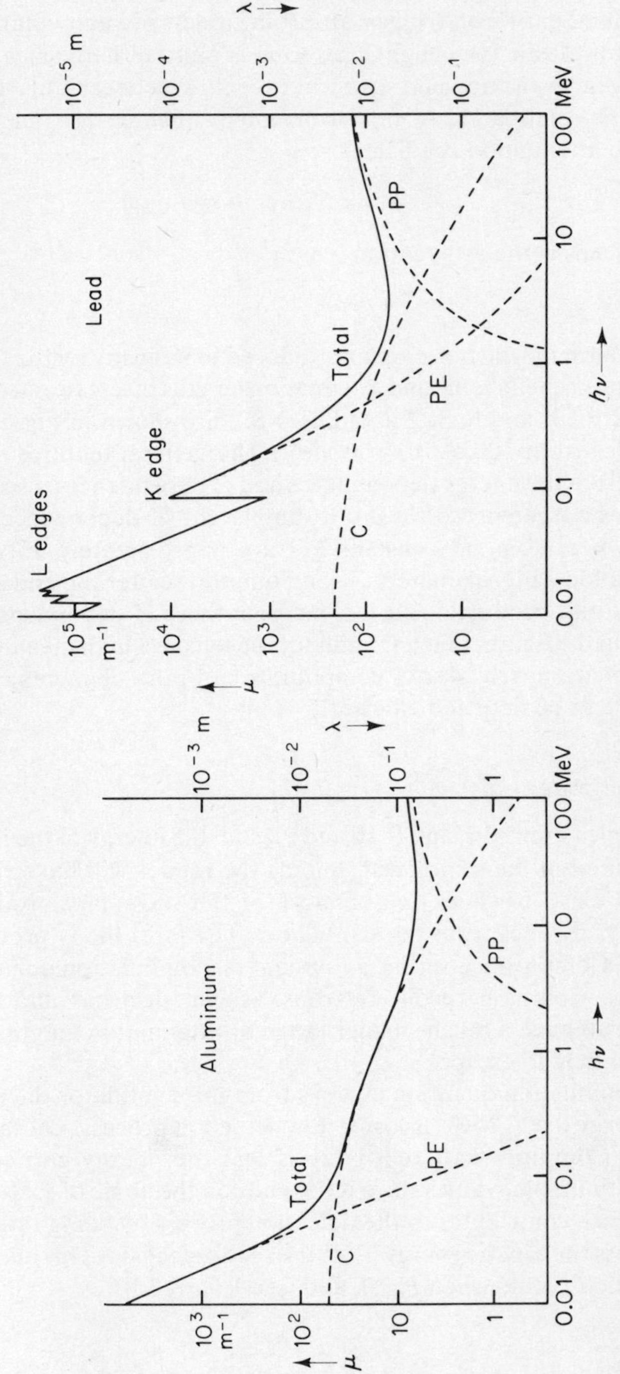

Fig. 9.12. γ-ray attenuation coefficients μ and attenuation lengths λ for aluminium and lead (from the data given by R. D. Evans, *The Atomic Nucleus*, McGraw-Hill, 1955).

which occurs when $\theta = 180°$. The probability distribution for T_e can be obtained by combining the relationship between T_e and θ (Eq. (8.20)) with the cross section for scattering through the angle θ (the Klein–Nishina formula, 8.22). This gives the differential cross section

$$d\sigma = \frac{\pi r_0^2}{\gamma^2}\left\{2+\beta^2\left[\frac{1}{\gamma^2}+\frac{T_e}{\beta h\nu}-\frac{2}{\gamma\beta}\right]\right\}\frac{dT_e}{m_0c^2} \tag{9.17}$$

for scattering in which the recoil energy is between T_e and $T_e + dT_e$. Here r_0 is defined by Eq. (8.4),

$$\gamma = \frac{h\nu}{m_0c^2},$$

and

$$\beta = \frac{T_e}{h\nu - T_e}.$$

Values of $d\sigma/dT_e$ are plotted in Fig. 9.13 for three different values of $h\nu$.

The presence of the sharp discontinuity at $(T_e)_{max}$ has some important consequences for the detection of γ-rays, as we shall soon see. The discontinuity is known as the *Compton edge*. It exists because the value of T_e is insensitive to the angle of scattering when this is large (see Fig. 8.3).

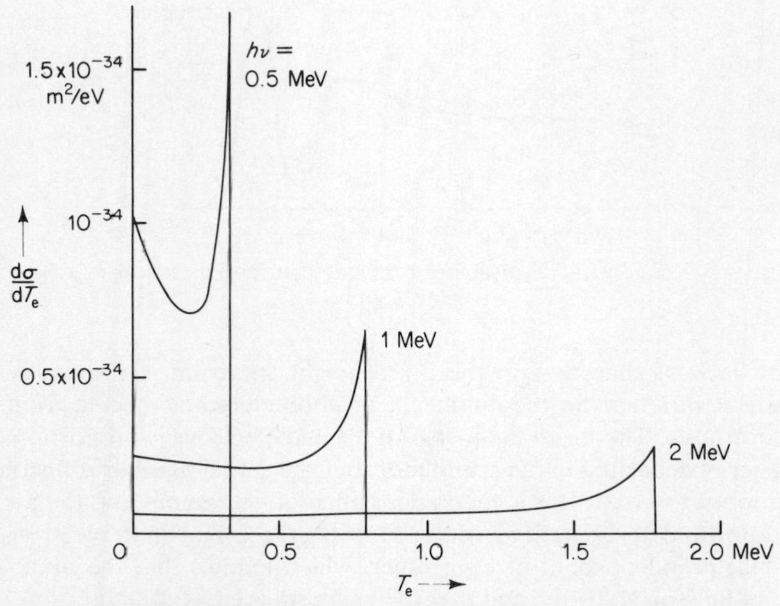

Fig. 9.13. Differential cross section for the production of Compton recoil electrons of energy T_e, for three different energies of the incident γ-ray.

Pulse height distributions

We now have enough information to consider the distribution of pulse heights that is obtained from the photomultiplier when a monoenergetic beam of γ-rays is being detected.

The simplest case is that in which the energy of the incident γ-ray is less than 1.022 MeV, so that the primary event can only be photoelectric ejection or Compton scattering. When it is photoelectric ejection, as illustrated in Fig. 9.10, no energy is lost from the scintillator (unless perhaps the primary event occurs very close to the surface of the scintillator), and so the height of the photomultiplier signal is proportional to $h\nu$. On the other hand when the initial event is Compton scattering, and the scattered γ-ray passes through the scintillator without further interaction, the distribution of signal heights depends on the shape of $d\sigma/dT_e$. But when the scattered γ-ray is itself absorbed (by a further Compton scattering or by a photoelectric event, as in Fig. 9.11(*a*)), the total signal height is again proportional to $h\nu$. The pulse height spectrum therefore consists of a single peak, corresponding to the full energy of the γ-ray, together with a continuous distribution of pulse heights, up to a height corresponding to the energy of the Compton edge. This type of spectrum is shown in Fig. 9.14.

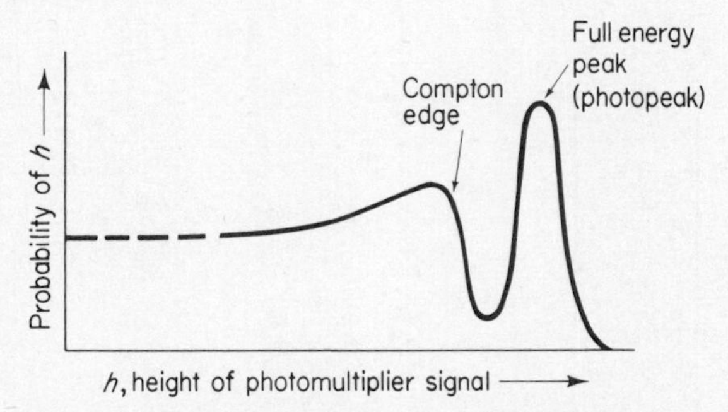

Fig. 9.14. Typical pulse height distribution for a 1 MeV γ-ray.

The lack of sharpness in the pulse height spectrum is caused by the statistical variations in the number n of photoelectrons ejected from the photocathode. The mean number $\bar{n}$ of ejected electrons is proportional to the energy deposited in the scintillator, but the actual number n fluctuates from event to event. It is a good approximation to assume that the atomic excitation and de-excitation events producing these final photoelectrons are statistically independent of each other, which implies that the number n follows Poisson statistics, and that the mean square deviation is

$$\sigma^2 \equiv \overline{(n-\bar{n})^2} = \bar{n}. \qquad (9.18)$$

When $\bar{n}$ is large the probability distribution for n has a Gaussian shape, centred at $\bar{n}$, with a width (full width at half maximum) given by (see Problem 9.3)

$$\Delta n = (8 \ln 2)^{1/2} \sigma = 2.35 \bar{n}^{1/2}. \tag{9.19}$$

The ratio of this width to the mean value is therefore

$$\frac{\Delta n}{\bar{n}} = \frac{2.35}{\bar{n}^{1/2}} = 2.35 \left(\frac{W}{h\nu} \right)^{1/2}, \tag{9.20}$$

where W is the mean energy required to produce one secondary photoelectron. A peak or edge corresponding to the photon energy $h\nu$ is therefore smeared out by an amount corresponding to the energy

$$\Delta E = 2.35 \left(\frac{W}{h\nu} \right)^{1/2} h\nu. \tag{9.21}$$

The ratio $\Delta E / h\nu$ is known as the *resolution* of the peak. For example, W typically has the value 1 keV (see the discussion below), and so the photopeak of Fig. 9.14, corresponding to $h\nu = 1$ MeV, would then have the resolution

$$2.35 (10^{-3})^{1/2} \simeq 7\%. \tag{9.22}$$

Other factors contribute to the widths of the observed peaks in the pulse height distribution. One of these is the finite value δ of the mean secondary emission ratio of the photomultiplier dynodes (see Section 9.2.1). This causes the spread $\Delta n / \bar{n}$ in the number of electrons collected at the anode of the photomultiplier to be larger than the spread in the number of photoelectrons initially ejected from the photocathode. The resolution of a peak corresponding to the energy E is increased to

$$\frac{\Delta E}{E} = 2.35 \left(\frac{W}{E} \right)^{1/2} \left(\frac{\delta}{\delta - 1} \right)^{1/2}.$$

The amplifying and recording system also may have a finite resolution r, giving an overall resolution

$$\frac{\Delta E}{E} = \left[(2.35)^2 \frac{W}{E} \frac{\delta}{\delta - 1} + r^2 \right]^{1/2}. \tag{9.23}$$

When the incident γ-ray has sufficient energy to cause pair production the energy of one or both of the resulting annihilation quanta may be lost, resulting in two discrete peaks. These are called the *single escape* and *double escape* peaks respectively. Alternatively, one or both of the annihilation quanta may be Compton scattered and then lost.

Scintillator materials

Having discussed the origin of the peaks and edges in the pulse height spectrum of a scintillator, we now consider the scintillator itself. Most materials produce some visible light when a γ-ray is absorbed in them, but the efficiency for this process is usually small, and the light that is produced is usually reabsorbed by the materials, so making them unsuitable as scintillators. The material of a scintillator must therefore be carefully chosen. Before describing the properties of two widely used scintillators, we consider the basic excitation and emission processes of the atoms or molecules of which a scintillator is composed.

An atom of the scintillator is excited, and perhaps also ionized, by the transfer of energy from an electron (photoelectron, Compton recoil electron, pair production electron or positron, or Auger electron) when it collides with, or passes nearby, the atom. When the atom is ionized the secondary electron that is released may also have enough kinetic energy to cause further excitation or ionization, although its range is very short and so this happens only in the immediate vicinity of the track of the primary electron. As a result of these primary and secondary processes the initial energy E of the primary electron is converted into the ionization energy I (for simplicity we suppose that all the atoms or molecules of the scintillator have the same value of I) of n_I atoms, and the mean excitation energy $\bar{\mathscr{E}}$ of n_E atoms or ions,

$$E = n_I I + n_E \bar{\mathscr{E}}. \tag{9.24}$$

We noted in the previous section that when the medium is a gas the mean energy W required to produce one ion is almost independent of E. The constancy of W applies also to liquids and solids, and so n_I is proportional to E. It is found that n_E is also usually approximately proportional to E. The energies I and $\mathscr{E}$ eventually reappear in a variety of forms, one of which is the energy of the visible and near ultraviolet photons in which we are presently interested, but whatever the details of the de-excitation process we would expect their rates to be independent of the way in which the atoms are initially excited. This would be false if the excited atoms are formed very close together and interact with each other, but this does not happen when the primary particle is an electron. We conclude therefore that the mean number of visible and ultraviolet photons is proportional to E, for all compositions of scintillator, as assumed in the discussion of the pulse height spectra.

An important requirement for the scintillator is that is must not re-absorb the de-excitation photons, for otherwise they would not be able to reach the photomultiplier. This condition is similar to the condition met in the design of certain laser media (see Chapter 6), and as in that problem it requires the use of special properties of certain materials. Two widely used classes of materials are the organic scintillators and the inorganic crystal scintillators.

Organic scintillators

As a specific example we start by discussing the properties of the scintillator material anthracene. Each molecule has the chemical composition $C_{14}H_{10}$, and consists of three hexagonal benzene rings (C_6H_6, ⬡) joined together in a line and sharing common C–C bonds (i.e. ⬡⬡⬡). It belongs to the class of aromatic organic molecules. As in the case of the dye molecules used in dye lasers (see Section 6.4.3) the anthracene molecule has a manifold of singlet excited states, many of them lying very close in energy to each other, and an essentially separate manifold of triplet excited states (see Fig. 6.20). The singlet excited states are illustrated in Fig. 9.15. Any of the

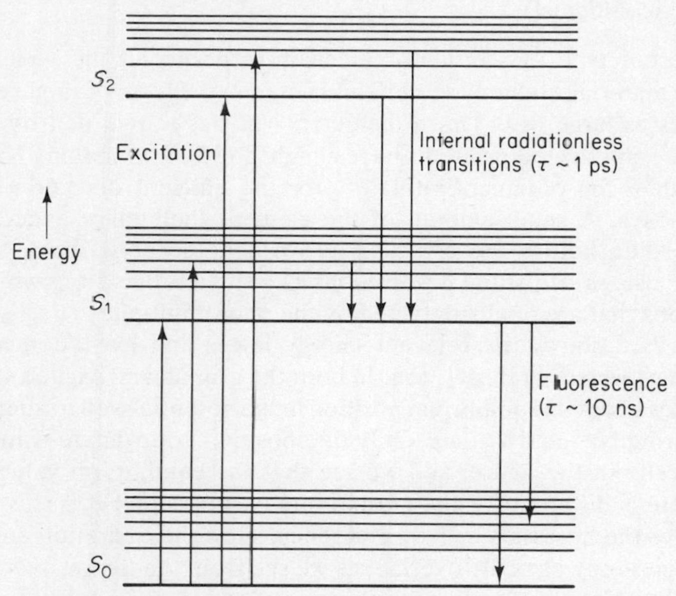

Fig. 9.15. Schematic illustration of the singlet excited states of the anthracene molecule.

sub-levels of the singlet states S_1, S_2 and higher, can be excited by a primary electron travelling through anthracene, and as in the case of the dye molecules, these sub-levels all decay very quickly (~1 ps) by internal radiationless transitions to the lowest vibrational sub-level of the S_1 state, as indicated in the figure. This sub-level then fluoresces (that is, it decays radiatively by an optically allowed transition) with a lifetime which is 27 ns for crystalline anthracene. Again as in the case of dye molecules, the emitted photons have energies which are usually lower, and therefore wavelengths which are longer, than those of absorbed photons (this can be seen in Fig. 6.21, and is also obvious from the lengths of the excitation and fluorescence arrows in Fig. 9.15). This gives the necessary self-transparency to the scintillator. The fluorescence photons have a mean energy of approximately 2.8 eV. An average of approximately 140 eV is expended in creating each one, giving a *fluorescence conversion efficiency* of approximately 2%.

Many other aromatic organic molecules are used as scintillators, either in the pure form or mixed with other compounds, and in the crystalline, plastic or liquid states. In general they have fluorescence conversion efficiencies of the same order as that of anthracene, but their decay lifetimes are often shorter (typically ~2 ns).

Inorganic scintillators

A different type of scintillator mechanism occurs in the widely used inorganic material, sodium iodide. This is used in the form of single crystals, sometimes as large as 0.3 m in diameter. NaI has a high density (3.67×10^3 kg/m^3) and the iodine atoms have a high Z (53), thus making NaI more suitable than the organic scintillators for the efficient detection of high energy γ-rays. A small amount of the element thallium is added to the sodium iodide before the crystal is grown. In the crystalline form the thallium exists as interstitial positive ions Tl^+, and it is these ions which emit the photons that are finally detected by the photomultiplier.

Figure 9.16 shows the relevant energy levels and excitation and de-excitation processes of the Tl^+ ion. In both the ground and excited state the ion vibrates about its equilibrium position in the potential well formed by the neighbouring Na^+ and I^- ions. Only one position co-ordinate is indicated schematically in the figure, and we see that the equilibrium value of the co-ordinate is different for the ground and excited state. It is this feature which gives the necessary self-transparency, since the excitation and emission energies only partially overlap, as we see from the figure.

Crystalline NaI is an insulator, with an energy gap between its valence and conduction bands, and much of the energy of the primary electron goes into exciting valence electrons into the conduction band. An electron excited in this way leaves behind a positive hole which is attracted by the Coulomb field of a conduction electron in such a way that the hole and electron sometimes bind together to form a localized system known as an *exciton*. The primary electron therefore gives rise to excitons along its path, and these are able to diffuse away. When they encounter a ground state Tl^+ ion they give up their energy to the ion, as indicated by the transitions labelled 'excitation' in the figure. The ions vibrate about their equilibrium positions with a thermal

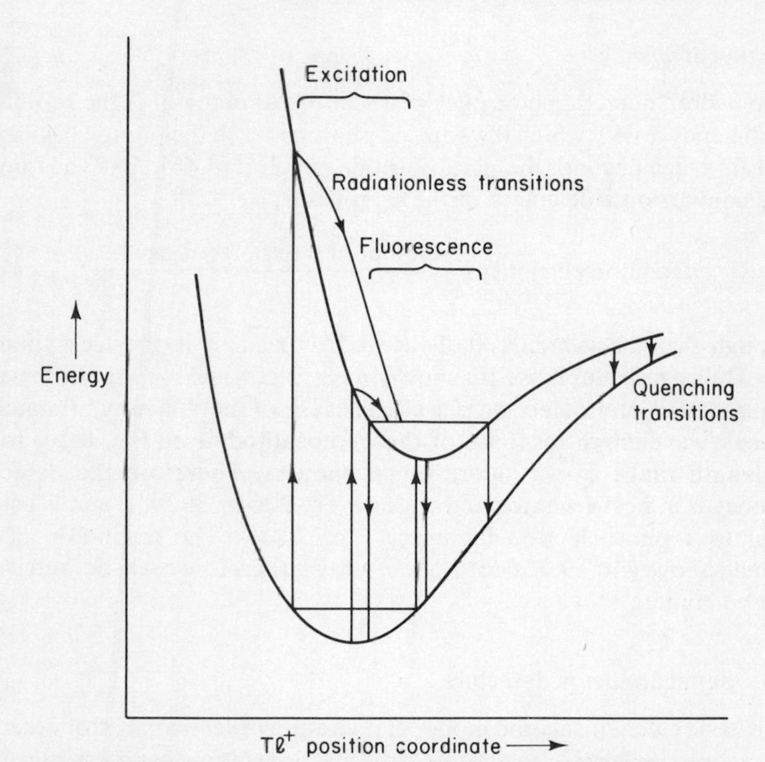

Fig. 9.16. The curves show the dependence of the energy of the Tl^+ ion in its ground and excited state, on its position in the crystal NaI(Tl). The upward and downward pointing arrows indicate excitation and de-excitation processes respectively. Internal radiationless transitions are also marked.

energy ($\sim kT$), and so the transitions occur for a range of positions of the ion, as shown in the figure (to be more exact we should consider the phonons of the crystal, but this is not essential for our present purposes). After this excitation the ions quickly decay non-radiatively to the bottom of the excited state potential well, from where they decay more slowly (with a lifetime of 230 ns) by photon emission, as indicated. Another decay mode, labelled as the quenching transition in the figure, occurs where the two potential curves are separated by $\lesssim kT$, since then the energy difference can be taken up as heat energy (phonons). The transition probability for quenching is smaller than for fluorescence. The average energy of the fluorescence photons is approximately 3 eV, and an average of approximately 33 eV is expended in producing each one, giving the exceptionally high fluorescence conversion efficiency of approximately 9%. A similar inorganic scintillator is CsI(Tl), but this has the significantly longer decay time of 1 μs.

Detection efficiency

The overall detection efficiency of a scintillator detector is the product of the efficiency α with which the emitted photons reach the photocathode, the quantum efficiency ε of the photocathode (see Eq. (8.46)), and the fluorescence conversion efficiency f of the scintillator,

$$\text{detection efficiency} = \frac{\text{number of photoelectrons}}{h\nu} = \alpha\varepsilon f. \qquad (9.25)$$

Although f can be as high as 0.09, as we have seen, α is often less than 0.5 (even when a reflector is used as shown in Fig. 9.10), and ε is usually less than or equal to 0.1 photoelectron per eV, because of the difficulty of matching the peak wavelength response of the photocathode (see Fig. 8.20) to the wavelength range of the fluorescence photons. Therefore the detection efficiency is at best 1 photoelectron for every 200 or 300 eV, and is usually nearer to 1 photoelectron for every 1 or 2 keV. The resolution of 7% obtained above (Eq. (9.22)) for 1 MeV γ-ray is therefore realistic, and it may often be higher.

9.3.3 Semiconductor detectors

The basic mechanisms and modes of operation of semiconductor detectors have many similarities to those of the gas ionization counters and scintillation detectors, and so we are able to use much of the information in the previous two sections. The discussion of the present section will be further shortened by treating in detail only the type of semiconductor detector that is most useful for γ-ray studies, namely the lithium-drifted germanium detector, usually abbreviated as the Ge(Li) detector.

We saw in Section 8.3.2 that when free electrons or holes are created in the region of a p–n junction, or are created elsewhere and diffuse to this region, they are quickly swept across the junction by the electric field that exists there. Use is made of this in the photodiode and the solar cell for example, in which the free electron–hole pairs are created by photon absorption. Fast charged particles, such as electrons, can also create electron–hole pairs as they travel through the semiconductor. This therefore provides a means of detecting γ-rays. The energy of the γ-ray is first converted to electron kinetic energy, as discussed for scintillators, this kinetic energy is then partially used to create electron–hole pairs, and finally the charged particles are swept across the junction and detected as in a gas ionization counter.

There are several advantages in detecting γ-rays in this way. For example the density of the semiconductor is much higher than that of a gas and so the stopping power is much higher than that of gas ionization counters. Another advantage is that the mean energy W needed to create a charged particle pair is an order of magnitude lower for a semiconductor (2.9 eV for Ge and 3.7 eV for Si) than for a gas (typically about 30 eV), and is two or three orders of magnitude lower than the mean energy needed to create a secondary photoelectron in a scintillation detector (~200 to 2000 eV). This means that the energy resolution of the semiconductor detector is much better than that of the other types of detector.

Although p–n junctions similar to those described in Section 8.3.2 for solar cells have been developed for use as X-ray (and also particle) detectors, this type of junction is not useful for γ-ray detection because the γ-rays have a high penetrating power and therefore have only a small probability of being stopped in the short depth of the junction. The depth can be increased by increasing the reverse bias applied across the junction, but this causes large space-charges to be built up, which give a non-uniform electric field inside the junction region. The practical limit to the bias is approximately 1000 V, giving a maximum depth of approximately 5 mm, which would stop only a few percent of γ-rays of a few MeV energy.

This problem has been solved by introducing a long region of intrinsic semiconductor between the p and n regions, as shown in Fig. 9.17. The description intrinsic usually implies that the semiconductor is pure, with no donor or acceptor impurities. When a potential difference is applied across this type of region there are no fixed charges (ionized donors or acceptors) inside it and so the field in the region is uniform. Because of this the depth of the field and the magnitude of the applied potential difference are not limited as they are for a p–n junction. Furthermore the uniform field is analogous to the electric field inside a gas ionization chamber, and any free

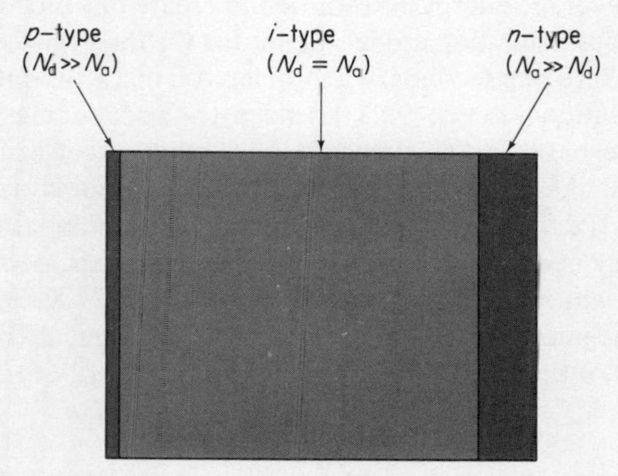

Fig. 9.17. A p–i–n junction for use as a semiconductor detector. N_d and N_a are the densities of donor and acceptor impurities respectively.

charged carriers (electrons or holes) formed inside it are quickly swept to the p or n regions, where they can be collected and registered.

Rather than trying to construct an intrinsic region free of all impurities, a technique is used for forming a region that contains impurities and yet is effectively still intrinsic. This is achieved by introducing into the region exactly equal numbers of donor and acceptor atoms, so that these atoms combine together to form stable molecules which remain electrically neutral and inactive when inside the semiconductor crystal lattice.

The practical details for making the Ge(Li) semiconductor illustrate this technique. The first stage is to prepare a thick slice of p-type germanium (the acceptor impurity is usually aluminium or gallium) of resistivity a few Ω m. A high concentration of lithium (a donor impurity) is then introduced by diffusion into the surface of one face, giving a thin n^+-type region. After this the slice is heated to approximately 60 °C and an electric field is applied across it. This causes the lithium atoms to ionize, and since Li^+ ions have an exceptionally small diameter they quickly drift under the influence of the electric field through the crystal structure of the main body of the slice. Here they encounter the acceptor atoms and form the molecules mentioned above. After cooling the slice and grinding away the uncompensated p-type region it then has the form shown in Fig. 9.17. It retains this form only if it is kept at a low temperature, since otherwise the free lithium in the n-type region would continue to diffuse; the low temperature also reduces the number of thermally excited free carriers, and hence decreases the noise level of the detector. For these reasons Ge(Li) detectors are kept permanently at the temperature of liquid nitrogen (77 K). Detectors made in this way have active regions up to approximately 20 mm deep and up to approximately 100 cm^3 in volume.

Because the average energy W required to create one electron–hole pair is two or three orders of magnitude smaller for Ge than the average energy required to produce one secondary photoelectron in a scintillation detector, the energy resolution (see Eq. (9.21)) is at least an order of magnitude better than that of the best NaI(Tl) detector. One other effect causes a further improvement in the energy resolution of semiconductor detectors. This arises from the fact that the average energy W is comparable with the minimum energy needed to create an electron–hole pair, namely the band gap energy E_g, which has the value 0.74 eV for Ge at 77 K. Because of this the spread of the number n of pairs about the mean value $\bar{n}$ is reduced, and the variance σ^2 of the distribution is given not by Eq. (9.18) but by

$$\sigma^2 = F\bar{n} \tag{9.26}$$

where F is less than unity. As an extreme example, if W were to be equal to E_g the number of pairs would always be exactly $h\nu/W$ (to the nearest integer below) and then σ^2 and F would be zero. The factor F is known as the *Fano factor.* Its calculated value is 0.05 for both Ge and Si, although experimentally measured values tend to be slightly larger than this.

The noise level of the detector and amplifier system gives rise to another contribution to the observed effective energy width ΔE of a peak in the pulse-height distribution. The magnitude of the contribution depends on the detector and amplifier, but not on the energy of the corresponding γ-ray. The contribution is independent of that given by Eq. (9.21), and so the two contributions may be combined in the same way that independent errors are combined, giving

$$\Delta E = [F(2.35)^2 WE + (\Delta E_0)^2]^{1/2} \tag{9.27}$$

where ΔE_0 is the noise contribution (typically a few keV). For example in the case of a 662 keV γ-ray (from the decay of ^{137}Cs) the first term gives a resolution of approximately 0.1% and a width of approximately 0.7 keV, while the term in ΔE_0 usually increases the resolution to approximately 0.3% and the total width to approximately 2 keV. This is to be compared with the resolution of 7% and width of approximately 50 keV obtained with a good NaI(Tl) detector. An example of a high resolution spectrum obtained with a Ge(Li) detector is shown in Fig. 9.18.

9.3.4 Cerenkov shower detectors

For very high energy photons ($\gtrsim$100 MeV) the usual method of detection is by means of a large block of transparent material containing a high proportion of atoms of high atomic number (for example, lead glass),

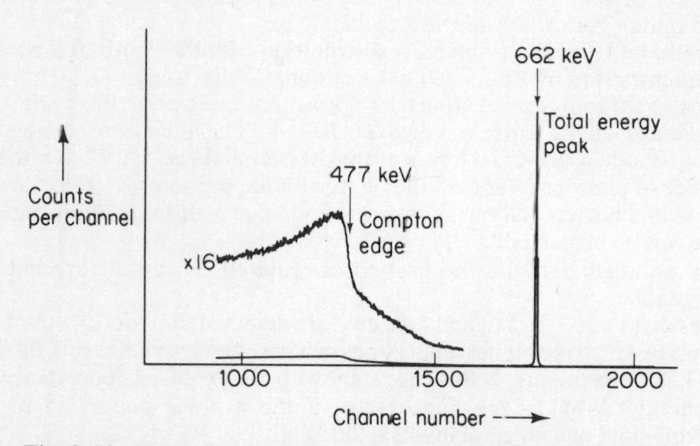

Fig. 9.18. Example of a pulse-height spectrum obtained with a Ge(Li) counter. The incident γ-rays (from the decay of ^{137}Cs) have an energy of 662 keV. The counter has a volume of 25 cm^3.

attached to a large area photomultiplier. A high-energy photon entering the block produces an electron–positron pair (see Section 8.2.5) within a short distance, and then the electron and positron lose energy, mostly by the bremsstrahlung process (see Section 3.1.6). The bremsstrahlung photons in turn give further electron–positron pairs, and in this way a large shower of electrons and positrons develops in the block. The total shower can be contained in a depth of about 50 cm of lead glass, even for photons of energy 100 GeV. As well as producing high-energy bremsstrahlung photons the charged particles of the shower give rise also to visible radiation through the Cerenkov effect (see Section 3.1.8). The mean number of Cerenkov photons is approximately proportional to the initial photon energy, and these photons are detected by the photomultiplier, as in the case of scintillation detectors (see Section 9.3.2). The typical energy resolution of the Cerenkov shower detector is about 8%.

PROBLEMS: CHAPTER 9

(Answers to selected problems are given in Appendix E.)

9.1 What can be deduced about the positions of the valence and conduction bands of an intrinsic semiconductor which has a long wavelength limit of 200 nm for photoelectron production, and of 1.77 μm when used as a photoconductive detector?

9.2 Estimate the minimum internal gain required of a proportional counter for the detection of single ultraviolet photons if the detector capacity is 10 pF and the signal amplifier has an effective noise resistance of 1 MΩ and a bandwidth of 1 MHz. If the same counter is used with no internal amplification to detect a flux of 10^4 ultraviolet photons per second, with a load resistance of 10^{10} Ω, what is the required bandwidth of the amplifier?

9.3 Show that a Gaussian probability distribution with the width (full width at half maximum) given by Eq. (9.19) has a variance equal to $\bar{n}$.

9.4 A weak light source irradiating the photocathode of a photomultiplier produces an average anode current signal of 10^{-9} A. The quantum efficiency of the photocathode is 0.1 and the gain of the dynode chain is 10^7. What is the average number of photons falling on the photocathode per second? What observation time would be necessary to confirm a 1% change in this rate, if tube and resistor noise can be neglected?

9.5 Draw parallels between the properties required of scintillator materials and laser media.

9.6 γ-rays of energy 1.0, 2.0, and 3.0 MeV are detected by a Ge(Li) counter, giving spectra in which the highest energy peaks have effective widths of 2.08, 2.70, and 3.20 keV respectively. Is this consistent with the expected dependence of width on energy? What is the Fano factor if the average energy to produce an electron–hole pair in germanium is 2.85 eV?

APPENDIX 

Some useful formulas in vector calculus

The two sets of co-ordinates used in the book are the cartesian co-ordinates (x, y, z) and the spherical polar co-ordinates (r, θ, ϕ). The relationships connecting the two sets are

$$\left.\begin{array}{lll} x = r\sin\theta\cos\phi, & y = r\sin\theta\sin\phi, & z = r\cos\theta, \\ r = (x^2+y^2+z^2)^{1/2}, & \theta = \cos^{-1}(z/r), & \phi = \tan^{-1}(y/x), \end{array}\right\} \tag{A.1}$$

and the relationships connecting the corresponding unit vectors are

$$\left.\begin{array}{l} \mathbf{i}_r = \mathbf{i}_x \sin\theta\cos\phi + \mathbf{i}_y \sin\theta\sin\phi + \mathbf{i}_z\cos\theta, \\ \mathbf{i}_\theta = \mathbf{i}_x\cos\theta\cos\phi + \mathbf{i}_y\cos\theta\sin\phi - \mathbf{i}_z\sin\theta, \\ \mathbf{i}_\phi = -\mathbf{i}_x\sin\phi + \mathbf{i}_y\cos\phi, \end{array}\right\} \tag{A.2}$$

and

$$\left.\begin{array}{l} \mathbf{i}_x = \mathbf{i}_r\sin\theta\cos\phi + \mathbf{i}_\theta\cos\theta\cos\phi - \mathbf{i}_\phi\sin\phi, \\ \mathbf{i}_y = \mathbf{i}_r\sin\theta\sin\phi + \mathbf{i}_\theta\cos\theta\sin\phi + \mathbf{i}_\phi\cos\phi, \\ \mathbf{i}_z = \mathbf{i}_r\cos\theta - \mathbf{i}_\theta\sin\theta. \end{array}\right\} \tag{A.3}$$

The operator grad (for gradient) operates only on scalar fields $f(\mathbf{r}, t)$. In cartesian co-ordinates it is defined as

$$\operatorname{grad} f = \mathbf{i}_x\frac{\partial f}{\partial x} + \mathbf{i}_y\frac{\partial f}{\partial y} + \mathbf{i}_z\frac{\partial f}{\partial z}. \tag{A.4}$$

This can also be written as

$$\operatorname{grad} f \equiv \nabla f,$$

where the operator ∇ is defined as

$$\nabla = \mathbf{i}_x \frac{\partial}{\partial x} + \mathbf{i}_y \frac{\partial}{\partial y} + \mathbf{i}_z \frac{\partial}{\partial z}. \tag{A.5}$$

In spherical polar co-ordinates,

$$\text{grad } f = \mathbf{i}_r \frac{\partial f}{\partial r} + \mathbf{i}_\theta \frac{1}{r} \frac{\partial f}{\partial \theta} + \mathbf{i}_\phi \frac{1}{r \sin \theta} \frac{\partial f}{\partial \phi}. \tag{A.6}$$

The operators div (for divergence) and curl or rot (for rotation) operate on vector fields $\mathbf{F}(\mathbf{r}, t)$. In cartesian co-ordinates,

$$\text{div } \mathbf{F} \equiv \nabla \cdot \mathbf{F} = \frac{\partial F_x}{\partial x} + \frac{\partial F_y}{\partial y} + \frac{\partial F_z}{\partial z}, \tag{A.7}$$

and

$$\begin{aligned} \text{curl } \mathbf{F} &\equiv \nabla \wedge \mathbf{F} \\ &= \mathbf{i}_x \left(\frac{\partial F_z}{\partial y} - \frac{\partial F_y}{\partial z} \right) + \mathbf{i}_y \left(\frac{\partial F_x}{\partial z} - \frac{\partial F_z}{\partial x} \right) + \mathbf{i}_z \left(\frac{\partial F_y}{\partial x} - \frac{\partial F_x}{\partial y} \right). \end{aligned} \tag{A.8}$$

In spherical polar co-ordinates,

$$\text{div } \mathbf{F} = \frac{1}{r^2} \frac{\partial}{\partial r} (r^2 F_r) + \frac{1}{r \sin \theta} \frac{\partial}{\partial \theta} (\sin \theta \, F_\theta) + \frac{1}{r \sin \theta} \frac{\partial F_\phi}{\partial \phi}, \tag{A.9}$$

and

$$\begin{aligned} \text{curl } \mathbf{F} = \mathbf{i}_r &\frac{1}{r \sin \theta} \left[\frac{\partial}{\partial \theta} (\sin \theta \, F_\phi) - \frac{\partial F_\theta}{\partial \phi} \right] \\ &+ \mathbf{i}_\theta \frac{1}{r} \left[\frac{1}{\sin \phi} \frac{\partial F_r}{\partial \phi} - \frac{\partial (r F_\phi)}{\partial r} \right] + \mathbf{i}_\phi \frac{1}{r} \left[\frac{\partial (r F_\theta)}{\partial r} - \frac{\partial F_r}{\partial \theta} \right]. \end{aligned} \tag{A.10}$$

Four combinations of the operators grad, div, and curl are used in the text. The most frequently used of these is

$$\text{div (grad } f) \equiv \nabla^2 f,$$

where the Laplacian operator ∇^2 is

$$\begin{aligned} \nabla^2 &= \frac{\partial^2}{\partial x^2} + \frac{\partial^2}{\partial y^2} + \frac{\partial^2}{\partial z^2} \\ &= \frac{1}{r^2} \frac{\partial}{\partial r} \left(r^2 \frac{\partial}{\partial r} \right) + \frac{1}{r^2 \sin \theta} \frac{\partial}{\partial \theta} \left(\sin \theta \frac{\partial}{\partial \theta} \right) + \frac{1}{r^2 \sin^2 \theta} \frac{\partial^2}{\partial \phi^2}. \end{aligned} \tag{A.11}$$

This operator can also be applied to a vector field, and then its meaning is

$$\nabla^2 \mathbf{F} = \mathbf{i}_x \nabla^2 F_x + \mathbf{i}_y \nabla^2 F_y + \mathbf{i}_z \nabla^2 F_z. \tag{A.12}$$

Two other combinations are equivalent to null operators. They are

$$\text{curl (grad } f) = 0 \tag{A.13}$$

(this is due to the fact that grad f, having the direction of steepest ascent, cannot return upon itself and therefore cannot have a non-zero rotation), and

$$\text{div}\,(\text{curl}\,\mathbf{F}) = 0. \tag{A.14}$$

The fourth combination is

$$\text{curl}\,(\text{curl}\,\mathbf{F}) = \text{grad}\,(\text{div}\,\mathbf{F}) - \nabla^2\mathbf{F}. \tag{A.15}$$

We have used this only on vector fields having a zero divergence, and so the new combination grad (div $\mathbf{F}$) vanishes in these circumstances. The remaining combinations in pairs of the operators grad, div, and curl (namely div div, curl div, grad curl, and grad grad) have no meaning.

Two integral relationships are used in the text. The first is the divergence theorem

$$\int_S \mathbf{F} \cdot \mathrm{d}\mathbf{S} \equiv \int_V (\text{div}\,\mathbf{F})\,\mathrm{d}V, \tag{A.16}$$

where S is the surface enclosing the volume V. The second is Stokes' theorem

$$\oint_L \mathbf{F} \cdot \mathrm{d}\mathbf{L} \equiv \int_S (\text{curl}\,\mathbf{F}) \cdot \mathrm{d}\mathbf{S}, \tag{A.17}$$

where L is the closed path at the perimeter of the surface S.

APPENDIX 

The birth of the photon

Although the details of the origin of the concept of the photon are not essential to an understanding of the present state of electromagnetic radiation theory, there are few physicists who fail to be fascinated by this and other aspects of the history of their subject. In this appendix Planck's struggle to explain the measured wavelength distribution of thermal radiation is briefly sketched, followed by an equally short account of Einstein's reasons for suggesting that the radiation field is quantized. Full accounts of the various aspects of the subject can be found in the books by Hermann, Kuhn, Whittaker, Kangro, and Stuewer (details of which are given in Appendix D).

When Planck started his studies of radiation theory Wien had already shown by thermodynamic reasoning that the energy density $u(\nu)$ at the frequency ν of radiation in thermal equilibrium at the temperature T must satisfy the displacement law

$$u(\nu) = \nu^3 f(\nu/T), \tag{B1}$$

where f is a function which is not given by these arguments alone. (The relationship (4.16) is also known as Wien's displacement law, but this is essentially only a particular deduction from the more general law (B1).) The form of u that was known to fit the limited experimental measurements available at that time was also due to Wien, and is known as the Wien distribution law,

$$u(\nu) = \frac{8\pi b\nu^3}{c^3} \mathrm{e}^{-a\nu/T}, \tag{B2}$$

where a and b are constants independent of ν and T. Planck attempted to provide a proof of this law by considering the energy and entropy of classical 'resonators' in equilibrium with the radiation. These resonators exist in the walls which enclose the radiation, and are able to emit and absorb radiation. He found it necessary to suppose that the entropy S of a resonator of frequency ν and energy U is given by

$$S = \frac{U}{a\nu} \ln\left(\frac{U}{\mathrm{e}b\nu}\right) \tag{B3}$$

(where e is the base of natural logarithms), but was not able to justify the uniqueness of this form.

Although there had already been indications from experimental measurements that the constants a and b might depend on ν in the region of low ν, Planck learned in 1900 that the accurate measurements of Lummer and Pringsheim definitely disagreed with the Wien distribution law in the wavelength region 12 to 18 μm. Observing that Eq. (B3) gives

$$\frac{\partial^2 S}{\partial U^2} = -\frac{\alpha}{U} \tag{B4}$$

(where $\alpha = \mathrm{e}b/a$), Planck modified this (*Annalen der Physik*, **4**, 553–563 and 711–727 (1901)) by putting

$$\frac{\partial^2 S}{\partial U^2} = -\frac{\alpha}{U(\beta + U)}, \tag{B5}$$

which he described as 'the simplest by far of all the expressions which yield S as a logarithmic function of U (a condition which probability theory suggests) and which besides coincides with the Wien law for small values of U'. Using the standard relationship

$$\frac{\partial S}{\partial U} = \frac{1}{T}, \tag{B6}$$

Eq. (B5) yields after integration

$$\frac{1}{T} = \frac{\alpha}{\beta} \ln\left(\frac{\beta + U}{U}\right), \tag{B7}$$

and hence

$$U = \frac{\beta}{\mathrm{e}^{\beta/\alpha T} - 1}$$

Then using the relationship (also derived classically by Planck)

$$u(\nu) = \frac{8\pi\nu^2}{c^3} U_\nu$$

between the energies at the frequency ν of radiation and elementary resonators in mutual equilibrium, the distribution

$$u(\nu)=\frac{8\pi\nu^2\beta}{c^3}\frac{1}{e^{\beta/\alpha T}-1}$$

is obtained. Since this must satisfy Wien's displacement law (B1), we find that

$$\beta/\alpha=\gamma\nu \tag{B8}$$

and hence

$$u(\nu)=\frac{8\pi\alpha\gamma\nu^3}{c^3}\frac{1}{e^{\gamma\nu/T}-1}, \tag{B9}$$

which is of course the Planck distribution law (4.12). Planck noted that this law was obeyed by the new experimental results.

In trying to justify the new law Planck concentrated on the form of the entropy of the elementary resonators. From Eqs. (B6), (B7), and (B8), and putting

$$h=\alpha\gamma,$$

we see that

$$S=\frac{h}{\gamma}\left[\left(1+\frac{U}{h\nu}\right)\ln\left(1+\frac{U}{h\nu}\right)-\frac{U}{h\nu}\ln\left(\frac{U}{h\nu}\right)\right]. \tag{B10}$$

Planck considered N resonators at the frequency ν, each having the mean energy U, and he divided the total energy NU into elements of width ε. The number of such elements is therefore

$$P=NU/\varepsilon.$$

Using the arguments that Boltzmann had employed for evaluating the entropy of a gas, Planck noted that the number of ways in which the P indistinguishable energy elements can be distributed amongst the N distinguishable resonators is

$$R=\frac{(N+P-1)!}{(N-1)!P!}. \tag{B12}$$

If N and P are sufficiently large Stirling's approximation can be used, and then Eqs. (B11) and (B12) yield

$$\ln R=N\left[\left(1+\frac{U}{\varepsilon}\right)\ln\left(1+\frac{U}{\varepsilon}\right)-\frac{U}{\varepsilon}\ln\frac{U}{\varepsilon}\right]. \tag{B13}$$

Comparing Eqs. (B10) and (B13), and using the fact that the entropy of a state which can be attained in W different ways is

$$S = k \ln W, \tag{B14}$$

Planck found that the energy elements must be proportional to ν,

$$\varepsilon = h\nu, \tag{B15}$$

where h is the universal constant that now bears his name.

Kuhn (see Appendix D) argues that Planck did not at this time, either in his papers or lectures, infer a discrete energy distribution for his resonators, as assumed by almost all later authors, but that he supposed only that the continuous energy U of an individual resonator has to be divided into ranges

$$nh\nu \leqslant U < (n+1)h\nu$$

for the purpose of evaluating the entropy. Planck's treatment of the radiation field itself was certainly purely classical, with no hint that the radiation energy might be quantized in the same units. It appears that Einstein was the first to suggest (*Annelen der Physik*, **20**, 199–206 (1906)) that the energy of an elementary resonator can take only values which are integral multiples of $h\nu$, and that during absorption and emission of radiation the energy of the radiator changes discontinuously by an integral multiple of $h\nu$.

Einstein was also the first to suggest (*Annalen der Physik*, **17**, 132–148 (1905)) that the radiation field itself is quantized. He considered the entropy of the field rather than that of the oscillators. Wien's form (B2) of the radiation law (valid at high frequencies) gives the radiant energy

$$E = \frac{8\pi h\nu^3 V\,\Delta\nu}{c^3}\,\mathrm{e}^{-h\nu/kT}$$

in the frequency interval $\Delta\nu$ in a cavity of volume V. Rearranging, this gives

$$\frac{1}{T} = -\frac{k}{h\nu}\ln\left(\frac{E}{a\nu^3 V\,\Delta\nu}\right),$$

where a is a constant. Equation (B6) then gives the entropy S in terms of E and V,

$$S = -\frac{kE}{h\nu}\left[\ln\left(\frac{E}{a\nu^3 V\,\Delta\nu}\right) - 1\right].$$

If the same radiant energy E, in the same frequency interval, is now confined in a larger cavity of volume V_0, the change in entropy is

$$S - S_0 = \frac{kE}{h\nu}\ln\left(\frac{V}{V_0}\right). \tag{B16}$$

But the relationship between the entropy of a system and the number of ways W in which the system can be attained is given by Eq. (B14), and so the difference in entropy of two states of a system is

$$S - S_0 = k \ln \frac{W}{W_0} = k \ln \frac{P}{P_0} \tag{B17}$$

where P and P_0 are the probabilities of the two states. Combining Eqs. (B16) and (B17) shows that the relative probability of finding the radiant energy E in the volumes V and V_0 is

$$\frac{P}{P_0} = \left(\frac{V}{V_0}\right)^{E/h\nu}. \tag{B18}$$

Now in the case of a tenuous gas of N independent molecules contained inside a cavity of volume V_0, the probability that any one molecule occupies a subspace of volume V is

$$p = \frac{V}{V_0}$$

and therefore the probability that all N molecules occupy this same subspace is

$$P = p^N = \left(\frac{V}{V_0}\right)^N. \tag{B19}$$

Einstein compared Eq. (B18) with Eq. (B19) and noted that as far as the probability and entropy of the radiation at high frequency is concerned, the radiation behaves as though its energy is subdivided into independent quanta of magnitude $h\nu$. This observation marked the birth of the photon. Einstein went on in the same paper to explain the photoelectric effect in terms of single photons giving up their energy to eject photoelectrons from metal surfaces. As outlined in Table 1.1, many years were to pass before the concept of the photon was generally accepted.

APPENDIX

Quantum mechanical treatment of the radiation field

The purpose of this appendix is to give a brief introduction to the quantum mechanical treatment of the electromagnetic field, so that we may gain a deeper insight into the photon and wave nature of the field. We must start by considering certain properties of the harmonic oscillator.

A particle of mass m moving in a one-dimensional potential $\frac{1}{2}bx^2$ has the total energy

$$U=\tfrac{1}{2}m\left(\frac{\mathrm{d}x}{\mathrm{d}t}\right)^2+\tfrac{1}{2}bx^2=\tfrac{1}{2}p^2/m+\tfrac{1}{2}bx^2. \tag{C1}$$

The wavefunctions $\psi_n(x)$ for the motion are solutions of the Schroedinger equation

$$H\psi_n(x)=E_n\psi_n(x) \tag{C2}$$

where the Hamiltonian H is given by

$$H=-\frac{\hbar^2}{2m}\frac{\mathrm{d}^2}{\mathrm{d}x^2}+\tfrac{1}{2}bx^2. \tag{C3}$$

The discrete energies that the oscillator can possess are

$$E_n=\hbar\omega(n+\tfrac{1}{2}) \tag{C4}$$

where $\omega(=\sqrt{b/m})$ is the angular frequency of the classical motion.

Let us now express H in terms of the *destruction operator* (otherwise known as the annihilation operator)

$$a=\frac{jp}{\alpha\hbar\sqrt{2}}+\frac{\alpha x}{\sqrt{2}}=\frac{1}{\alpha\sqrt{2}}\frac{\mathrm{d}}{\mathrm{d}x}+\frac{\alpha x}{\sqrt{2}} \tag{C5}$$

and the *creation operator*

$$a^{\dagger}=-\frac{jp}{\alpha\hbar\sqrt{2}}+\frac{\alpha x}{\sqrt{2}}=-\frac{1}{\alpha\sqrt{2}}\frac{\mathrm{d}}{\mathrm{d}x}+\frac{\alpha x}{\sqrt{2}}, \tag{C6}$$

where p is the momentum operator and

$$\alpha=\left(\frac{m\omega}{\hbar}\right)^{1/2}. \tag{C7}$$

When taking the products of the operators a and $a^{\dagger}$ we must be careful to distinguish between the operations

$$(x)\left(\frac{\mathrm{d}}{\mathrm{d}x}\right)f(x)=xf'(x)$$

and

$$\left(\frac{\mathrm{d}}{\mathrm{d}x}\right)(x)f(x)=f(x)+xf'(x).$$

We find that

$$a^{\dagger}a=-\frac{1}{2\alpha^2}\left(\frac{\mathrm{d}^2}{\mathrm{d}x^2}-\alpha^4x^2\right)-\frac{1}{2},$$

$$aa^{\dagger}=-\frac{1}{2\alpha^2}\left(\frac{\mathrm{d}^2}{\mathrm{d}x^2}-\alpha^4x^2\right)+\frac{1}{2},$$

and so

$$aa^{\dagger}-a^{\dagger}a=1.$$

The Hamiltonian is therefore

$$H=\hbar\omega(a^{\dagger}a+\tfrac{1}{2})=\hbar\omega(aa^{\dagger}-\tfrac{1}{2}).$$

If we now go further and look at the product of H with $a^{\dagger}$, we find that

$$\begin{aligned}Ha^{\dagger}&=\hbar\omega(a^{\dagger}aa^{\dagger}+\tfrac{1}{2}a^{\dagger})=\hbar\omega[a^{\dagger}(1+a^{\dagger}a)+\tfrac{1}{2}a^{\dagger}]\\&=a^{\dagger}H+\hbar\omega a^{\dagger}.\end{aligned} \tag{C8}$$

The importance of this relationship becomes apparent when we apply $Ha^{\dagger}$ to one of the wavefunctions ψ_n,

$$Ha^{\dagger}\psi_n=(a^{\dagger}H+\hbar\omega a^{\dagger})\psi_n=a^{\dagger}(E_n+\hbar\omega)\psi_n,$$

and then rewrite this result as

$$H(a^\dagger \psi_n) = (E_n + \hbar\omega)(a^\dagger \psi_n). \tag{C9}$$

We see that the function $(a^\dagger \psi_n)$ is also a wavefunction of the motion, and that the energy associated with it is $E_n + \hbar\omega$. This implies that if the oscillator can have the energy E_n, then it can also have all the energies $E_n + m\hbar\omega$, where m is any positive integer.

As a check on this we can apply $a^\dagger$ to the ground state wavefunction

$$\psi_0(x) = \alpha^{1/2}\pi^{-1/4} e^{-\frac{1}{2}\alpha^2 x^2},$$

obtaining

$$a^\dagger \psi_0(x) = \alpha^{3/2}\pi^{-1/4} 2^{1/2} x\, e^{-\frac{1}{2}\alpha^2 x^2} = \psi_1(x),$$

the wavefunction of the first excited state. The effect of $a^\dagger$ is to create one further quantum of energy: hence its name. More generally, it can be shown that

$$a^\dagger \psi_n = (n+1)^{1/2} \psi_{n+1}. \tag{C10}$$

Similarly the destruction operator a reduces the number of energy quanta by one,

$$a\psi_n = n^{1/2} \psi_{n-1}. \tag{C11}$$

In the special case $n = 0$, Eq. (C11) becomes

$$a\psi_0 = 0.$$

This result is useful because it shows that

$$E_0\psi_0 = H\psi_0 = \hbar\omega(a^\dagger a + \tfrac{1}{2})\psi_0 = \tfrac{1}{2}\hbar\omega\psi_0,$$

so confirming that the zero-point energy is $\frac{1}{2}\hbar\omega$, and that the energy spectrum is given correctly by Eq. (C4).

A useful operator that we shall need shortly is the *number operator*

$$\hat{n} = a^\dagger a. \tag{C12}$$

It has this name because when it is applied to the eigenfunction ψ_n it gives

$$\hat{n}\psi_n = a^\dagger n^{1/2} \psi_{n-1} = n\psi_n. \tag{C13}$$

To relate these properties of the harmonic oscillator wavefunctions and operators to the quantization of the electromagnetic field let us consider the energy in a normal mode of a cavity. The simplest mode to start with is that formed between two perfectly conducting parallel sheets of infinite extent, separated by an integral number of half wavelengths. If the plates are normal to the z-direction and the electric field is in the x-direction, the electric and magnetic fields of the standing wave between the plates are

$$\mathbf{E} = \mathbf{i}_x \frac{E_0}{2}[\cos(\omega t - kz + \phi) - \cos(\omega t + kz + \phi)]$$
$$= \mathbf{i}_x E_0 \sin(kz)\sin(\omega t + \phi), \tag{C14}$$

and

$$\mathbf{B} = \mathbf{i}_y \frac{E_0}{c}\cos(kz)\cos(\omega t + \phi). \tag{C15}$$

The magnetic vector potential for this mode is

$$\mathbf{A} = \mathbf{i}_x \left(\frac{E_0}{kc}\right)\sin(kz)\cos(\omega t + \phi). \tag{C16}$$

The fields **E** and **B** are out of phase by 90°; when E^2 is a maximum, B^2 is a minimum, and vice-versa. Averaging over the z direction we find that the total energy in a volume V is

$$U = \tfrac{1}{2}(\varepsilon_0 E^2 + \mu_0^{-1} B^2)$$
$$= \frac{V}{2}[\varepsilon_0 E_0^2 \sin^2(\omega t + \phi) + \varepsilon_0 E_0^2 \cos^2(\omega t + \phi)] = \tfrac{1}{2} V \varepsilon_0 E_0^2. \tag{C17}$$

The total energy is shared equally between the electric and magnetic fields, and the way in which it oscillates between the two forms is analogous to the oscillation between the potential energy and kinetic energy in the harmonic oscillator.

Another way to see the analogy with the harmonic oscillator is to express U in terms of **A**. We find that

$$U = \tfrac{1}{2} V \varepsilon_0 \left[\left(\frac{\mathrm{d}A_x}{\mathrm{d}t}\right)^2 + \omega^2 A_x^2\right] \tag{C18}$$

where

$$A_x = \frac{E_0}{\omega}\cos(\omega t + \phi).$$

Comparing Eqs. (C1) and (C18) we see that suitable analogues of the position x and momentum p of the harmonic oscillator are

$$x \to \left(\frac{V\varepsilon_0}{m}\right)^{1/2} A_x = \left(\frac{V\varepsilon_0}{m}\right)^{1/2} \frac{E_0}{\omega}\cos(\omega t + \phi),$$

$$p \to (V\varepsilon_0 m)^{1/2} \frac{\mathrm{d}A_x}{\mathrm{d}t} = (V\varepsilon_0 m)^{1/2} E_0 \sin(\omega t + \phi).$$

Going one step further this analogy leads to the destruction and creation operators

$$a = \frac{jp}{\alpha\hbar\sqrt{2}} + \frac{\alpha x}{\sqrt{2}} \rightarrow \left(\frac{V\varepsilon_0}{2\hbar\omega}\right)^{1/2} E_0[j \sin(\omega t + \phi) + \cos(\omega t + \phi)]$$

$$= \left(\frac{V\varepsilon_0}{2\hbar\omega}\right)^{1/2} E_0 \, e^{j(\omega t + \phi)},$$

$$a^\dagger = a^*,$$

and to the relationship

$$A_x = \left(\frac{\hbar}{2V\varepsilon_0\omega}\right)^{1/2} (a + a^\dagger).$$

This analogy implies that the total electromagnetic energy in the mode has the discrete energy spectrum

$$U = \hbar\omega(n + \tfrac{1}{2})$$

regardless of the volume V of the cavity in which the mode exists, so that the energy can be changed only in units of $\hbar\omega$.

Now let us look at the quantization of the electromagnetic field more rigorously. In making the transition from the classical to the quantum mechanical formulations it is usual to define a and $a^\dagger$ in terms of free plane waves rather than standing waves (so that they can be applied to a wider range of problems). For a plane wave of wave-vector $\mathbf{k}$ and polarization direction $\mathbf{i}_p$ (where $\mathbf{k} \cdot \mathbf{i}_p = 0$) the vector potential can be written as

$$\mathbf{A} = \mathbf{i}_p\{A_k \, e^{j(\omega t - \mathbf{k} \cdot \mathbf{r})} + A_k^* \, e^{-j(\omega t - \mathbf{k} \cdot \mathbf{r})}\}. \tag{C19}$$

The destruction and creation operators are now associated with A_k and A_k^* respectively, and are defined by

$$a = \left(\frac{2\varepsilon_0\omega V}{\hbar}\right)^{1/2} A_k \tag{C20}$$

and

$$a^\dagger = \left(\frac{2\varepsilon_0\omega V}{\hbar}\right)^{1/2} A_k^*. \tag{C21}$$

Then

$$\hat{\mathbf{A}} = \mathbf{i}_p\left(\frac{\hbar}{2\varepsilon_0\omega V}\right)^{1/2} \{a \, e^{j(\omega t - \mathbf{k} \cdot \mathbf{r})} + a^\dagger \, e^{-j(\omega t - \mathbf{k} \cdot \mathbf{r})}\}. \tag{C22}$$

The symbol $\hat{\mathbf{A}}$ is used to emphasize that this vector potential is now an operator. It has a physically significant meaning only when it is applied to a function, such as one of the harmonic oscillator functions ψ_n. We may ask, what is the argument of this function, and in what space does it and the operations a and $a^\dagger$ exist? In fact the only quantities that are measured physically are the expectation values

$$\bar{X} = \int \psi^*(x) X(x) \psi(x)\, \mathrm{d}x \qquad \text{(C23)}$$

corresponding to the operators X (where X represents the quantity being measured). We shall see below that these expectation values are evaluated without having to specify the space of the functions $\psi(x)$ and operators $X(x)$. The electric and magnetic field operators associated with $\hat{\mathbf{A}}$ are (from Eqs. (2.38) and (3.8), taking $\phi = 0$),

$$\hat{\mathbf{E}} = \mathbf{i}_p \left(\frac{\hbar\omega}{2\varepsilon_0 V}\right)^{1/2} \{-aj\, \mathrm{e}^{\mathrm{j}(\omega t - \mathbf{k}\cdot\mathbf{r})} + a^\dagger j\, \mathrm{e}^{-\mathrm{j}(\omega t - \mathbf{k}\cdot\mathbf{r})}\} \qquad \text{(C24)}$$

and

$$\hat{\mathbf{B}} = (\mathbf{k} \wedge \mathbf{i}_p) \left(\frac{\hbar}{2\varepsilon_0 \omega V}\right)^{1/2} \{-aj\, \mathrm{e}^{j(\omega t - \mathbf{k}\cdot\mathbf{r})} + a^\dagger j\, \mathrm{e}^{-\mathrm{j}(\omega t - \mathbf{k}\cdot\mathbf{r})}\}. \qquad \text{(C25)}$$

As a first example of the use of these operator forms, let us find the electric and magnetic energies in the field described by the function ψ_n. Using the abbreviation

$$e = \mathrm{e}^{\mathrm{j}(\omega t - \mathbf{k}\cdot\mathbf{r})},$$

the result of applying $|\hat{\mathbf{E}}|^2$ to ψ_n is

$$(\hat{\mathbf{E}}^* \cdot \hat{\mathbf{E}})\psi_n = \frac{\hbar\omega}{2\varepsilon_0 V} \{a^\dagger je^* - aje\}\{-aje + a^\dagger je^*\}\psi_n$$

$$= \frac{\hbar\omega}{2\varepsilon_0 V} \{a^\dagger je^* - aje\}\{-jen^{1/2}\psi_{n-1} + je^*(n+1)^{1/2}\psi_{n+1}\}$$

$$= \frac{\hbar\omega}{2\varepsilon_0 V} \{n\psi_n + (n+1)\psi_n + \text{terms in } \psi_{n-2} \text{ and } \psi_{n+2}\}.$$

Therefore the energy in the electric field is

$$\frac{V}{2} \int \psi_n^* (\varepsilon_0 |\hat{\mathbf{E}}|^2) \psi_n\, \mathrm{d}\tau = \tfrac{1}{2}\hbar\omega(n + \tfrac{1}{2}).$$

The same value is found for the energy in the magnetic field, and so the total energy is $\hbar\omega(n + \frac{1}{2})$. More generally, the Hamiltonian of the field can be written as

$$\hat{H} = \frac{V}{2}(\varepsilon_0 |\hat{\mathbf{E}}|^2 + \mu_0^{-1} |\hat{\mathbf{B}}|^2) = \sum_{\mathbf{k},p} (a_{\mathbf{k},p}^\dagger a_{\mathbf{k},p} + \tfrac{1}{2})\hbar\omega_k \qquad \text{(C26)}$$

where the labels **k** and p specify the wave-vector and direction of polarization of the fundamental modes of the field. Each mode then has the total energy $\hbar\omega_k(n_k+\frac{1}{2})$.

As a second example we consider an important state $\psi(\alpha)$ called the *coherent state* of the electromagnetic field. We shall see that it is the quantum mechanical state which most closely approaches the classical electromagnetic wave. It is an eigenstate of the destruction operator,

$$a\psi(\alpha)=\alpha\psi(\alpha),$$

but it is not a pure state having a definite number n of quanta. Instead it is a superposition of pure states ψ_n, but having a form which ensures that there is a definite mean number $\bar{n}$ of quanta in the state. In fact the coherent state is defined as

$$\psi(\alpha)=\exp\left(-\tfrac{1}{2}|\alpha|^2\right)\sum_{n=0}^{\infty}\frac{\alpha^n}{(n!)^{1/2}}\psi_n \tag{C27}$$

when the mean number is $\bar{n}=|\alpha|^2$. The exponential factor ensures that the coherent state function is normalized to unity. To check that this definition gives the mean number $|\alpha|^2$ of quanta, we use Eqs. (C12) and (C13) to evaluate

$$\bar{n}=\int\psi^*(\alpha)\hat{n}\psi(\alpha)\,d\tau=e^{-|\alpha|^2}\int\psi^*(\alpha)\sum_n\frac{\alpha^n n}{(n!)^{1/2}}\psi_n\,d\tau$$

$$=e^{-|\alpha|^2}\sum_{n=0}^{\infty}\frac{|\alpha|^{2n}n}{n!}=e^{-|\alpha|^2}|\alpha|^2\sum_{m=0}^{\infty}\frac{|\alpha|^{2m}}{m!}=|\alpha|^2. \tag{C28}$$

To investigate the fluctuations in the number of quanta in the coherent state we now evaluate

$$\overline{n^2}=\int\psi^*(\alpha)\hat{n}^2\psi(\alpha)\,d\tau=e^{-|\alpha|^2}\sum_n\frac{|\alpha|^{2n}n^2}{n!}$$

$$=e^{-|\alpha|^2}\sum_n\frac{|\alpha|^{2n}[n(n-1)+n]}{n!}=|\alpha|^4+|\alpha|^2.$$

Therefore the root-mean-square fluctuation is

$$\Delta n=(\overline{n^2}-\bar{n}^2)^{1/2}=|\alpha|=\bar{n}^{1/2}. \tag{C29}$$

$\therefore\quad \Delta n/\bar{n}=(\bar{n})^{-1/2}$.

As $\bar{n}$ becomes larger the relative fluctuation $\Delta n/\bar{n}$ becomes smaller and less important.

To find the electric field strength of the coherent state we need to evaluate

$$\int \psi^*(\alpha)\hat{E}\psi(\alpha)\,\mathrm{d}\tau$$
$$= \left(\frac{\hbar\omega}{2\varepsilon_0 V}\right)^{1/2} \mathrm{e}^{-|\alpha|^2} \int \psi^*(\alpha) \sum_n \frac{\alpha^n}{(n!)^{1/2}} \{-jen^{1/2}\psi_{n-1}$$
$$+ je^*(n+1)^{1/2}\psi_{n+1}\}\,\mathrm{d}\tau$$
$$= \left(\frac{\hbar\omega}{2\varepsilon_0 V}\right)^{1/2} \{-je\alpha + je^*\alpha^*\} = 2\left(\frac{\hbar\omega}{2\varepsilon_0 V}\right)^{1/2} |\alpha| \sin(\omega t - \mathbf{k}\cdot\mathbf{r} + \theta)$$
$$= \left(\frac{2\bar{n}\hbar\omega}{\varepsilon_0 V}\right)^{1/2} \sin(\omega t - \mathbf{k}\cdot\mathbf{r} + \theta), \tag{C30}$$

where α is here given the general form

$$\alpha = |\alpha|\,\mathrm{e}^{\mathrm{j}\theta}.$$

On the other hand $\overline{E^2}$ is given by

$$\int \psi^*(\alpha)\hat{E}^2\psi(\alpha)\,\mathrm{d}\tau = \left(\frac{\hbar\omega}{2\varepsilon_0 V}\right)[4|\alpha|^2 \sin^2(\omega t - \mathbf{k}\cdot\mathbf{r} + \theta) + 1],$$

so that the root-mean-square fluctuation in the field is

$$\Delta E = (\overline{E^2} - \bar{E}^2)^{1/2} = \left(\frac{\hbar\omega}{2\varepsilon_0 V}\right)^{1/2}. \tag{C31}$$

We see that the relative fluctuation $\Delta E/\bar{E}$ is also proportional to $(\bar{n})^{-1/2}$.

The phase of the coherent state is less easy to establish. Although the classical electromagnetic wave can have a definite measurable phase, the corresponding phase of a quantum-mechanical wavefunction is in general an arbitrary and unobservable quantity. It is possible however to define a phase operator $\hat{\phi}$ which relates to the classical phase ϕ in the limit of very large $\bar{n}$. It is given by

$$a = (\hat{n} + 1)^{1/2}\,\mathrm{e}^{\mathrm{j}\hat{\phi}}.$$

It is then possible to show that in the limit of large $\bar{n}$ the root-mean-square fluctuations in n, $\sin\phi$, and $\cos\phi$ are related by

$$\Delta n\; \Delta(\cos\phi) = \tfrac{1}{2}|\sin\theta|, \tag{C32}$$

$$\Delta n\; \Delta(\sin\phi) = \tfrac{1}{2}|\cos\theta|. \tag{C33}$$

In these circumstances $\Delta(\cos\phi)$ and $\Delta(\sin\phi)$ are small, and it follows also that

$$\phi = \theta$$

and

$$\Delta n\ \Delta\phi = \tfrac{1}{2}. \tag{C34}$$

It can be shown more generally that for any state of the radiation field

$$\Delta n\ \Delta\phi \geqslant \tfrac{1}{2}. \tag{C35}$$

This is the *uncertainty relationship* connecting the fluctuations in n and ϕ. It is analogous to the relationship

$$\Delta p\ \Delta x \geqslant \tfrac{1}{2}\hbar \tag{C36}$$

connecting the uncertainties in the momentum and position of a particle, and the relationship

$$\Delta E\ \Delta t \geqslant \tfrac{1}{2}\hbar \tag{C37}$$

connecting the uncertainties in the energy and duration of a state. Relationships (C35) and (C37) may be regarded as equivalent, in the sense that for a field of frequency ν the uncertainty Δn implies

$$\Delta E = h\nu\ \Delta n$$

while the uncertainty $\Delta\phi$ implies

$$\Delta t = \nu^{-1}\frac{\Delta\phi}{2\pi},$$

together giving

$$\Delta E\ \Delta t = \hbar\ \Delta n\ \Delta\phi.$$

The product $\Delta n\ \Delta\phi$ has its smallest value (i.e. $\frac{1}{2}$) only for the coherent state $\psi(\alpha)$. This is therefore the state of the field which most closely resembles the classical (non-fluctuating) electromagnetic wave. The fluctuations $\Delta\phi$ and $\Delta E/\bar{E}$ for the coherent state are illustrated in Fig. C1 for the two cases $\bar{n} = 4$ and 400. In the second of these the uncertainties in the field strength are represented approximately by the thickness of the line. Clearly this corresponds closely to the classical wave $E_0 \sin(\omega t - \mathbf{k}\cdot\mathbf{r})$. On the other hand when $\bar{n} = 4$ the uncertainties are

$$\Delta E \sim E_0/2\bar{n}^{1/2} = \tfrac{1}{4}E_0$$

and

$$\Delta\phi \sim \frac{1}{2\Delta n} = \tfrac{1}{4}.$$

It is of course possible to define the field amplitude exactly, by taking the pure state ψ_n of the radiation field, since this has a constant number n of photons. But then the phase ϕ would be completely unspecified, as we see

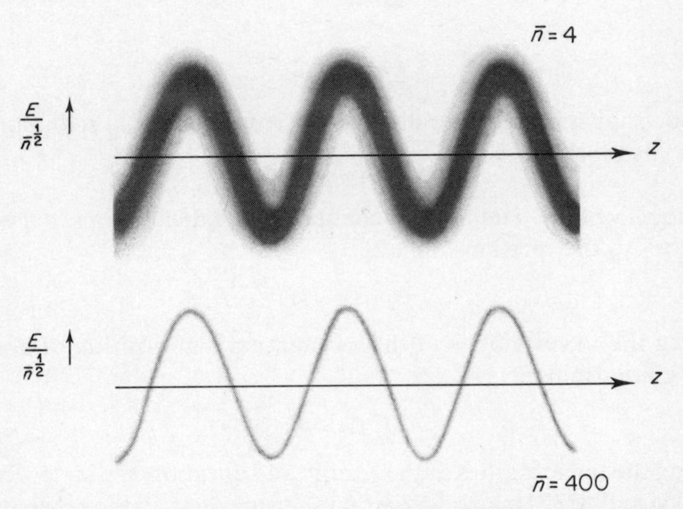

Fig. C1. Schematic representation of the electric field strength of the coherent state, for $\bar{n} = 4$ and 400, showing the uncertainties implied by the relationship $\Delta n\ \Delta\phi = \frac{1}{2}$. The depth of shading represents the relative probability of finding the field strength E.

from the uncertainty relationship (C35). Conversely the phase could be accurately defined, but then the amplitude would be completely unknown!

As an example of the nearly classical wave let us consider a radio transmitter which radiates 10 kW of power at a wavelength of 300 m. At a distance of 10 km the power received per unit area, assuming that the aerial is not highly directional, is

$$\bar{N} \sim \frac{10^4}{4\pi(10^4)^2} \sim 10^{-5}\ \mathrm{W/m^2}.$$

Each photon of the field has the extremely small energy

$$\hbar\omega \simeq 7 \times 10^{-28}\ \mathrm{J},$$

ans so if a receiving aerial has an effective area of $10\ \mathrm{m^2}$ the number of photons collected in one period of the wave is very large,

$$\bar{n} = \frac{\bar{N}A}{\hbar\omega\nu} \sim 10^{17}.$$

The resulting uncertainties in the phase and amplitude of the field are therefore so small that it can be considered to all intents and purposes as a classical field. This is true also of typical laboratory sources of visible light. Laser beams usually have even larger values of $\bar{n}$, giving beams of almost constant amplitude and phase.

We might ask if photons are the only elementary particles which can add together to simulate a classical wave. A necessary condition is that the normal modes of the field must be able to be occupied by more than one particle. This therefore excludes electrons or any other fermions, since only bosons (particles having a spin $n\hbar$ where n is zero or an integer) can give multiple occupancy of a mode. In principle therefore sufficient numbers of any identical bosons (e.g. neutral pions) could be made to simulate a sine wave, although it is difficult to imagine how this could be achieved in practice. The photon is the only massless boson and hence the only one that can have a sufficiently low energy for it to be produced in large enough numbers.

As a final example of the use of the field quantization techniques let us briefly consider the processes of absorption and emission of photons. When a photon is absorbed by an atom the state of the radiation field changes from ψ_n to ψ_{n-1} while the state of the atom changes from $\psi(1)$ to $\psi(2)$ say. In the case of an optically allowed transition the transition probability B_{12} can be shown to be proportional to

$$\left|\int \psi_{n-1}^{*}\psi(2)^{*}\hat{\mathbf{D}}\cdot\hat{\mathbf{E}}\psi_n\psi(1)\,\mathrm{d}\tau\right|^2$$

where $\hat{\mathbf{D}}$ is an operator corresponding to the electric dipole moment that appears in the semi-quantal treatment. This transition probability can be separated into two parts. The first concerns the atomic wavefunctions only, and gives $|\mathbf{D}_{12}|^2$ where $\mathbf{D}_{12}$ is the transition electric dipole moment obtained previously (Eq. (4.43)). The second part depends only on the field properties, and with the help of Eq. (C24) can be written as

$$B_{12}\propto\left(\frac{\hbar\omega}{2\varepsilon_0 V}\right)\left|\int \psi_{n-1}^{*}(ae-a^{\dagger}e^{*})\psi_n\,\mathrm{d}\tau\right|^2$$

$$=\left(\frac{\hbar\omega}{2\varepsilon_0 V}\right)\left|\int \psi_{n-1}^{*}[en^{1/2}\psi_{n-1}-e^{*}(n+1)^{1/2}\psi_{n+1}]\,\mathrm{d}\tau\right|^2=\left(\frac{\hbar\omega}{2\varepsilon_0 V}\right)n. \tag{C38}$$

This shows that the transition probability for absorption is proportional to the number of photons in the field, as assumed in Chapter 4. Note that the absorption probability is *not* proportional to E^2, the square of the electric field strength, since this is proportional to the total energy in the field at the absorption frequency, namely $\hbar\omega(n+\frac{1}{2})$. The zero-point energy clearly does not influence the absorption process.

Emission of a photon by the atom in state 2 increases the number of photons in the field from n to $n+1$. The transition rate is therefore proportional to

$$\left|\int \psi_{n+1}^{*}(ae-a^{\dagger}e^{*})\psi_n\,\mathrm{d}\tau\right|^2=(n+1) \tag{C39}$$

We see from this that emission can occur even when the field initially contains no photons; this is of course the process of spontaneous emission. If the state ψ_0 were a null state of some sort, with zero amplitude, instead of the state of zero-point fluctuation, the spontaneous decay rate would be zero. We can therefore regard the spontaneous decay as being in this sense caused by the zero-point fluctuations. The remaining part of the emission rate is of course the rate for stimulated emission. It is proportional to n, as assumed in Chapters 4 and 6.

One further point about the stimulated emission rate, of the greatest importance for the operation of lasers (see Chapter 6), is that the extra photon goes into the *same mode* as the n photons causing the stimulated emission. A radiation field consists in general of more than one occupied mode, and its state is therefore a superposition of the states $\psi_{n_k}(\mathbf{k})$, corresponding to n_k photons in mode $\mathbf{k}$. Several of these modes may have frequencies within the natural width of the atomic transition, but because

$$\int \psi^*_{n_k+1}(\mathbf{k})a^\dagger\psi_{n_{k'}}(\mathbf{k}')\,\mathrm{d}\tau$$

is non-zero only if $\mathbf{k}=\mathbf{k}'$, the only photons that can cause emission into mode $\mathbf{k}$ are those already in mode $\mathbf{k}$. The stimulated photon therefore has the same direction and momentum as the stimulating photons (more precisely, it occupies the same cell of phase space, as discussed in Chapter 5).

APPENDIX 

Additional reading

CHAPTER 1

1. E. Whittaker, *A History of the Theories of Aether and Electricity* (Thomas Nelson and Sons Ltd., London, 1951). A source book for the historial details.
2. H. Kangro, *Early History of Planck's Radiation Law* (Taylor and Francis, 1976).
3. R. H. Stuewer, *The Compton Effect* (Science History Publications, New York, 1975). A full and penetrating account of the history of wave–particle duality, both before and after the discovery of the Compton effect.

CHAPTER 2

4. B. I. Bleaney and B. Bleaney, *Electricity and Magnetism* (Oxford University Press, 1976).
5. I. S. Grant and W. R. Phillips, *Electromagnetism* (John Wiley and Sons, 1975). Both these books contain a comprehensive treatment of Maxwell's equations and electromagnetic waves.

CHAPTER 3

6. J. D. Jackson, *Classical Electrodynamics* (John Wiley and Sons, 2nd ed., 1975). A comprehensive and advanced treatment of the classical aspects.
7. A. Corney, *Atomic and Laser Spectroscopy* (Clarendon Press, Oxford, 1977). Further information on multipole fields. This book is also useful additional reading for Chapters 4, 6, and 7.

CHAPTER 4

8. M. Garbuny, *Optical Physics* (Academic Press, 1965). Further material on atomic spectra and on the interaction of radiation with matter (and so is useful reading for Chapter 8 also).
9. R. Loudon, *The Quantum Theory of Light* (Clarendon Press, Oxford, 1973). A comprehensive quantal treatment. Useful additional reading for Chapter 5 also. See also Corney[7].

CHAPTER 5

10. M. Born and E. Wolf, *Principles of Optics* (Pergamon Press, 5th ed., 1979). Contains a more advanced and detailed treatment of coherence properties (and also of diffraction, polarization, and all aspects of wave optics).
11. G. W. Stroke, *Coherent Optics and Holography* (Academic Press, 1966). An advanced classical treatment of coherence and holography.
    See also Loudon[9].

CHAPTER 6 AND 7

12. A. Yariv, *Quantum Electronics* (John Wiley and Sons Inc., New York, 1967). A more advanced and detailed treatment of lasers, scattering, and non-linear optics.
13. M. J. Beesley, *Lasers and their Applications* (Taylor and Francis Ltd., London, 1972). Contains further information on practical details and the uses of lasers.
14. A. L. Schawlow (editor), *Lasers and Light* (a collection of Scientific American articles, published by W. H. Freeman and Co., San Francisco, 1969). Contains further material at an elementary level, on a variety of subjects, including Chemical and Biological Effects of Light, Vision, Colour, Infrared Astronomy, Photon Echoes, and Holography.
15. W. R. Bennett, *The Physics of Gas Lasers* (Gordon and Breach, 1977).
    See also Corney[7].

CHAPTER 8

16. W. Heitler, *The Quantum Theory of Radiation* (3rd ed., Clarendon Press, Oxford, 1954). A full and advanced treatment.
17. W. E. Burcham, *Nuclear Physics* (Longmans, 1963). Contains material on the interaction with matter, and detection of X and γ-rays.
    See also Garbuny[8].

CHAPTER 9

18. R. A. Smith, F. F. Jones, and R. P. Chasmar, *The Detection and Measurement of Infra-red Radiation* (Clarendon Press, Oxford, 1968). Detailed treatment of infra-red detectors.
19. J. E. Carroll, *Physical Models for Semi-conductor Devices* (Edward Arnold, 1974). Contains further details on photo-conductivity, solar cells, and semi-conductor lasers.
20. J. B. A. England, *Techniques in Nuclear Structure Physics*, Part I (Macmillan, 1974). For more detailed discussion of X and γ-ray detectors.

Appendix B

21. A. Hermann, *The Genesis of Quantum Theory, 1899–1913* (*The MIT Press*, 1971).
22. T. S. Kuhn, *Black-body Theory and the Quantum Discontinuity 1894–1912* (Clarendon Press, Oxford, 1978).

See also Whittaker[1], Kangro[2] and Stuewer[3].

Appendix C

See London[9] and Yariv[12].

APPENDIX 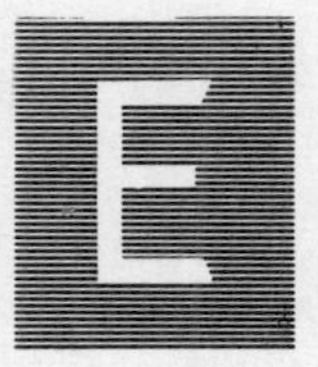

Answers to selected problems

2.1 $\oint \mathbf{H}\cdot d\mathbf{l} = d\phi/dt$, where $\phi = \int \mathbf{D}\cdot d\mathbf{S}$, assuming that there are no free currents.

2.3 Take $N \sim 300\ \mathrm{W\,m^{-2}}$ (at a temperate latitude). Then $\bar{E} \sim 340\ \mathrm{V\,m^{-1}}$ (somewhat larger than the usual vertical DC field, $\simeq 100$ V/m), and $\bar{B} \sim 1.1\times 10^{-6}\ \mathrm{W\,m^{-2}}$ ($=0.011$ gauss, which is ~ 50 times smaller than the earth's DC magnetic field).

2.5 $n_1 \sin\theta = n_2 \sin\theta''$, $\theta''_B = \pi/2 - \theta_B$, $\sin 2\theta_B = \sin 2\theta''_B$.

$$\therefore \quad \frac{d\theta''}{d\theta} = \frac{n_1 \cos\theta}{n_2 \cos\theta''} = \frac{n_1^2}{n_2^2}, \quad \text{when } \theta = \theta_B.$$

For $\theta = \theta_B + \Delta\theta$,

$$\sin 2\theta - \sin 2\theta'' = \left(2\cos 2\theta - 2\cos 2\theta'' \frac{n_1^2}{n_2^2}\right)\Delta\theta = \frac{2\cos 2\theta\,(n_1^2 + n_2^2)}{n_2^2}\,\Delta\theta.$$

$$\therefore \quad R^2 = \Delta\theta^2 \frac{(n_1^2+n_2^2)}{n_2^4}\frac{1}{\tan^2 2\theta}. \quad \tan 2\theta = \frac{2\tan\theta}{1-\tan^2\theta} = \frac{2n_1 n_2}{n_1^2 - n_2^2}.$$

$$\therefore \quad R^2 = \frac{(n_1^4 - n_2^4)^2}{4n_1^2 n_2^6}\,\Delta\theta^2 = 0.362\,\Delta\theta^2.$$

$$\therefore \quad \text{If } R^2 < 0.01,\ \Delta\theta < 0.167^0 = 10'.$$

2.7 Momentum in pulse $= E/c = m\Delta v$.

$$\therefore \quad \Delta v = 10^{-3}\ \mathrm{m/s}.$$

Angular momentum in pulse $= E/\omega = I\,\Delta\omega'$, where $\omega = 2\pi c/\lambda$, ω' = angular velocity of sphere.

$$\therefore \quad \Delta\omega' = 0.096\ \mathrm{rad/s}.$$

2.9 At distance r from the sun, radiation force $=\dfrac{1.35}{c}\times 10^3\left(\dfrac{R}{r}\right)^2 \pi a^2$,

where $R = 1.5\times 10^{11}$ m, a = radius of sphere.

Gravitational force $=\dfrac{4\pi a^3\rho}{3}\dfrac{GM_s}{r^2}$. But $\dfrac{GM_s}{r^2} = R\omega^2$,

where $2\pi/\omega = 1$ year.

$$\therefore \quad a = \frac{3\times 1.35\times 10^3}{4c\rho R\omega^2} = 5.7\times 10^{-7}\text{ m}.$$

3.1 Away from the sheet, in free space, the field consists of plane waves travelling away from the sheet. We know that Maxwell's equations are satisfied for these plane waves. Take a close path, in the y–z plane, enclosing the sheet, with $\Delta y = l$, $\Delta z \to 0$. $\mathbf{D}$ is the same on both sides of the sheet.

$$\int \text{curl}\,\mathbf{H}\cdot d\mathbf{S} = \int \mathbf{H}\cdot d\mathbf{l} = -\frac{E_0 2l}{\mu_0 c} = I_0 l.$$

Instantaneous power in field, at $z = \pm 0$, is

$$2E_0^2/(\mu_0 c) = \mu_0 c I_0^2/2.$$

Power supplied to current $= -\mathbf{I}\cdot\mathbf{E} = \mu_0 c I_0^2/2$.

3.3 Suppose that the dipoles 1 and 2 are in the z direction, with their centres at $\pm \mathbf{i}_x l/2$. Then

$$\mathbf{E}_{1,2} \sim \mathbf{i}_\theta \frac{\omega^2}{r_{1,2}} \sin\theta \cos(\omega t - kr_{1,2} + \alpha_{1,2}),$$

where $r_{1,2} = r \mp \mathbf{i}_r\cdot\mathbf{i}_x l/2 = r \mp \tfrac{1}{2}l \sin\theta\cos\phi$, and $\alpha_1 = 0$, $\alpha_2 = \pi$.

$$\therefore \quad \mathbf{E}_{1,2} \sim \mathbf{i}_\theta \frac{\omega^2}{r}\sin\theta\,[\mp\cos(\omega t - kr) - \tfrac{1}{2}kl \sin\theta\cos\phi \sin(\omega t - kr)],$$

neglecting terms in $1/r^2$.

$$\therefore \quad \mathbf{E}_1 + \mathbf{E}_2 \sim -\mathbf{i}_\theta \frac{\omega^2 \sin^2\theta\, kl \cos\phi \sin(\omega t - kr)}{rc} \sim \omega^3.$$

3.4 $\mathbf{i}_s = \mathbf{s}/s$, where $\mathbf{s} = \mathbf{r} - \mathbf{x}$.

$$\frac{\partial \mathbf{s}}{\partial t} = -\frac{\partial \mathbf{x}}{\partial t} = -\mathbf{v}, \quad \frac{\partial^2 \mathbf{s}}{\partial t^2} = -\mathbf{a}.$$

$$s = (r^2 + x^2 - 2\mathbf{r}\cdot\mathbf{x})^{1/2}, \quad \frac{\partial s}{\partial t} = \frac{xv - \mathbf{r}\cdot\mathbf{v}}{s}.$$

$$\therefore \quad \frac{\partial \mathbf{i}_s}{\partial t} = -\frac{\mathbf{v}}{s} - \frac{\mathbf{s}(xv - \mathbf{r}\cdot\mathbf{v})}{s^3}.$$

$$\therefore \quad \frac{\partial^2 \mathbf{i}_s}{\partial t^2} = -\frac{\mathbf{a}}{s} + \frac{\mathbf{v}(xv - \mathbf{r}\cdot\mathbf{v})}{s^3} + \frac{\mathbf{v}(xv - \mathbf{r}\cdot\mathbf{v})}{s^3} + \frac{3\mathbf{s}(xv - \mathbf{r}\cdot\mathbf{v})^2}{s^5} - \frac{\mathbf{s}(v^2 + xa - \mathbf{v}^2 - \mathbf{r}\cdot\mathbf{a})}{s^3}$$

$$= -\frac{\mathbf{a}}{s} + \frac{\mathbf{s}(\mathbf{r}\cdot\mathbf{a})}{s^3} + \text{terms in } \frac{1}{s^2} \text{ etc (taking } x \ll s)$$

$$= \frac{-\mathbf{a} + \mathbf{i}_s(\mathbf{i}_s\cdot\mathbf{a})}{s} + O\left(\frac{1}{s^2}\right) = \frac{\mathbf{i}_s \wedge (\mathbf{i}_s \wedge \mathbf{a})}{s} + O\left(\frac{1}{s^2}\right).$$

3.5 $v/c = 1/(n \cos\chi) = 0.872 \pm 0.022.$

$$\therefore \quad \frac{m}{m_0} = \left(1 - \frac{v^2}{c^2}\right)^{-1/2} = 2.04 \pm 0.19.$$

$$\therefore \quad \text{kinetic energy} = 969 \pm 180 \text{ MeV}.$$

3.7 Group velocity $= v_g = \mathrm{d}\omega/\mathrm{d}k. \quad k = 2\pi n/\lambda,\ \omega = 2\pi c/\lambda.$

$$\therefore \quad v_g = \frac{\mathrm{d}\omega}{\mathrm{d}\lambda}\Big/\frac{\mathrm{d}k}{\mathrm{d}\lambda} = -\frac{2\pi c}{\lambda}\left(-\frac{2\pi n}{\lambda^2} + \frac{2\pi}{\lambda}\frac{\mathrm{d}n}{\mathrm{d}\lambda}\right)^{-1} = \frac{c}{n}\left(1 - \frac{\lambda}{n}\frac{\mathrm{d}n}{\mathrm{d}\lambda}\right)^{-1}.$$

$$n = a + b/\lambda^2 = 1.000275.$$

$$v_g = \frac{c}{n}\left(1 + \frac{2b}{n\lambda^2}\right)^{-1} = c(a + 3b/\lambda^2)^{-1} = \frac{c}{1.000285}.$$

$$\therefore \quad \Delta t = \Delta\left(\frac{l}{v}\right) = \frac{10^4}{3 \times 10^8} \times 0.00285 = 9.5 \text{ ns}.$$

3.8 Plasma frequency $= 9 \times 10^7$ Hz (corresponding to $\lambda = 3.3$ m). Therefore there is a radio black-out for $\lambda > 3.3$ m.

3.9 Take density $\sim 2 \times 10^3$ kg/m^3. All the electrons (6 per atom) have a binding energy less than 2.5 keV. $\quad \therefore \quad N \sim 6.0 \times 10^{29}$ m^{-3}.

$\therefore \quad n = 1 - 5.89 \times 10^{-5} = \cos\alpha$, where $\alpha =$ glancing angle.

$\therefore \quad \frac{1}{2}\alpha^2 = 5.89 \times 10^{-5}$, $\alpha = 1.09 \times 10^{-2}$ rad $= 37'$.

3.10 Put

$$x = \text{Re}(\varepsilon) = 1 + \frac{a(\omega_0^2 - \omega^2)}{c},$$

$$y = \text{Im}(\varepsilon) = -\frac{ab}{c}$$

where

$$c = (\omega_0^2 - \omega^2)^2 + b^2.$$

Then

$$(x-1)^2 + \left(y + \frac{a}{2b}\right)^2 = \left\{a^2(\omega_0^2 - \omega^2)^2 + \left[-ab + \frac{a}{2b}c\right]^2\right\}c^{-2} = \frac{a^2}{4b^2}$$

If $a = 1$, $b = 0.2$, $\omega_0 = 1$ (see Fig. on next page):

ω	0.6	0.8	0.9	0.95	1.00	1.05	1.1	1.2	1.4	1.6
Re (ε)	2.42	3.12	3.50	2.97	1.00	1.03	−1.50	−0.88	0.00	0.37
Im (ε)	−0.44	−1.18	−2.63	−4.04	−5.00	−3.96	−2.38	−0.86	−0.21	−0.08
Re ($\varepsilon^{1/2}$)	1.56	1.80	1.98	2.00	1.75	1.24	0.81	0.42	0.32	0.61
−Im ($\varepsilon^{1/2}$)	0.14	0.33	0.66	1.01	1.43	1.60	1.47	1.03	0.32	0.07

When plotted, Re ($\varepsilon^{1/2}$) and − Im ($\varepsilon^{1/2}$) are as shown in Fig. 3.23. ε always lies in the bottom two quadrants, and therefore the square root of interest always lies in the bottom right quadrant, with a positive value of n.

3.11 The half-wave condition at which the maximum intensity is passed is $\Delta n \cdot l/\lambda = m + \frac{1}{2}$. Therefore $m = 100$. At $\lambda = (1+\alpha) \times 500$ nm, $\Delta n = 0.1(1 - 0.1\alpha)$.

$$\therefore \quad \frac{\Delta n \cdot d}{\lambda} = 100.5(1 - 0.1\alpha)(1+\alpha)^{-1} = 100.5(1 - 1.1\alpha) = 100.5 - 100.55\alpha.$$

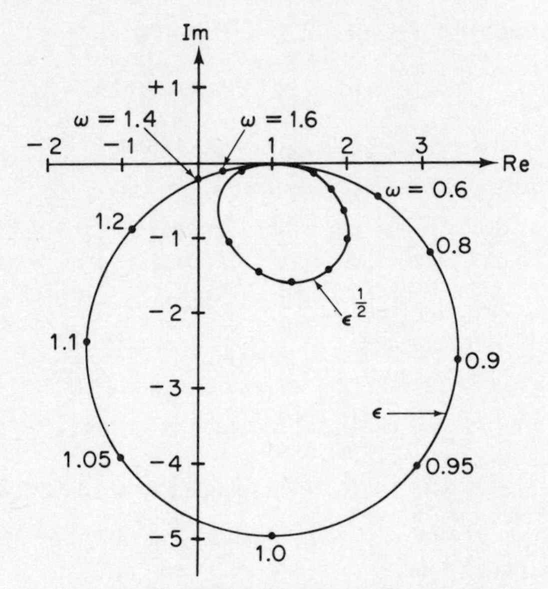

Figure for Answer 3.10.

Intensity is reduced by factor $\cos^2(\pi/4) = 0.5$ when $110.55\alpha = \pm 1/8$.

$$\therefore \quad \alpha = \pm 1.13 \times 10^{-3}, \quad 2\Delta\lambda = 1.13 \text{ nm}.$$

The range of wavelengths is 1.13 nm (FWHM).

4.1 Probability $= p = \exp(-\hbar\omega/kT)[1 - \exp(-\hbar\omega/kT)] = 10^{-4}$. Either $h\omega \gg kT$, or $\hbar\omega \ll kT$. In the first case, $p \simeq \exp(-\hbar\omega/kT)$, $T = 7.87$ K. In the second case, $p \simeq \hbar\omega/kT$, $T = 7.25 \times 10^5$ K.

4.3 Mean temperature T_0 is given by $\sigma T_0^4 A = P$,

$$\therefore \quad T_0 = 2434 \text{ K}.$$

Put $T = T_0 + S$, $T_0 = 2434$ K, $\bar{P} = 100$ W, $P = 2\bar{P}\sin^2 \omega t$.

$$\therefore \quad C\dot{S} = (P - \bar{P}) - \bar{P}(T^4 - T_0^4)/T_0^4 = -\bar{P}\cos 2\omega t - 4\bar{P}S/T_0.$$

$$\therefore \quad \dot{S} + aS = -b\cos 2\omega t, \text{ where } a = 9.34 \text{ s}^{-1}, b = 5.68 \times 10^3 \text{ Ks}^{-1}.$$

$$\therefore \quad S = -b\frac{a\cos 2\omega t + 2\omega \sin 2\omega t}{a^2 + 4\omega^2},$$

which has amplitude $b/(a^2 + 4\omega^2)^{1/2} = 9.04$ K.

4.4 $B = \dfrac{Ac^3}{8\pi h\nu^3} = \dfrac{c^3}{8\pi h\nu^3\tau} = 5.7 \times 10^{34}$ m/kg.

4.5 Ratio $= [\exp(\varepsilon/kT) - 1]^{-1}$.
(i) 25.3, (ii) 6×10^{-68}.

4.6 Degeneracies are 3 for upper state, 1 for lower state.

$$\therefore \quad \text{(Eq. 4.36) } \tau = \frac{3m\varepsilon_0 c^3}{2\pi e^2 \nu^2 f} = 17 \text{ ns}.$$

4.7 (i) $0^{\pm} \leftrightarrow 1^{\mp}$, $\frac{1}{2}^{\pm} \leftrightarrow \frac{1}{2}^{\mp}$
(ii) $0^{\pm} \leftrightarrow 1^{\pm}$, $\frac{1}{2}^{\pm} \leftrightarrow \frac{1}{2}^{\pm}$
(iii) $0^{\pm} \leftrightarrow 2^{\pm}$
(iv) $\frac{1}{2}^{\pm} \leftrightarrow 1\frac{1}{2}^{\pm}$
(v) not possible
(vi) $\frac{1}{2}^{\pm} \leftrightarrow 2\frac{1}{2}^{\mp}$
(vii) not possible

5.1 From Fig. 5.9(*b*), $\theta = 1.22\lambda/d = 2.03 \times 10^{-7}$ rad $= 0.04$ sec of arc.

5.2 $a(\omega) \propto \int e^{-j\omega t} E(t)\, dt \propto \int e^{-a(t-t_1)^2+b}\, dt$, where

$$-a(t-t_1)^2 + b = -((t-t_0)^2/4T^2) - j(\omega - \bar{\omega})t.$$

$$\therefore \quad b = -(\omega-\bar{\omega})^2 T^2 - j(\omega-\bar{\omega})t_0, \qquad |a(\omega)| \propto e^{-(\omega-\bar{\omega})^2 T^2}.$$

RMS deviation of e^{-ax^2} is $(2a)^{-1/2}$ (see answer to Q9.3)

$$\therefore \quad \Delta t \Delta\omega = (T)(4T^2)^{-1/2} = \tfrac{1}{2}.$$

5.3 Put separation of sources $= d$, distance from source to screen $= D$, distance along screen $= x$.
(*a*) Putting $a = kd/D$, $E(x) \propto \cos(\omega t - ax) + \cos(\omega t) + \cos(\omega t + ax) = \cos \omega t[2\cos(ax) + 1]$, $I(x) \propto [2\cos(kdx/D) + 1]^2$, has zeros at $kdx/D = 2\pi/3 + 2\pi n$, $4\pi/3 + 2\pi n$, has maxima at $kdx/D = n\pi$.
(*b*) With different zero in x, $I(x) \propto [\cos(\omega t - ax/2) + \cos(\omega t + ax/2)]^2 + \cos^2(\omega t) \propto 4\cos^2(ax/2) + 1 = 2\cos(ax) + 3$, visibility $= (5-3)/(5+3) = 0.25$.
(*c*) Same visibility, different spacing.

5.5 $\delta = 1/(e^{h\nu/kT} - 1)$.
(*a*) $\delta = 6.2 \times 10^{-6}$, (*b*) $\delta = 7.0 \times 10^5$.

6.1 In laser beam, $W = 1.27 \times 10^3$ W/m². Stefan's law: $W = \sigma T^4 = 5.67 \times 10^{-8} \times (5800)^4 = 6.42 \times 10^7$ W/m².

$$\therefore \quad \text{required factor} = \left(\frac{6.42 \times 10^7}{1.27 \times 10^3}\right)^{1/2} = 224.$$

6.2 (i) $t_c = \dfrac{L}{c[\frac{1}{2}(1 - R_1 R_2)]} = 3.3 \times 10^{-7}$ s.

(ii) $\Delta\nu_{nat} = \dfrac{1}{2\pi\tau} = 1.57 \times 10^7$ Hz,

$$\Delta\nu_{dop} = 7.16 \times 10^{-7} \nu_0 \left(\frac{T}{A}\right)^{1/2} = 1.55 \times 10^9 \text{ Hz}.$$

$\therefore$ Doppler width dominates.

(iii) $G = e^{\kappa L}$, $\kappa = \dfrac{c^2 g(0)\Delta n}{8\pi\nu^2\tau} - \dfrac{1}{ct_c}$, $g(0) = \dfrac{0.939}{\Delta\nu_{dop}}$.

$\therefore \quad \kappa = 9.27 \times 10^{-2} - 10^{-2} = 8.27 \times 10^{-2}$, $G = 1.086$.

(iv) $\Delta\nu_n = \dfrac{c}{2L} = 1.5 \times 10^8$ Hz.

(v) From (iii), $G > 1$ for $|\nu - \nu_0| \lesssim \Delta\nu_{dop}$.
$\therefore$ range of ν is $\sim 3 \times 10^9$, which includes ~ 20 axial modes.

6.3 $\kappa_\nu \propto \dfrac{g(0)\Delta n}{\nu^2 (B_{ij})^{-1}} \propto \dfrac{\lambda^2 B_{ij}\Delta n}{\Delta\nu_d} \propto \lambda^3 B_{ij}\Delta n.$

$$\therefore \quad \frac{\kappa(632.8\ \text{nm})}{\kappa(3.39\ \mu\text{m})} = 9.4\times10^{-3}\,\frac{\Delta n(632.8\ \text{nm})}{\Delta n(3.39\ \mu\text{m})}.$$

$\therefore$ Longer wavelength line needs smaller inversion density.

6.5 Need $\kappa_\nu > 10^{-1}$, $\quad \therefore \quad \dfrac{c^2\Delta n}{8\pi\nu^2\tau\Delta\nu} \geqslant 10^{-1}.$

But $\tau\Delta\nu = \dfrac{1}{2\pi}$, $\quad \therefore \quad \Delta n \geqslant \dfrac{4\times10^{-1}}{\lambda^2} = 1.6\times10^{16}\ \text{m}^{-3}.$

Minimum power required to maintain Δn is

$$\sim \frac{\Delta n\times h\nu}{\tau} = \frac{\Delta n\times 4\times10^{-17}\ \text{J}}{\lambda^2\times4.5\times10^4\ \text{s}} \geqslant 8\times10^{11}\ \text{W m}^{-3}.$$

$\therefore$ Minimum power in $10^{-7}\ \text{m}^3$ is ~ 100 kW.

6.7 $F = \dfrac{d^2}{4L\lambda}$, $\quad d \geqslant (12L\lambda)^{1/2} = 1.95$ mm and 4.51 mm.

6.8 $w_0 = 1.92\times10^{-4}$ m. $\quad z_0 = 0.194$ m. $\quad w$ (at 0.25 m) $= 3.14\times10^{-4}$ m, w (at 10 m) $= 9.93\times10^{-3}$ m.

6.9 $\lambda = d(\sin\theta + \sin\theta)$. $\quad \therefore \quad \sin\theta = \frac{1}{2}$, $\theta = 30°$.

$$\therefore \quad \frac{d\theta}{dt} = \frac{d\lambda}{dt}\Big/(2d\cos\theta) = 1.15\times10^{-4}\ \text{rad/s}.$$

6.11 Beam passes symmetrically through each prism. If α = angle of prism, n = refractive index, $\theta = \theta_B$ = angle of incidence, then $\sin\theta = n\sin(\alpha/2)$, $\tan\theta = n$. Both conditions are satisfied if

$$\tan^2\theta = n^2 = \frac{n^2\sin^2(\alpha/2)}{1-n^2\sin^2(\alpha/2)},$$

$$\tan(\alpha/2) = 1/n.$$

7.1 Suppose lens is at $z = -z_1$, waist is at $z = 0$. Then

$$R = w_0[1+(z_1/z_0)^2]^{1/2} = w_0\left[1+\left(\frac{z_1\lambda}{\pi w_0}\right)^2\right]^{1/2},$$

$$\therefore \quad z_1 = \frac{\pi w_0^2}{\lambda}\left[\left(\frac{R}{w_0}\right)^2 - 1\right]^{1/2},$$

which has a maximum value when $w_0 = R/\sqrt{2}$.

7.2 When $L = L_0$ the vector phasors representing the wavelets form a closed circle. When $L \neq L_0$, angle of arc of circle is $\phi = 2\pi L/L_0$, length of chord $= s \propto 2\sin(\phi/2)$, intensity $\propto s^2 \propto \sin^2(L\pi/L_0)$.

8.1 For a 1 MeV γ-ray, $\lambda = 1.240\times10^{-12}$ m. $\lambda'_{max} - \lambda = 2h/m_0c = 4.852\times10^{-12}$ m

$$\therefore \quad \lambda_{max} = 6.092\times10^{-12}\ \text{m}, \quad h\nu'_{min} = 0.204\ \text{MeV},$$

$(E_e)_{max} = 1.0 - 0.204 = 0.796$ MeV. $(E_e)_{min} = 0$.

8.2 Each of the 3 particles has

$$E = m_0c^2 + \tfrac{1}{3}(h\nu - 2m_0c^2),$$

and hence

$$p = \frac{1}{c}(E^2 - m_0^2c^4)^{1/2} = \frac{1}{3c}(h^2\nu^2 + 2h\nu m_0c^2 - 8m_0^2c^4)^{1/2}.$$

The total momentum is that of the photon,

$$\therefore \quad p = \frac{h\nu}{3c}, \text{ which gives } h\nu = 4m_0c^2.$$

8.3 K-shell photoelectron energy = 912 keV. Auger energies = $E_K - 2E_L$ = 56.2 keV for $K^{-1} \rightarrow L_1^{-2}$, 57.6 keV for $K^{-1} \rightarrow L_2^{-2}$, and 62.0 keV for $K^{-1} \rightarrow L_3^{-2}$.

8.5 Compton scattering dominates. $\sigma \sim \frac{1}{2}\sigma_{\text{Thomson}}$, from Fig. 8.5(a).

$$\mu = N_a\sigma, \text{ where } N_a \text{ number of electrons/m}^3$$

$$= \frac{\rho \times \rho_{\text{water}} \times Z}{m_p \times A} \sim \frac{\rho \times 10^3}{2m_p}.$$

$$\therefore \quad \mu \sim \frac{1}{2}\frac{\rho \times 10^3}{2m_p}\frac{8\pi r_0^2}{3} \sim 10\rho \text{ m}^{-1}.$$

8.6 Single line is due to photoelectrons. Assume ionization energy is negligible. $\therefore$ $E = \gamma$-ray energy. Maximum energy of Compton scattered electrons is

$$h\nu\frac{2\gamma}{1+2\gamma} = 0.75h\nu,$$

$$\therefore \quad \gamma = 1.5, \qquad E = h\nu = 1.5m_0c^2 = 0.767 \text{ MeV}.$$

9.1

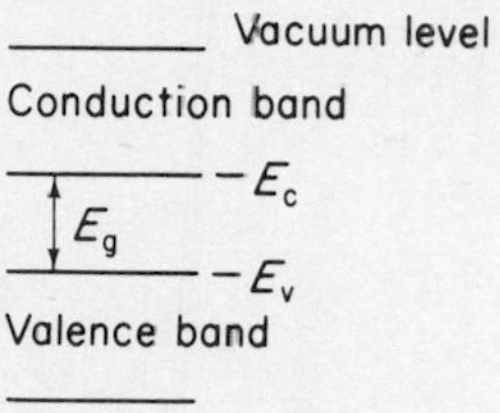

Vacuum level

Conduction band

$-E_c$

E_g

$-E_v$

Valence band

200 nm $\rightarrow E_v = 6.2$ eV.

1.77 μm $\rightarrow E_g = 0.7$ eV.

$\therefore$ $E_c = 5.5$ eV.

9.2 (i) Signal (without gain) = $e/C = 1.6 \times 10^{-8}$ V, RMS noise voltage = $(4kTR\Delta\nu)^{1/2} = 1.3 \times 10^{-4}$ V. Therefore required gain is $\gtrsim 10^4$.

(ii) Mean signal level = $neR = 1.6 \times 10^{-5}$ V, RMS noise level = $1.3 \times 10^{-5}(\Delta\nu)^{1/2}$ V,

$\therefore$ need $\Delta\nu \lesssim 1$ Hz.

9.3 The Gaussian probability distribution $P(x) = e^{-\alpha x^2}$ has the variance

$$\sigma^2 = \overline{x^2} = (\int x^2 e^{-\alpha x^2} dx)/(\int e^{-\alpha x^2} dx) = 1/2\alpha.$$

The probability distribution has half its maximum value when $\alpha x^2 = \ln 2$. Therefore the width is

$$\Delta x = 2\left(\frac{\ln 2}{\alpha}\right)^{1/2} = (8 \ln 2)^{1/2}\sigma = 2.35\sigma.$$

9.4 $\bar{n} = \dfrac{10^{-9} \times 10^{-7} \times 10}{1.6 \times 10^{-19}} = 6.25 \times 10^3.$

Fluctuations in the current are caused by fluctuations (shot noise) in the number of photoelectrons. Suppose that there are N photoelectrons in time t. Then $\bar{N} = 6.25 \times 10^2\, t$, $\sigma(N) = N^{1/2}$. Require $\sigma(N) \leqslant 0.01\, \bar{N}$, therefore $t \geqslant 16$ s.

9.6 $(\Delta E)^2 = (2.35)^2 \mathrm{FWE} + (\Delta E_0)^2 = aE + b.$

$$E \;= 10^6 \qquad 2 \times 10^6 \qquad 3 \times 10^6 \quad (\mathrm{eV})$$

$$(\Delta E)^2 = 4.33 \times 10^6 \quad 7.29 \times 10^6 \quad 10.24 \times 10^6 \quad (\mathrm{eV})^2$$

Relationship is linear, as expected (with $a = 2.96$ eV, $b = 1.37 \times 10^6\, (\mathrm{eV})^2$).

$$\therefore \quad F = \frac{a}{(2.35)^2 W} = 0.188.$$

Index